Advances in Intelligent Systems and Computing

Volume 907

Series editor

Janusz Kacprzyk, Systems Research Institute, Polish Academy of Sciences,
Warsaw, Poland
e-mail: kacprzyk@ibspan.waw.pl

The series "Advances in Intelligent Systems and Computing" contains publications on theory, applications, and design methods of Intelligent Systems and Intelligent Computing. Virtually all disciplines such as engineering, natural sciences, computer and information science, ICT, economics, business, e-commerce, environment, healthcare, life science are covered. The list of topics spans all the areas of modern intelligent systems and computing such as: computational intelligence, soft computing including neural networks, fuzzy systems, evolutionary computing and the fusion of these paradigms, social intelligence, ambient intelligence, computational neuroscience, artificial life, virtual worlds and society, cognitive science and systems, Perception and Vision, DNA and immune based systems, self-organizing and adaptive systems, e-Learning and teaching, human-centered and human-centric computing, recommender systems, intelligent control, robotics and mechatronics including human-machine teaming, knowledge-based paradigms, learning paradigms, machine ethics, intelligent data analysis, knowledge management, intelligent agents, intelligent decision making and support, intelligent network security, trust management, interactive entertainment, Web intelligence and multimedia.

The publications within "Advances in Intelligent Systems and Computing" are primarily proceedings of important conferences, symposia and congresses. They cover significant recent developments in the field, both of a foundational and applicable character. An important characteristic feature of the series is the short publication time and world-wide distribution. This permits a rapid and broad dissemination of research results.

More information about this series at http://www.springer.com/series/11156

Zhanna Anikina

Editor

Going Global through Social Sciences and Humanities: A Systems and ICT Perspective

Proceedings of the 2nd International Conference "Going Global through Social Sciences and Humanities", 27–28 February 2019, Tomsk, Russia

 Springer

Editor
Zhanna Anikina
Research Centre Kairos
Tomsk, Russia

ISSN 2194-5357   ISSN 2194-5365   (electronic)
Advances in Intelligent Systems and Computing
ISBN 978-3-030-11472-5   ISBN 978-3-030-11473-2   (eBook)
https://doi.org/10.1007/978-3-030-11473-2

Library of Congress Control Number: 2018967432

This Springer imprint is published by the registered company Springer Nature Switzerland AG
The registered company address is: Gewerbestrasse 11, 6330 Cham, Switzerland

Preface

Dear Reader,

This is a book of contributions submitted to the 2nd International Conference *Going Global through Social Sciences and Humanities* (GGSSH 2019) held on 27–28 February 2019. The conference was organized by Research Centre Kairos (Tomsk, Russia) and held in an online format to make it available to a diverse audience of academics. The event met over 100 participants. The conference aimed to focus on a variety of issues such as interdisciplinary pedagogy, language teaching and learning, cultural studies and linguistics. A special concern was given to global academic integration and professional development for research.

The following topics were discussed in detail:

- Educational environments
- Language teaching and learning
- Lifelong learning
- Cultural studies
- Linguistic studies
- Practices for global research

The event was important for several reasons. First, it became the platform for academics to discuss current trends and prospects in a variety of issues within Social Sciences and Humanities. In addition, the foundations of doing research globally as well as other relevant topics were considered.

Second, the event became a part of the project *What it takes to be a researcher* carried out in Research Centre Kairos. The project aimed to support Russian academics to function in an international English-speaking academic context. One of the strands within the activities was launching writing support programmes with the focus on various aspects of academic writing. Although we received positive feedback about our support, such an approach turned to be weak to provide sustainable change in Russian academics' scholarly publishing. Our evaluation of the initiatives we had undertaken affected the direction of our current projects and future plans. What we concluded is that we failed to take writing for publication as a social process existing in a wider context. For that reason, participation in GGSSH

2019 resulting in paper publication became an important step contributing to Russian academics' publishing experience.

Third, the event was important for Research Centre Kairos, since it became the source of Russian academics' perceptions and assumptions about key challenges in doing research globally. This knowledge will develop a better understanding of the factors that may contribute to the design of professional development schemes to enhance Russian academics' performance in research practices on a global scale. This is our next big job to be done. Along with that, the project promises to be of much benefit to non-English-speaking academic community beyond Russia challenged to communicate their research through the medium of English.

We would like to thank *Advances in Intelligent Systems and Computing* for their support in our important work. We are also grateful to the academics who took part in GGSSH 2019. They made the event a worthwhile platform for development.

We look forward to seeing you next year.

Zhanna Anikina

LKTI 2017

Program Committee

Chair

Zhanna Anikina   Research Centre Kairos, Tomsk, Russia

International Advisory Board

Terry Lamb	University of Westminster, London, UK
Katya Levchenko	Turnitin, Utrecht, Netherlands
Joffrey Planchard	Nature Research, London, UK
Joe Sykes	Akita International University, Akita, Japan

Organizing Committee

Marina Bovtenko	Novosibirsk State Technical University, Novosibirsk, Russia
Ksenia Girfanova	Tomsk State University of Architecture and Building, Tomsk, Russia
Liubov Goncharova	Research Centre Kairos, Tomsk, Russia
Ekaterina Daminova	Tomsk Polytechnic University, Tomsk, Russia
Nikolay Kachalov	Tomsk Polytechnic University, Tomsk, Russia
Anna Kloyster	Omsk State Technical University, Omsk, Russia
Olga Kvashnina	Tomsk Polytechnic University, Tomsk, Russia
Tatiana Martseva	Tomsk State Pedagogical University, Tomsk, Russia
Innokentiy Novgorodov	North-Eastern Federal University, Yakutsk, Russia
Olga Obdalova	Tomsk State University, Tomsk, Russia
Irina Sharapova	Research Centre Kairos, Tomsk, Russia

Contents

Educational Environments

Training a Pre-service Foreign Language Teacher Within the Linguo-Informational Educational Environment

Konstantin Bezukladnikov[1]($\boxtimes$), Boris Kruze[1], and Boris Zhigalev[2]

[1] Perm State Humanitarian Pedagogical University,
Perm 614990, Russian Federation
konstantin.bezukladnikov@gmail.com, bkruze@gmail.com
[2] Linguistics University of Nizhny Novgorod, Nizhny Novgorod 603000,
Russian Federation
zhigalev@lunn.ru

Abstract. The paper describes the linguo-informational educational environment for training a pre-service teacher of foreign languages. Such training is aimed to be performed within higher educational establishments of professional pedagogical education. Such project assumes training of pre-service teacher of a foreign language with attributes, adequate to the range of their functions, as well as the system of strategies given in the linguo-informational context. Linguo-informational context of intercultural communication that exists within linguo-informational boundaries, implies the principles of synergy, multi-disciplinary integration, non-linearity and the openness of the system of linguo-informational training of pre-service teachers of foreign languages. This leads to the development of coherent content and tools which provide the implementation of these principles. Building linguo-informational educational environment is an effective way to structure the educational process, where it functions according to the properties of linguo-informational context of intercultural communication, laws and regulations of national systems of education as well as the requirements to the professional competence of a teacher.

Keywords: Methodical design · Educational environment · Principle · Integration · Synergism · Linguo-informational training · Foreign language teacher

1 Introduction

Current papers on pre-service teacher training has clearly stated the unanimous methodological position of the scientists according to which cross-cultural communication cannot be considered apart from the information environment as a component of modern polyculture. It results in thorough studies on the development and creation of a specific linguo-informational context for the training of pre-service foreign language teachers as the unity of cultural, psychological and linguistic characteristics of interaction [1].

© Springer Nature Switzerland AG 2019
Z. Anikina (Ed.): GGSSH 2019, AISC 907, pp. 3–14, 2019.
https://doi.org/10.1007/978-3-030-11473-2_1

The researchers agree on the complex nature of this phenomenon. At the same time they point out that the linguo-informational environment is subject to rapid development which complicates studying of this phenomenon synchronically. This means that the scientific conclusions and solutions on the results of the linguo-informational training of pre-service foreign language teachers system design must have universal character which would provide adequate ideas for the process of such a training apart from the possible substantial changes [18].

Methodical design of the linguo-informational training system for a pre-service foreign language teacher at the substantial and technological levels assumes a ready product with the features corresponding to the range of its possible applications and functional purposes, and at the same time it represents the algorithmic description of a way of actions in the given context. The features of the linguo-informational context of cross-cultural communication include so-called linguo-informational environment, its methodological boundaries, that widely apply such synergy principles as interdisciplinary integration, nonlinearity and openness of the linguo-informational training system of a future foreign language teacher. It results in the necessity both at the substantial and technological levels to design a set of tools for monitoring such features and to implement the principles mentioned into training practice [2]. Similar questions in education has led the Russian scientists to intensive development of spatial approach in the recent years. This approach is the basis for the transfer from the reproductive subject - object characteristics to synergetic, interdisciplinary and competence-based ones where a person is perceived as an individual. Who is independently conceiving and actively generating own linguo-informational meanings.

2 Methods and Materials

The objective of the research, reflected in this article is the methodical design of linguo-informational educational environment for pre-service foreign language teacher training.

The specific objective defined the scientific material – the philosophical, psychological and pedagogical, linguistic and didactic literature devoted to the study and analysis of the instruments applied in the educational environment design. The research objectives determined the use of research methods: theoretical analysis of philosophical, normative, psychological and pedagogical, methodological and linguistic literature, teaching experience on the research problem (including international); analysis of regulatory documents relating to the study of domestic and foreign educational standards of higher professional education, training programs, domestic and foreign educational systems; methodological design, modeling, planning, forecasting; empirical methods (pedagogical observation, conversations, interviews, reflective and criterial analysis products of linguo-informational activity of the pre-service teachers of a foreign language on the basis of content of personal web portals); interpretative research methods (analysis, comparison, ranking and scaling, synthesis of theoretical research, practical experience, empirical data); experimental methods (experimental teaching); statistical methods of experimental data processing.

3   Discussion of Reflective Practice

The basis for determining linguo-informational context of intercultural communication as a factor for building linguo-informational training systems were set up by the studies which describe the problems in education within the framework of postnonclassic scientific paradigm and the systematic approach as well as philosophical and psychological studies [3]. They are approved by the idea of integration and non-linearity of the cognitive processes associated with the revolution in the system of information exchange. For this study, the provisions that postnonclassic science involves the development of personality were followed by the idea of self-organization, and synergy [1].

To understand the essence of linguo-informational training of the pre-service teachers of foreign languages it is very important to study the field of developing multicultural identity [1] of the pre-service teacher, communicative competence, providing the opportunity to participate in cross-cultural communication [4], professionally oriented foreign language teaching [18].

Significant update of methodological positions related to teaching intercultural communication and multicultural identity were widely introduced in the US and Europe [16, 23]. These studies were the basis for designing the competence model as the target value for higher education, including foreign language [5].

The problem of linguo-informational context used for training specialists in the field of foreign languages teaching was studied from various angles: approaches to harnessing the potential of information technology in the context of a multicultural development [1, 6]; particularly the application of information technologies in foreign language education [7], the role and place of information technology in foreign languages teaching [22, 25].

Considerable impact was made by the numerous studies devoted to the definition and classification of specific linguistic features of intercultural communication in information space [8–10, 29], the relation of signs with other forms of intercultural communication. Scientific interest is the outcome of the psychological, sociological, pedagogical, methodical, and technical aspects of linguo-informational context of intercultural communication [15], since such a review provides a holistic view on the tendencies of training the teacher of a foreign language.

The researchers stated the complex nature of this phenomenon as it requires a multifaceted study. At the same time linguo-informational environment is subject to rapid development, which makes it difficult to study this phenomenon in synchronous mode [11, 27, 28]. This means that the conclusions and decisions in designing linguo-informational training of the pre-service teachers of a foreign language should be universal, which would ensure adequate representation of the training course, regardless of the newly occurring substantive changes.

It can be argued that to date there has been completely developed scientific basis, indicating the systemic nature of the problem of designing a system of linguo-informational training of the pre-service teachers of a foreign language, its theoretical and practical relevance and at the same time noting methodological difficulties in designing such a system. Despite the significance of the research in the field of methodology of foreign languages teaching and efficiency of modern methodological approaches to learning, versatility and undoubted importance of the research on the problems of formation of the communicative competence and information competence of the teacher of a foreign language, linguo-informational training of the pre-service teachers of a foreign language in high school so far has not received sufficient theoretical and practical development.

The theoretical and practical relevance of the research is determined by the needs of the modern methods and practice of students training - pre-service teachers of foreign languages as the development and conceptualization of linguo-informational training system at the university.

The design of linguo-informational educational environment allows to prove and model a wide range of individual educational routes for each student on different education levels. Such educational environment can serve as an effective means for linguo-informational integration of pre-service foreign language teacher training process.

Spatial approach is a view on the linguo-informational system of a pre-service foreign language teacher training from a position of the educational environment definition – "fields of opportunities for education" setting certain borders in which the realization of an infinite set of options for mutual-and self-education is potentially admissible [3].

The system of linguo-informational pre-service foreign language teacher training on the basis of the educational environment concept receives its territorial measurement, and is structured as the route projected by a student. Thus the "narrow" linguo-informational educational environment is an environment of a separate room for instruction. The linguo-informational context of pre-service foreign language teacher training assumes creating a wider linguo-informational educational environment which, by its nature, is not limited and includes the whole world. For the methodical support it is necessary to project a set of students' self-realization tools where they can not only show their independence, but also feel it at full [4].

We formulate the basic principles of an educational environment which is potent of forming the personality of pre-service foreign language teachers who harmoniously identify themselves in the linguo-informational context of cross-cultural communication [27].

The principle of expansion of educational environment– constant and systematic increase in the opportunities for linguo-informational education and self-education for pre-service foreign language teacher;

The principle of linguo-informational educational environment enrichment – purposeful filling of a student's social life by cultural values, providing knowledge of these values in various forms of cultural representation. It reveals in behavior of other people and in own student's manifestations; amplification – enrichment of the content, forms and methods of specifically infant modes – games, practical and graphic activities and communication of students with each other [28].

The principle of linguo-informational educational environment openness increase – increase in opportunities for pre-service foreign language teachers in the choice of the instructional content, ways, methods, forms and places of linguo-informational training, in accumulation and reflection of linguo-informational educational experience of rule-making and transfer of external attributes of the information culture to the internal valuable relations. It also includes recognition of spontaneity of a pre-service foreign language teacher linguo-informational development as a process during which there might be internal contradictions which act as internal drivers to actualize the creative potential, stimulate creative transformation of the experience gained [12, 13].

Such an educational environment was created within the Program of the strategic development of Perm State Humanitarian Pedagogical University. According to the program the educational environment of a higher education institution is constructed at qualitatively new stage in the innovative educational programs and systems of teacher training development, has big social importance and potential of application. The program is focused on the solution of the problems related to the improvement of higher education on the other metodological and technological basis which forms competence and new personal qualities adequate to the continuous processes of the changing world and society [14, 15].

Consideration of foreign language education issues has led scientists to intensive development in recent years. Thus, spatial approach [16, 17] which gives grounds for the transfer of education to the reproductive nature of the subject-object to a synergistic, multi-disciplinary and competence, in which the person is understood as self-thinking and actively generating own linguo-informational meanings.

Designing linguo-informational educational space allows to justify and simulate a wide range of individual educational routes for each student at various levels of education. This educational space can serve as an effective means of linguo-informational integration process of training pre-service teacher of a foreign language.

Spatial approach is a view of linguo-informational training of the pre-service teacher of a foreign language from the standpoint of determining the educational environment - the "field of opportunities for education" that defines certain boundaries within which a potentially infinite set of acceptable implementation options for inter-action arrises [19, 20].

According to Shendrik if the environment is something external, predetermined, acting in terms of the limitations, which sets certain limits, the space or environment is a place for the existence and development i.e. the opportunities for self-realization [21, 22]. Educational environment is a reflection of beingness in relation to any entity in time and space [20].

The most important are those characteristics of space as, firstly, the length - "zoom", the value of the territory in which the specified and admissible educational problem exists, secondly, the initial position or purposes of participating in educational processes, thirdly, the space quality that makes the time and the rhythm of the education processes organization, and finally, the methodology, which is built on the basis of the educational process.

Educational potential of linguo-informational space, which we would rather call environment, is provided by linguo-informational needs of students, and therefore constantly expands, often moving from one reality to another [17].

In relation to the process of pre-service foreign language teacher linguo-informational training the above mentioned thoughts mean the necessity to provide conditions for the creation of linguo-informational educational environment.

Such environment is based on the model of network interaction for the professional poly-lingual and poly-cultural training at departments, laboratories, leading schools of the university, in cooperation with schools and partner higher education institutions concerning information and project technologies. The most important tool in the realization of that models was the telecommunication system for the international professional interaction e.g. on-line round tables, forums, conferences, master classes, workshops.

The publishing department of the university was transformed into a media center supporting the program by sharing the university innovative experience gained in all fields of activity through network projects, while participating in the state priority projects and programs of the education development. The main directions in the realization of the task were: the organization of experimental regional platforms for the development of innovative experience and creation of an open access network of materials and results of their activity. The media center, on the one hand, spreads both the results of the research activity of the university faculty and the samples of the most successful and valuable students' scientific practice. On the other hand, the center coordinates the involvement of the representatives of different social groups to the projects carried out on the basis of the university by means of different media (including interaction with TV channels and radio stations, development and promotion of the web portal).

Process of teaching disciplines at the university within the Program is supported with various presentation and service equipment, such as audience scoring, systems of presentations, electronic boards, systems of videoconferences, conference rooms with the possibility of broadcasting the events to other university buildings and in the Internet; systems of multi-language support that allow convenient perception. Transition to wireless networks in all the educational and administrative buildings and hostels providing a free access of employees and students to internal and external resources of the Perm State Humanitarian Pedagogical University promotes effective involvement of all the participants in the linguo-informational environment.

The university library with traditional printed editions and electronic resources became the center of information infrastructure for linguo-informational training of pre-service foreign language teacher. The special attention is paid to the acquisition of foreign scientific collections on the problems of a linguo-didactics and foreign language education in a digital format. Remote access allows to vary classroom and out-of-class forms of linguo-informational training.

The next important tool of linguo-informational educational environment of pre-service foreign language teachers training is the university Internet platform providing higher education institutions, scientific research institutions, teachers, graduate students, undergraduates and other interested parties with the access to educational, methodical, research, administrative and other kinds of activity at a more advanced level.

The main task of the portal is to form the innovative educational environment promoting the development of education and research by means of improving the information support of educational, methodical and scientific processes on the basis of information and communication technologies.

Objective achieved provides the solution of the following tasks:

1. Collecting, systemization and formation of the qualitative, public, free information resources supporting the process of linguo-informational training of pre-service foreign language teachers.
2. Simplification of the students' access to leading Russian and international information resources on foreign language education.
3. Technological information solutions for individual participation in creation and wide use of new qualitative scientific and educational products in a foreign language.
4. Providing constantly operating feedback between authors and users of foreign language, educational and scientific resources.
5. Creation of the mechanisms providing the effective implementation of geographically distributed foreign language educational and research projects.

The portal contains the qualitative, public information resources supporting the process of linguo-informational training of pre-service foreign language teachers according to the basic requirements of the Ministry of Education and Science and educational associations.

Information infrastructure of the university meets all the linguo-informational requirements and provides access to foreign language information. All the buildings and hostels are connected in an internal corporate network, with the possibility of broadband wireless Internet access from any point; equipped with computer and office tools, and also technical means of instruction. Traditional seminar audiences and computer classes were replaced by universal audiences adapted for various types of education and research activity.

The first and the only on the territory of the Commonwealth of Independent States the International Baccalaureate Provider Center was established on the basis of Perm State Humanitarian Pedagogical University. The goals of the Center are to ensure the application of the professional competences developed by students in the process of instruction in the linguo-informational educational environment, realization of its principles of expansion, enrichment and openness.

The main functions of the Provider Center in the context of linguo-informational educational environment is the organization of the international conferences and workshops of professional development. Experts from different countries of Europe, Africa, the Middle East are involved in these events [23–25]. It creates an authentic context for cross-cultural communication of pre-service foreign language teachers on the issues of linguo-didactics and foreign language education [26, 27].

Another task of the Provider Center within the linguo-informational educational environment is the development of the project list within the Program. The students take active part in the organization, planning and implementation of these projects. The list of the projects developed by the Provider Center assumes a certain variety, each student is provided with an opportunity to apply the competences developed in the process of instruction at this or that discipline or the program [21].

The individual educational route of each student includes project work - the final qualification project summarizing the results of research, educational and practical activity, project work at the interdisciplinary humanitarian level in the linguo-informational context. Project work in the center evolves in cooperation with famous scientists, graduates and students [29].

Thus, students are paid for the participation in research and linguo-informational development. It is promoted by "Perm State Humanitarian Pedagogical University Students' Final Qualification Papers Contest", organization of on-line student's scientific conferences, seminars, round tables [29], students's work in the organizing committees of the scientific events held.

The results of the projects in which the students of the Faculty of Foreign Languages participated, supported by the Programme of Strategic Development of Perm State Humanitarian Pedagogical University, were the development of a web portal of linguo-informational training of the pre-service teacher of a foreign language on bachelor and master's levels. The portal is based on Web 2.0-3.0 technologies. The main idea of the portal is based on the technology of web portfolio of the pre-service teacher of a foreign language.

The purpose of the portal is monitoring and evaluating the process and the outcome of multicultural development of the pre-service teachers of foreign languages in linguo-informational context. Thus, it helps students in planning, self-evaluation, reflection and correction of development in linguo-informational context, providing conditions for the formation of linguo-didactic and linguo-informational competence of future foreign language teachers within constant cross-cultural communication in linguo-informational context. The stakeholders in the field of multicultural training of future foreign language teachers are students, teachers, parents, educational institutions of higher linguistic teacher education, the city, the region, the potential employer, potential consumers of foreign language education service – all of them are involved in this work.

Linguo-didactic content of the portal is organized by academic disciplines, which are able to form the competences required in the field of foreign languages teaching. They are listed in chronological order to reflect the process of competence formation for all years and at all levels of study in higher school and after graduation in the framework of continuous education throughout their lives. The linguo-informational portal consists of several sections:

- Introduction, News Feed, which publishes posts for future masters of the current and planned training activities (tasks, timing, control activities);
- Competence profile of the pre-service teachers of foreign languages, which includes the competences of the pre-service teacher of a foreign language;
- Virtual class where the faculty member uploads the necessary course materials, organizes discussion and fulfillment of specified tasks, dossier, where students store, systemize and conceptualize the results of the course; career prospects.

Students' work with the portal begins with an introduction to learn about the aims and objectives of use, features and technology of work with the sections, and the theoretical foundations. This part allows students to obtain a holistic view of their training at every level of education, to know the criteria for evaluation of results, and formulate a primary goal of their education.

Work in the portal for students begins with obtaining a personal login and password required to provide individual work. After receiving the personal login and password in the free form, they represent the most general information about themselves, their beliefs of the nature of foreign languages teachers, including presentation on the future self-realization in multicultural education, personal tasks in this activity. "Profile" is stored in the portal, and the student has the opportunity on later stages of development to compare their initial presentation with the resultant training. For the faculty section "Profile" is the most general information about the personality of the pre-service teachers of foreign languages, their interests and preferences, which is useful in further selection of individual development trajectory.

Section Forum is designed for remote advising, the pre-service teacher of a foreign language may ask the faculty member and other students to help resolve some problems. The faculty member, in turn, has the opportunity to provide students with additional materials for self-study (for example, in the course of working with the web-quest, assignments and so on), recommendations for improvement in critical essays, abstract lesson plans and other products of students' linguo-informational activity.

In the future, students master the content according to the curriculum subjects. This is facilitated by a section of the Portal "Virtual Classroom", which provides all the necessary information related to the study of specific disciplines.

The most important substantive content of this section is the so-called navigators which perform the principles of concentration and openness of linguo-informational educational environment.

Navigator is a type of a visual diagram of the individual educational environment that routes the development of educational programs in general and their individual components. Navigator is a projection, it gives the names, addresses and directions of activity of the pre-service teacher of a foreign language during the course of studies in the educational program, discipline, course, unit, the acquisition of competencies. Navigators include the electronic versions of the curriculum, resource directories. General Navigation for the development of educational programs contains information about various academic disciplines, contributing to the organization of effective and non-linear content selection, structuring the content, text processing in modules and sections. Navigators may have different formats, which correspond to the logic of enrichment of linguo-informational educational environment. The navigator can have the directory structure, followed by an interactive readme module. Each links to other modules so that the user has the option of switching between modules. Nonlinear principle allows to come back to the material studied, this process is based on the relationship between process and object, between old and new material. The electronic navigation is not just a list of terms and explanation given by the end of the textbook – it's an original dynamic help system. In the process of own linguo-informational development a pre-service teacher of a foreign language independently controls the change of modules and modes of activity. To do this, a navigator provides all sorts of elements. Linguo-informational navigators can adapt to the needs of a student working with it; it allows varying the depth and complexity of the studied material, focusing on the specific needs of a particular student, generating additional illustrative material, providing graphical and geometrical interpretation of the concepts being studied and obtained by a student. Such educational environment provides the need to show student progress in the form of electronic records, diaries, etc. This function is successfully implemented by the pre-service teacher's electronic journal, which is used, for example, as a navigator for teaching practice. In this case the emphasis shifts, instead of a prepared navigator students are encouraged to independently design navigation themselves. This feature certainly contributes to the principles of expansion, enrichment and openness of linguo-informational educational environment. Thus, the Navigator for the pre-service teacher of a foreign language becomes a sort of guide in linguo-informational educational environment. The requirements to the competence level of the pre-service teachers of a foreign language are presented in the portal as professional profile based on the competences, which are formed on an interdisciplinary level. Students see the results of their development, estimate the level of progress presented in the "competency profile", thereby forming a profile of their professional competence. Evaluation here may change as the student achieves higher levels of competence, as the portal provides it interactively. The data automatically correlate with the official student assessment for the relevant academic subject that promotes the development of skills of critical analysis of their own activities. The discrepancy between the level of self-evaluation with a formal assessment is a cause for a scientific and methodological discussion, during which the student is able, on the basis of the relevant traces of their activities, to present a more convincing rationale and the manifestation of their competences in the field of multicultural education.

For the faculty it is a cause for a sharper focus during the exam. Students store, structured and analyzed results of their creative teaching, research and professional work, which they first of all illustrate during routine tests, and, secondly, use in their future work in the field of multicultural education. For preparation and application of materials in the "profile" section, pre-service teachers of foreign languages also need to acquire computer skills, skills of digitizing analogue documents, creating electronic materials, use information technologies. This is incentive for students to master a variety of formats and necessary communicative tools for intercultural communication in the linguo-informational context. In turn, it is the guarantee that pre-service teachers of foreign languages would necessarily develop linguo-informational competence.

"Career" section describes the career prospects and professional development. The portal, which is an important part of the students training in the Faculty of Foreign Languages, is stored on the server of the University and is a basis of self-presentation at the State Qualification Commitee. Organization of education process with the linguo-informational web portal has a potential for e-learning, since it is housed on the university server and is accessible from any computer with access to the Internet. During State Qualification Committee meeting the students present their professional profile of a teacher of a foreign language, as well as demonstrate their level of linguo-informational competence. When working with the Portal pre-service teachers of a foreign language create and organize a discussion on the results of their activity.

4   Conclusion

Summarizing the above, we point out that the linguo-informational educational environment designed provides a nonlinear, variable linguo-informational context for the training of pre-service foreign language teachers. This environment is based on the principles of expansion, enrichment, and openness. It is an integrative basis for the design of nonlinear and open content and technologies for the pre-service foreign language teacher in the context of linguo-informational training.

References

1. A Basis for Practice: The Diploma Years Programme. International Baccalaureate Organization. Peterson House, London (2015)
2. Bezukladnikov, K.E., Kruze, B.A.: An outline of an ESP teacher training course. World Appl. Sci. J. **20**, 103–106 (2012)
3. Bezukladnikov, K.E., Novosyolov, M.N., Kruze, B.A.: The international teacher's foreign language professional communicative competency development. Procedia - Soc. Behav. Sci. **154**, 329–332 (2014)
4. Bezukladnikov, K.E., Shamov, A.N., Novoselov, M.N.: Modeling of educational process aimed at forming foreign language professional lexical competence. World Appl. Sci. J. **22** (7), 903–910 (2013)
5. Bezukladnikov, K.E., Zhigalev, B.A.: Writing as the aim and means in teaching a foreign language: problems of assessment. Life Sci. J. **11** (2014). http://www.lifesciencesite.com/lsj/life1111s/154_26188life1111s14_685_689.pdf. Accessed 10 Oct 2018
6. Bezukladnikov, K.E., Zhigalev, B.A., Vikulina, M.A.: Pedagogical measuring of education quality. Life Sci. J. **7** (2014). http://www.lifesciencesite.com/lsj/life1107s/076_24661life1107s14_356_359.pdf. Accessed 10 Oct 2018
7. Bezukladnlkov, K.E., Kruze, B.A., Mosina, M.A.: Interactive approach to ESP teaching and learning. World Appl. Sci. J. **24**(2), 201–206 (2013)
8. Bezukladnikov, K.E., Kruze, B.A.: Modern education technologies for pre-service foreign language teachers. Procedia - Soc. Behav. Sci. **200**, 393–397 (2015)
9. Bezukladnikov, K.E., Novoselov, M.N., Kruze, B.A.: Peculiarities of pre-service foreign language teachers' professional communicative competency development. Yazyk I Kultura [Language and Culture] **38**, 152–171 (2017)
10. Bezukladnikov, K.E., Nazarova, A.V.: Foreign monologue teaching: an integrated basis. Yazyk I Kultura [Language and Culture] **39**, 135–154 (2017)
11. Bezukladnikov, K., Kruze, B.A., Vakhrusheva, O.V.: Cadets' foreign language teaching: the context of self-organization. Yazyk I Kultura [Language and Culture] **41**, 217–241 (2018)
12. Gural, S.K., Smokotin, V.M.: The language of worldwide communication and linguistic and cultural globalization. Lang. Cult. **1**(25), 4–13 (2014)
13. Campbell, C.: Learning-Based Teaching. Oxford University Press, Oxford (2013)
14. Deller, Sh: Teaching Other Subjects Through English. Oxford University Press, Oxford (2014)
15. Dürscheid, C.: Medienkommunikation im Kontinuum von Mündlichkeit und Schriftlichkeit. Theoretische und empirische Probleme. Zeitschrift für angewandte Linguistik **38**, 37–56 (2013)
16. Gower, R.: Teaching Practice Handbook. Macmillan Heinemann English Language Teaching, London (2014)
17. Holec, H.: Autonomy and Foreign Language Learning. Pergamon Press, Oxford (2015)
18. Hutchinson, T.: English for Specific Purposes. Cambridge University Press, Cambridge (2013)
19. Hymes, D.: Foundations in Sociolinguistics Edition. Routledge, London (2013)
20. Kimball, M.A.: The Web Portfolio Guide: Creating Electronic Portfolios for the Web. Longman, Harlow (2013)

21. Kruze, B., Oskolkova, V., Ozhegova, E.: The Competence – based approach in the russian federation: the definition of the notion and structure of the professional competence of a pre-service teacher. World Appl. Sci. J. (WASJ) **20**(20), 20–23 (2012)
22. Markel, M.: Technical Communication. Situation and Strategies. Bedford/St. Martin's, New York (2012)
23. Montgomery, K.: Creating E-portfolios Using PowerPoint: A Guide for Educators. Sage, New York (2004)
24. Parry, S.B.: The quest for competencies: competency studies can help you make HR decision, but the results are only as good as the study. Training **33**, 48–56 (1996)
25. Petrides, L.A.: Cases Studies on Information Technology in Higher Education: Implications for Policy and Practice. Idea Group Publishing, New York (2000)
26. Petrova, G.I., Smokotin, V.M., Gural, S.K., Budenkova, V.Y.: The Idea of a University, its Spiritual-Humanitarian Values and Content. Procedia - Soc. Behav. Sci. **154**, 245–249 (2014)
27. Gural, S.: A new approach to foreign language discourse teaching as a super-complex, self-developing system. Procedia - Soc. Behav. Sci. **154**, 3–7 (2014)
28. Zhigalev, B., Bezukladnikov, K., Kruze, B.: Criteria related assessment and reflection as a way to increase motivation for foreign language acquisition in school and university. Yazyk I Kultura [Language and Culture] **37**, 153–166 (2017)
29. Zhigalev, B.A., Bezukladnikov, K.E., Kruze, B.A.: The variability of assimilative processes in texts with different stylistic coloring. Yazyk I Kultura [Language and Culture] **41**, 72–87 (2018)

Developing Professional Competences with Interactive Teaching Methods at Tertiary Education

Galina I. Egorova[1] ⓘ, Natalya I. Loseva[1] ⓘ, Andrey N. Egorov[1] ⓘ,
Elena L. Belyak[1] ⓘ, and Olga M. Demidova[2(✉)] ⓘ

[1] Industrial University of Tyumen, Tobolsk 626150, Russian Federation
egorovagi@list.ru, {lni99,elenabelyak}@yandex.ru,
egorovandre@mail.ru
[2] Tomsk Polytechnic University, Tomsk 634050, Russian Federation
berezovskaya-olechka@yandex.ru

Abstract. Within the Federal Educational Standards 3+ and stricter employers' requirements, there is an issue of new engineering thinking for Bachelor students' development. A training complex based on "subject-to-subject" interaction should be developed. This dialectal dependence can be solved with the systematic organization of specialists' and bachelors' professional competences development, where practical application, team-work, and peer learning are dominant.

The study intends to enhance the level of professional competences development of students in engineering universities. The results are: (1) project learning technology based on "subject-to-subject" interaction via group problems solution which provides development of professional knowledge and skills; (2) relevant methods and stages of project fulfillment tailored for a certain industry has been grounded; (3) it has been proved that group project fulfillment is necessary and important for learners' immersion into the future profession; (4) a method of "Sparring-partnership", used at practical classes and lectures, has been developed; (5) the conditions for ecological group projects have been stated as key ones for Tyumen region in terms of "Innovations in oil and gas refining cluster. Technology. Ecological compatibility. Safety"; (6) methodology of interactive methods application aimed at professional competences development via combination of different activities has been suggested.

We applied theoretical, empirical, and experimental methods.

Key results: aspects of engineering education modernity, importance, and its social and cultural role have been described; interactive methods (Sparring-partnership method, project method) have been developed and incorporated into the disciplines of the basic and variable part of the main professional educational programme (MPEP).

Keywords: Professional competences · Interactive methods ·
Project methods · "Sparring-partnership" method

© Springer Nature Switzerland AG 2019
Z. Anikina (Ed.): GGSSH 2019, AISC 907, pp. 15–22, 2019.
https://doi.org/10.1007/978-3-030-11473-2_2

1 Introduction

Bachelor's Degree students must be professionally competent as technologies are constantly developing. That means that a bachelor student should know chemical processes of functionally important chemical substances, materials on their bases, such as: fuel, polymers, plastics, rubber, synthetic fabrics paints and lacquers, solvents and extracting agents, freon and sprays [1].

However, professional competences cannot develop on their own. Main role here is played by students themselves whereas a teacher does not stand aside as well. Educational system of higher education engineering school moves in spirals. Taking into account historical traditions, experience and the role of a engineering school in specialists' training, some modern aspects should be mentioned. The following issues are the most relevant. First, special education technologies and methods should be used to improve the training. Second, the training program contents in a higher engineering school should be reconsidered with the account of engineering innovations and personal development parameters. Third, traditional training cannot be limited to simple acquisition of knowledge as a future engineer should develop a productive way of thinking characterized by novelty, search, problem statement and determined by personal qualities [2]. All these are important for engineering education. Modern pedagogy and professional education allow to use modern technologies and methods of training which develop professional competences rationally [3]. In the present study the efficiency of interactive methods is investigated.

2 Methods and Materials

Nowadays there is no unified definition and opinion on the application of the interactive methods in a higher education engineering school. We determine interactive methods as ways of effective and rational interaction among students, and between students and teachers. It should be noted that interactive methods include elements of natural education technology based on the collaborative learning. Here it is recommended to use project technologies as to organize the education process based on "subject-to-subject" relations, peer learning. Let us describe some peculiarities and examples of our work [4].

Within the "Monomers" module in the course of "Chemistry and technology of organic matters" we used a project technology while studying "Butadiene" production. The aim of the project is to describe the basis of the monomer production technology at LLC "SIBUR Tobolsk". The organization is rather simple: the students are coordinated on the project stages. The preliminary stage allows to choose the direction of development. Here, the project themes are suggested by the leader of the project group, but students may suggest their themes as well. The project development stage includes design specifications (project aim is production capacity increase; product characterization rates; participants list (major 18.03.01 Chemical technology); project timeline (1.10.17–30.02.18).

The technological stage deals with the following issues: technology specifications, raw materials, production, process typology and its main parameters, equipment characteristics and peculiarities of its application in the process, process flow diagram.

Control and assessment stage is linked with the assessment period within every semester. Students have a certain day (Thursday) to do the task; lecture rooms attendance is scheduled (not less than 6 h). The leader of the project group consults his/her team, the development direction is adjusted, new solutions are suggested, and possible obstacles in the project are discussed.

Final stage: the students record the results, make a presentation, discuss and reflect on their work. The projects are presented (with the results report) for the teachers committee - project groups leaders, here project groups give a final presentation on the project final stage, every participant reports his/her contribution in the group project during 5–7 min.

The peculiarity of the group project training is important for bachelor degree students when working on engineering tasks. Moreover this technology can be applied in course projects, if the task is given to a group [4, 7].

Regional component of the educational program is significant in professional competences development on the basis of project technologies. Let us describe the peculiarities of the group project performed by the students of CHTG – 14 (General Chemical Technology). When working on the "Butadiene" project the students prove its importance for three workshops: CGFU (central gas fractional unit), DBO-2/3, DBO-10 (single stage butane dehydrogenation), repair and maintenance service, energy service, and automation service.

Describing peculiarities of the main technological processes the students study main equipment, process, automation, raw materials, and product peculiarities, understand leading technology principles. A reactor block is used for regenerated air heating, single-stage dehydrogenation of a butane-butylene fraction in vacuum on the fixed-bed catalyst, heat recovery of gases regeneration and evacuation.

The engineering process is monitored via the HIS human interface station by instrumentation and controls data, analysis data of the line chromatograph, the quality audit laboratory, and material balance. The raw material is normal butane fraction, hot butane-butylene fraction, "Gudry" the dehydrogenating process catalyst. Schematically butadiene is produced by two separate installations – DBO-2/3, 2 reactor and compression blocks refer to DBO-2. In DBO-3 section butane-butylene-butadiene fraction is extracted from the contact gas. Extractions from the fraction and butadiene purification are performed in the DBO-10 workshop. Process pipeline length is 708 km. The process is controlled with microprocessors of new generation.

DBO-3 section is used for butane-butylene-butadiene fraction extraction from the contact gas supplied from the butane dehydration section - DBO-3. This section includes the following units: hydrocarbon absorption and desorption; hydrocarbon fraction distillation with butane-butylene-butadiene fraction extraction; carbon C_4 reabsorption; absorbent stabilization; heavy hydrocarbon debutanization.

Additionally the students learn that the section includes additional auxiliary equipment: oil seal and anti-freeze pump cooling; separation of gases from relief valves and manual outgassing lines; equipment for rejected product; pipelines and pumps emptying, mixing of exhaust gases with the power gas, waste water collecting, precipitation collection; discharging of industrial water from heat-exchange equipment.

Revealed scientific principles of production are especially interesting for the students (production process continuity; reactive agent surface extend; counter flow; optimal temperature and pressure application; catalysts; mechanization and automation) [4].

While characterizing the released product and its kinds the students describe physical and chemical features of the liquefied hydrocarbon gas, butadiene, ethane-propane fraction, unabsorbed gas, BBFs (butane-butylene-butadiene fractions), C_4 fractions, of distillation residues, hexane, acetylene fraction. They describe further utilization of the products with the account of the contemporary innovations of the Russian enterprises: propane is used in ethylene, propylene, polyethylene production, in everyday life; butane is used in butadiene production, as fuel material in pyrolysis ovens; isobutane is used in the production of isoprenics, as a monomer; isopentane in isoprenics production; pentane is raw material for isomer and artificial rubber production; hexane is a main component for benzins, solvents; butadiene is used in the production of monomer butadiene rubber, thermoplastic rubber.

Analyzing the products, students make investments proposals, suggesting possible sums of money to be invested for the expansion of the product mix. Estimating main facilities for the specified capacity, students assess the necessity of the hardening unit for raw material heating and evaporation, contact gas cooling process; collection of the rejected products; reactor blocks; functions of the line chromatographs.

3 Results

Immersing students into the future profession is a large advantage of the group project [5]. Here, different abilities develop, such as: abilities to work in a team, managerial abilities, and abilities to fulfill different activities (as leaders and as team members). Students learn other professions required in this sphere of production. Main professions are: industrial engineers, oil refining engineers; process unit operators; compressors and pumps engineers; instrumentation and process equipment fitters; water purification operators; laboratory assistants and samplers, etc. Learning new professions motivates students for additional training and the second higher education.

Some group projects were of ecological character having a key theme for Tyumen region, i.e. "Innovations in oil-gas refining cluster. Technology. Ecological compatibility. Safety". Project aim was to form new way of thinking and integral vision of the technology, ecological compatibility, safety of oil and gas chemical production, to show strong relation between engineering process quality and its impact on the society, economy and human activity. This aim was achieved when bachelors' learnt oil and gas refining processes with the help of dialectal interrelation of notions "technology-ecological compatibility-safety-production efficiency growth" [4, 7, p. 121].

We designed the projects in compliance with the following FSES principles: practical application, availability, cognitivity, and science and manufacture interrelation regarding the ecology and safety issues. The first project showed the most important issues in oil and gas processing together with ecology and safety issues.

Students formulated and discussed theoretical and practical peculiarities of oil refining: oil stabilization by means of separation and fractionation processes; oil dehydration and desalination; gave information about Oil-in-Water Emulsions, ways of demulsifying; industrial electro dehydrators; described oil treatment schemes; safety and ecological compatibility of oil treatment required for transportation.

The second project "Isobutylene separation and concentration unit. Technology. Ecological compatibility. Safety" reveals innovative methods of concentrated iso-butylene production; processes of vacuum dehydration on the fixed and fluid bed of the catalyst; peculiarities of isobutylene separation technology: based on the interaction with carboxylic acids and spirits. The students described the process flow diagram; proved the ecological safety and compatibility of the installation for isobutylene extraction and concentration.

The third project "Gas separation installation. Technology. Ecological compatibility. Safety" showed peculiarities of gas fractionating installations, purification process with gas fractioning units (GFU); stabilizers; proved the GFU operation mode in compliance with safety and ecological compatibility of the installation for isobutylene extracting and concentrating.

The fourth project "Propylene production. Technology. Ecological compatibility. Safety" theoretical and practical bases for monomer production were shown, world techniques of propylene production were described: Catofin; Oleflex; STAR; Snamprogetti/Yarsintez. The project described peculiarities of propylene extraction from different products during various processes and justified the role of propane dehydrogenation installation. Safety and ecological compatibility of the installation for propylene production were revealed as well.

These projects were performed by students of "Chemical technology", "Energy and resources saving processes in chemical engineering, petro chemical and biotechnologies" that provided collective solution to the challenges these companies experienced and facilitated new engineering thinking. In our work we applied an interactive method "Sparring - partnership", during practical classes and lectures [6]. For example, this method was used at the class on "Technological criterion of chemical processes efficiency estimation in a chemical reactor" in the discipline "Chemical reactors".

Methods peculiarities were: division of the group into two teams, each team had a student who prepared questions and tasks for the sparring partner from the other team in his theme. One student had a theme "Technological criteria - reagent substance conversion, product yield, and connection between them", the second student had a theme "Technological criteria of the process efficiency estimation in a chemical reactor: selectivity of the production process, consumption indexes". Before the class the teacher discussed the tasks with each student. Sparring partners answered the questions and did the tasks in turns during the class. At the end of the lesson the results were given with the emphasis on the comparative analysis of qualitative and quantitative characteristics of a chemical reactor operation.

Further we would like to give the example of a lecture in the form of a sparring-partnership. The theme was "Mother substances for the organic synthesis" in the discipline "Chemistry and technology of the organic substances". Methodology is similar to the one of the practical class. Example questions and tasks: Are oil, gas and coal renewable materials for the organic synthesis? Why can paraffin and olefin, syntactic gas be referred to the primary components for petrochemical synthesis production technology? Draw the scheme of hydrocarbon production from primary components: oil, gas, coal. Give comparative characteristics of production methods of petrochemical synthesis. Give oil composition, its classification; prove the essence of oil treating before the refining. Define the straight distillation of petroleum.

"Sparring-partnership" lecture on the theme "Paraffin" in discipline "Chemistry and technology of the organic substances" included the following questions and tasks: sources and properties of the saturated hydrocarbons used for organic chemical production; ways and technologies of wax precipitation; choose the most appropriate method for the methane hydrocarbon separation; draw the nods of the process flow diagrams; choose the most rational method of the methane hydrocarbon separation; draw the process flow diagram of C_1–C_5 hydrocarbon separation according to the description; describe the properties of the higher paraffin and methods of their extraction from raw materials; show the peculiarities of the Parex-method; describe the isomerizing reaction mechanism of n-paraffin; explain the choice of process conditions; compare the n-paraffin extraction methods.

Discussion, heuristic and information-driven methods have been used for the development of the professional competences. Discussion elements, opinion's criticism, contradictions in chemical and ecological materials were used in different forms of training. These methods develop high motivation via new motives for action, being the methods and ways of achieving the aim, which is important within new FSES. We used several types of professional situations. Problem situation "Development of new resources from wastes for oil-gas processing industry" (a description of a real problem situation). The purpose is to find the optimal solution to the problem of new resources development for oil-gas processing industry from wastes.

The students took into account certain groups of technologies: process, base and breakthrough technologies. We used this system for waste disposal efficiency estimation in the Fuel and Energy sector. We considered waste disposal as a part of industrial production, i.e. those which are by-products of the manufacturing process. Basic technologies of wastes disposal are the ones that are being used at the present moment.

Breakthrough technologies of wastes disposal are those that have not been used before or used insignificantly. They can improve production conditions, providing the best competitive advantages. The students described methods of wastes disposal efficiency via certain system of RES indexes (Resource capacity, Ecological, Security), estimating resources from the point of view of ecological compatibility and safety, which are necessary for deeper insight.

The technique included oil-gas refining companies monitoring according to the following criteria: availability, sufficiency (frequencies of mention in reports on wastes), objectivity, specificity of indexes. Basing on the reports, more than 20 oil-gas refining companies were listed. The priority was given to the companies having higher frequency of mentions in reports.

The situation for assessment "Assess the long-term key trends development possibilities in waste-free oil-gas refining industry basing on its current structure". The purpose: to analyze critically solutions made, give a motivated conclusion and solution. The assessment was systemized and analyzed with the account of current trends in wastes disposals. A comparative analysis of key trends in a long-term run for different companies has been done; basing on the obtained results global trends in waste-free oil-gas refining industry development have been determined, basing on their analysis, students created a route map till 2050. Illustrative situation "Wastes disposal – a reference point for future". The aim is to estimate the situation in general, analyze its solution, set questions, agree or disagree.

The students described the illustrative situation combining forecasting methods and strategizing the amount of wastes. Here it was important to include innovative waste disposal methods. The students determined different levels with a wide range of investigation fields within the scope of such basic trends as oil-gas refining, oil and gas production: high level of associated gas disposal; C2+ carbon fraction production.

Noting from the point of view of chemical engineering that a natural and associated gases are valuable hydrocarbons alongside with oil, the students paraphrased D.I. Mendeleev words that gas flaring results not only in air pollution but also in huge waste of money. The students proved that associated gas disposal together with the production of other commodities depends on enhancing the level of socio-cultural awareness, economical efficiency and stable development of the economy in the Tyumen region.

4 Conclusion

Nowadays the higher education engineering institution as socio-cultural phenomenon is constantly developing. In a social context it is a dynamically evolving system which has a significant impact on the social life. Here we notice interdependence: as a socio-cultural phenomenon the higher education engineering institution emerged to satisfy human needs in production and in true engineering knowledge about the world, engineering, civilization, influencing the development of all spheres of life.

A higher education engineering institution can be considered as a socio-cultural phenomenon because when we talk about its origin, its boundaries reach the boundaries of "culture", "professional culture", and "professional competences". From the other point of view an engineering institution is a stable training base of specialists, bachelors, masters of new generation. Thus, interactive methods of training become more important and necessary.

References

1. Egorova, G., Loseva, N., Belyak, E.: Otsenka sformirovannosti professional'nyh kompetentsiy bakalavrov tehnicheskih vuzov. Problemy pedagogicheskoi innovatiki v professional'nom obrazovanii [Estimating the level of bachelors' professional competences development]. In: Proceedings of 18th International Research and Science conference, pp. 43–45. Russian State Pedagogical University, Saint- Petersburg (2017). (in Russian)
2. Egorova, G., Egorov, A., Loseva, N., Belyak, E., Demidova, O.: Siberian arts and crafts as basis for development of cultural traditions and innovations of bachelors. Adv. Intell. Syst. Comput. **677**, 93–100 (2018)
3. Karpova, Yu.A.: Innovatsii, intelekt, obrazovaniye [Innovations, intellection, education]. MSFU, Moscow (1998). (in Russian)

4. Vedeneev, A.: Modelirovanie elementov myshleniya [Modeling elements of thinking]. Nauka, Moscow (1988). (in Russian)
5. Vekker, L.: Psyhologicheskie process [Psychological processes], vol. 2. Myshlenie i Intellect, Leningrad (1976). (in Russian)
6. Rezchikova, E.: Didakticheskie osnovaniya formirovaniya meta-kompetentsiy [Didactic grounds of meth competences formation]. In: Proceedings of IV Conference "TRIZ Journal. Application Practice of Methodological Instruments" (2012). http://www.metodolog.ru/node/1618. Accessed 12 Aug 2018. (in Russian)
7. Loseva, N.: Interaktivnye metody v prepodavanii himicheskih distsiplin v universitete [Interactive methods in teaching chemical disciplines in a University]. In: Proceedings of International Research and Practice Conference "Education and Science Development in the Modern World", pp. 120–122. AR-Consalt, Moscow (2014)

Providing Professional Self-realization for Technical University Students when Teaching the Humanities

Maria A. Fedorova[1](✉) ⓘ, Margarita V. Tsyguleva[2] ⓘ,
Helena V. Tsoupikova[2] ⓘ, and Irina N. Efimenko[2] ⓘ

[1] Omsk State Technical University, Pr. Mira 11, 644050 Omsk, Russian
Federation
sidorova_ma79@mail.ru
[2] Omsk State Automobile and Highway University, Pr. Mira 5, 644080 Omsk,
Russian Federation
m.v.tsyguleva@gmail.com, chisel43@yandex.ru,
efimenko_1951@bk.ru

Abstract. Development of professional competence is impossible without professional self-realization of a student at the University. This article discusses features and conditions of professional self-realization of a Technical University student during their research work (in the humanities) and a model of professional self-realization. The paper is both of theoretical and practical importance. Factors and basic components of student professional self-realization are defined. In the practical part of the study, they are applied to the scientific research training of students. The key concept of study, that is an individual educational path, is projected onto the student scientific activity and is used in its version - individual scientific-educational path. This concept is the basis for a soft model of research and educational process differentiation, as well as for the personality-centered education technologies piloted by the authors in a technical university, based on the example of the humanities. The methods used in the research are empirical (observation), polling methods (surveys), modeling, and the method of competences evaluation.

Keywords: Self-realization in education · Scientific research ·
Human congruity · Engineering education · Creative thinking ·
Research education

1 Introduction

In our opinion, formation and development of a professional competence is impossible without professional self-realization of a university student. In this article we consider the characteristics and conditions for professional self-realization of a technical university student during research (in humanities) and offer a model for professional self-realization.

The concept of self-realization is an international one. It is usually understood as education for a person. The term itself is rather new for education as it is a psychological one. However, even in psychology it only became widely used in 1990s. Still, a

© Springer Nature Switzerland AG 2019
Z. Anikina (Ed.): GGSSH 2019, AISC 907, pp. 23–30, 2019.
https://doi.org/10.1007/978-3-030-11473-2_3

lot of researches carry out their investigation using psychological terms and ideas. For instance, Ireyefoju takes into account Plato's human psychology and implements its main achievements in the Nigerian educational context showing that even without using this term scholars have paid much attention to realization of personal potential throughout the world history [9].

In a Russian educational tradition this term became up-to-date in 2000s, but as early as 1979 Ukeje described education for self-realization as one aimed at developing independent, self-reliant, free and responsible individual citizens who are capable of contributing to the development of society [22].

Nowadays a lot of Russian scholars try to analyze the possibilities of providing students self-realization and predict the outcomes that it may have for a person and a society as a whole [1–5, 11–21].

2 Theory

From a psychological point of view, self-realization is a combination of "instrumental style and motivationally-semantic characteristics, ensuring the permanence of aspirations and commitment to personal self-fulfillment in various spheres of life in the process of ontogenesis" [17, p. 38]. According to a number of philosophical concepts, self-realization can be understood as "intellectual and practical activities aimed at insight into a personality, his search for the meaning of life, running higher purpose, the discovery and development of person's own potential on the path to a better" [4, p. 3].

From the point of view of pedagogy, self-realization is the realization of students' personal potential. We agree with Khutorskoy, who notes that "self-realization is a learning challenge in terms of student mission" [14, p. 39]. Among several types of self-realization, defined by researchers, one of the most important is a professional one. Professional self-realization is associated with achievement of personal and socially important purposes by researchers [11] and awareness of their role in production and its implementation [18].

Let us quote Kalashnikova who thinks that "as integral components of professional self-realization should be recognized, the conscious analysis of the personal professional activities, reflection and construction of the system of professional contexts, openness to professional innovations, pushing the limits of the standard life, the desire to pursue professional activities of personal intentions and lifestyle." [11, p. 117–118]. Besides, Kalashnikova describes 3 stages of professional self-realization:

- Period of starting and adaptation to professional activities;
- Integration period;
- Approval period.

The influence of student self-realization on their professional success has been studied for educational programs in the field of Tourism [1, 5], Psychology [21], Pedagogy [12], and the humanities [2]. With respect to engineering training programs, self-realization of college students in mono-city is considered in the research of Bakhmat [3] while technical university students self-development - in the work of Prima [19]. We have not found any dedicated studies on the influence of students self-realization on the quality of their professional activity, but we believe that to be a successful scientific and educational activity for university students to contribute to the development of "higher level" competencies that employers, according to the results of our surveys obtained in 2016 [7], consider the most significant for technical university graduates.

Kienko and Morozova revealed the psychological components of self-realization in the profession: "satisfaction with labour, communication, professionally important thinking skills, emotional-volitional, professionally important qualities, dominance of professional orientation onto the client, and professional success" [15, p. 251]. Grigorieva and Povarenkov have identified key indicators of professional self-realization: satisfaction with professional activities, psycho-emotional state, and positive self-evaluation and sense of meanings realization [8, p. 173].

However, there is another point of view. Klimanov considers professional self-realization as a component of professional success. It is quite obvious that professional self-realization, being one of the spheres of human realization, is associated with professional tasks solving [16].

Self-realization involves the formation of a certain type of thinking and a set of competencies. So, self-realization in technical universities is associated with the formation of the engineering way of thinking. Engineering thinking includes the ability to set and solve engineering, scientific and technical tasks. Development of scientific and technical progress requires creative thinking skills not only to apply ready-made solutions and algorithms, but also offer the new ones. That is, formation of a technical mindset is possible with the involvement of students in engineering and technical creativity, including research and the ability to deal with professional tasks

Sviderskaya, exploring a teacher professional self-realization, finds that it, being "the process of identifying, understanding and implementing of the teacher's potential positive professional opportunities" [20, p. 55], includes (taking into account specificity of pedagogical activity) such components as: inner motivation; the focus of professional activity; focused processes of self-development, self-improvement, self-training and self-education; research activities; reflection, analysis and evaluation of the results of personal activities; planning the further process of self-realization based on the obtained results, adjusting the goals, objectives and methods of the planned action. It seems that such a definition of a teacher professional self-realization could also be applied to a professional self-realization. Nowadays, engineers carry out all seven kinds of pedagogical activities, defined by Sviderskaya (educational, methodological, organizational, reflexive-creative, communicative, research and innovative) [20]. The difference is functional specificity.

In this research we understand the *professional self-realization* of students during research activities as the identification and realization of students' personal potential on individual research education paths which contribute to formation of professional competence and competitiveness.

3   Methodology

Among the questions to answer before creating a self-realization model are the following:

- Is it possible to predict the success of professional self-realization?
- Can a need for professional self-realization be somehow generated or formed?
- What organizational and pedagogical conditions we need to create in a university and in the working place?

Based on the above questions, we have analyzed professional self-development, self-improvement, self-education and self-education activities of students through implementation of their research activities; reflection, analysis and evaluation of the results of personal activities in science; planning the further self-realization process on the basis of the results obtained, adjusting the goals, objectives and methods of the planned action.

The analysis results, as well as the study of scientific environment in technical universities, resulted in the following model of students self-realization, including factors, principles, components and aspects, one of which being soft differentiation.

4   Modeling

Professional self-realization factors (indicators) are the following:

- Professional self-awareness development;
- Formation of a certain type of professional thinking;
- Development of professional competence;
- Realization of a personal potential.

Components of the professional competence are the professional activity and professional mobility of a person.

Methodological framework of the given research is the human congruity approach, namely, the **principle of human congruity in education**; **the principle of training productivity; the principle of educational environment; the principle of student educational potential growth**; justification the need to lay the basis for "educational goal-setting along with social **order for human potential** [14]; **the idea** of competence as "a complex of personal qualities of a student (learner), depending on the experience of their activities in certain social and personally meaningful sphere" [14, p. 165]. These principles form the basis for the proposed study of the pedagogical technologies in a professional self-realization of technical university students in the process of their scientific research.

In our opinion, the main way to achieve professional self-realization is an individual path of learning way, the way to implement the student's personal potential during research activities.

Structural model of research education for technical university students, taking into account the principles of professional self-realization:

- Characteristics and principles of students research education;
- Goal-setting (formulation of the purpose and tasks);
- The content of science education (pedagogical conditions);
- Technologies;
- Methods and forms of students research education with respect to their professional self-realization;
- Planned results of research education.

Key aspects of this model are the following:

- Flexible differentiation, taking into account the personal preferences of students and the ability to select individual scientific and educational paths.
- Modular structure of research education, providing the flexibility of differentiation (selection of modules and interchangeability). Designing a modular system for science education of students through various forms, technologies, methods and research activities.
- Possibility of students' participation in comprehensive interdisciplinary research projects.
- Immersion into scientific research since the beginning of training in high school, including organization of the chairs of employers in universities.

In addition, based on the concept of individual path, we offer development of a soft model of differentiated research education for technical university students, which includes the following:

- Typology of employers and departments within the enterprise, depending on the availability and scope of research and development activities and related needs in the research competencies of the students.
- Availability and contents of variable and invariable content for a technical university graduate research competencies, depending on different requirements of employers.
- Consideration of students' individual characteristics from the viewpoint of science education, such as interest in research, motivation, defining the "cognitive personality types" and the need to provide mentally comfortable cognitive conditions for them [13, pp. 30–31].
- Variation of forms and methods for student research education and individual educational paths development [6].

5 Discussion and Conclusion

Concerning the above theory and modeling, the implementation of the pedagogic self-realization model includes:

- Forms of students research activities organization;
- Forms of pedagogical coaching of students scientific research activity;
- Ways of involving students into research activities, including rating technologies;

- Forms of introspection of personal potential realization in research activities (portfolio);
- Methods of scientific communication development in technical universities;
- Technologies of diagnostics for a student research competency development level as the main result of professional self-realization in the process of research education.

In particular, the individual research education paths by implementing the Portfolio of students research activities and the program "Student-Researcher" have been already approved in two engineering universities of Omsk (Omsk State Technical University (OmSTU) and Omsk State Automobile and Highway University. The productivity of both technologies, in addition to quantitative results, is also confirmed by the fact that they are still successfully introduced into the research systems of the universities. A special case of implementing the project methodology is the work of the Laboratory for Development of International Scientific Communication at the Department of Foreign Languages, OmSTU.

References

1. Aristarkhova, S.A.: Pedagogicheskaya podderzhka tvorcheskoj samorealizacii studentov vuza fizicheskoj kul'tury na zanyatiyah po russkomu yazyku i kul'ture rechi [Pedagogical support of creative self-realization of students of a physical culture high school at the work of Russian language and culture of speech]. Ph.D. thesis. Moscow State academy of Physical Culture (2015). (in Russian)
2. Astemirova, O.A.: Razvitie potrebnosti samorealizacii v professional'noj deyatel'nosti u studentov gumanitarnyh vuzov sredstvami social'nogo vospitaniya [Development of a need in self-realization in profession students of humanitarian higher schools by means of social education]. Ph.D. thesis. Moscow City Psychology and pedagogy University (2012). (in Russian)
3. Bakhmat, S.A.: Professional'naya samorealizaciya studentov kolledzha v usloviyah monoprofil'nogo goroda [Professional self-realization of college students in conditions of monocity]. Second. Vocat. Educ. **9**, 15–19 (2009). (in Russian)
4. Baryshnikova, T.I.: Pedagogicheskie usloviya razvitiya sposobnosti k samorealizacii u studentov vuza (Na materiale izucheniya inostrannogo yazyka) [Pedagogical conditions in developing the ability to self-realization among students of the University (on the material of learning a foreign language)]. Ph.D. thesis. Khabarovsk State pedagogical University (2002). (in Russian)
5. Belova, O.V.: Social'no-psihologicheskie osobennosti professional'nogo samoopredeleniya studentov pedagogicheskih vuzov [Socio-psychological features of professional self-determination of students of pedagogical universities]. Ph.D. thesis. Moscow City Psychology and pedagogy University (2013). (in Russian)
6. Fedorova, M.A., Tsyguleva, M.V., Vinnikova, T.A., Sishchuk, J.M.: Distance education opportunities in teaching a foreign language to people with limited health possibilities. Astra Salvensis **6**(1), 631–637 (2018)

7. Fedorova, M.A.: K voprosu o professional'noj samorealizacii studentov tekhnicheskogo vuza v sisteme nauchno-issledovatel'skoj podgotovki [Revisiting the issue of professional self-realization of Technical University students in the research training system]. Eidos 4 (2017), http://eidos.ru/journal/2017/400/. Accessed 10 Aug 2018. (in Russian)

8. Grigorieva, A.A., Povarenkov, YuP: Problema sootnosheniya i vzaimodejstviya professional'noj samoaktualizacii i professional'noj samorealizacii lichnosti [The problem of correlation and interaction of professional self-actualization and professional self-realization of personality]. Yarosl. Pedag. Bull. **3**, 166–174 (2015). (in Russian)

9. Ireyefoju, P.J.: Constructing education for self-realization on the basis of Plato's Human Psychology: the Nigerian experience. Int. Lett. Soc. Humanist. Sci. **48**, 192–197 (2015)

10. Jiménez, G.G.: Education is love and self-realization. Revista de Comunicación de la SEECI XX **39**, 162–183 (2016)

11. Kalashnikova, S.V.: Professional'noe samoopredelenie i professional'naya samorealizaciya prepodavatelej vuza MVD Rossii [Professional self-awareness and professional self-realization of Russian Ministry of Interior Affairs university teachers]. Bulletin of the Kaliningrad branch of the St. Petersburg University of the Ministry of Interior Affairs of Russia 36, 116–119 (2014). (in Russian)

12. Kalinina, N.V.: Samorazvitie studentov pedagogicheskogo kolledzha v usloviyah uchebnoj deyatel'nosti [Self-development of students of a pedagogical college in conditions of educational activity]. PhD thesis. Irkutsk State Pedagogical University (2003). (in Russian)

13. Karpov, A.O.: Social'nye paradigmy i paradigmal'no-differencirovannaya sistema obrazovaniya [Social paradigms and paradigmatic differentiated system of education]. Voprosy Philosophii **3**, 30–31 (2013). (in Russian)

14. Khutorskoy, A.V.: Didaktika [Didactics]. Tutorial for universities. Piter, Saint-Petersburg (2017). (in Russian)

15. Kienko, E.V., Morozova, I.S.: Professional'nye aspekty samorealizacii lichnosti [Professional aspects of personal self-realization]. Philos. Educ. **35**(2), 247–253 (2011). (in Russian)

16. Klimanov, A.A.: EHtapy stanovleniya teorii professional'noj samorealizacii kak komponenta professional'noj uspeshnosti [Stages of the formation of professional self-realization theory as a component of professional success]. Pedagog. Mod. **2**(22), 74–80 (2016). (in Russian)

17. Kudinov, S.I.: Funkcional'no-stilevoj podhod v issledovanii samorealizacii lichnosti [Functional-style approach in the study of personal self-realization]. In: Regional Proceedings on Science. Education. Practice, pp. 37–41. Ufa Eastern University, Ufa (2007). (in Russian)

18. Manakina, E.M.: Professional'naya samorealizaciya studentov v kontekste reformirovaniya sistemy rossijskogo professional'nogo obrazovaniya [Students' professional self-realization in the context of reforming the system of Russian vocational education]. Second. Vocat. Educ. **6**, 5–10 (2017). (in Russian)

19. Prima, A.K.: Student tekhnicheskogo vuza kak sub"ekt samorazvitiya [Technical university student as a subject of self-development]. Izvestiya UFU. Tech. Sci. **10**(123), 246–251 (2011). (in Russian)

20. Sviderkaya, S.P.: Nauchno-issledovatel'skaya deyatel'nost' kak uslovie sovershenstvovaniya professional'noj samorealizacii pedagoga [Research activity as a condition for improving the professional self-realization of a teacher]. Ph.D. thesis. Immanuel Kant Baltic Federal University (2016). (in Russian)

21. Tsvetkova, O.A.: Samorealizaciya i destruktivnost' lichnosti [Self-realization and destructiveness of personality]. In: International proceedings on Omsk social and humanities readings, pp. 340–341, Omsk State Technical Univesity, Omsk (2015). (in Russian)
22. Ukeje, B.O.: The role of education in a changing society. In: Ukeje, B.O. (ed.) Foundations of Education, pp. 369–389. Ethiope Publishing Corporation, Benin (1979)

Peculiarities of Psychological Culture Development for Gifted Senior Pupils in Rural Settings

Olga G. Kholodkova$^{(\boxtimes)}$ ⓘ and Galina L. Parfenova$^{(\boxtimes)}$ ⓘ

Altai State Pedagogical University, Barnaul 656031, Russian Federation
Holodkova.fnk@mail.ru, parfyonova@yandex.ru

Abstract. The main purpose of the article is to consider psychological culture of a gifted personality as a resource for achievements and social self-realization. The work introduces the outcomes of empirical research concerning differences in developmental components of psychological culture of gifted senior pupils and their peers with average and low performance in rural settings. In addition, the authors analyze the peculiarities of a cognitive and value-notional component of personal psychological culture identified in gifted senior pupils. The article emphasizes the importance of the formation of psychological culture for a gifted person as its systemic characteristics and an integral part of the basic culture. The review of modern problems of research of giftedness and psychological culture of personality is made. The basic components of psychological culture are determined, they are empirically studied in gifted students. The empirical data characterizing the level of psychological literacy, readiness for self-development and life orientations of gifted high school students are generalized. Statistical confirmation of sufficient level of development of cognitive and value-semantic components of psychological culture of the gifted senior pupils of rural school is resulted that allow to consider them as real internal resources of achievements and social self-realization of the gifted personality.

Keywords: Psychological culture · Giftedness · Rural senior pupils · Life-Purpose orientation

1 Introduction

The urgency to consider the issue of psychological culture development peculiarities of gifted rural senior pupils is unquestionable due to radical social, economic, cultural, political, informational and technological changes in modern society that require identification and development of gifted and talented youth as the key human resource that is able to provide social, cultural, spiritual and moral transformation.

Rural schools, that are often centers for education, culture, moral and social support for local community, possess significant resources for revealing and developing gifted students' potential. However, quite frequently due to a number of reasons rural educational establishments do not carry out any activities, aimed at identifying gifted children, peculiarities and conditions of their development, or the strategies used are

© Springer Nature Switzerland AG 2019
Z. Anikina (Ed.): GGSSH 2019, AISC 907, pp. 31–44, 2019.
https://doi.org/10.1007/978-3-030-11473-2_4

not effective to support a talented person and do not allow them to cultivate potential and hidden giftedness of children and youths. Undoubtedly, the senior school age, when a person is fully capable to recognize and realize skills, external and internal resources in self-development and achievements, is a significant stage in social, personal and professional self-determination. To implement the potential of gifted seniors, it is necessary to understand their psychological peculiarities, resources and barriers for development.

Under given circumstances, it is impossible to solve the problem of social self-realization for talented youths without researching their psychological culture as a system pattern and a component of basic culture. Psychological culture is one of the features of social maturity and personal competence, one of the factors of harmonious development and psychological well-being that effectively allows a personality to determine and realize oneself in the society and profession, and contributes to successful social adaptation, life and progress satisfaction.

Research of peculiarities of gifted rural seniors' psychological culture is of utmost importance because local authorities do not always possess social, cultural, educational and economic facilities that can promote a talented child with a proper support and assistance in educational area when a senior student is on the threshold of adulthood.

Psychological culture of gifted senior pupils in a rural school is an internal resource for development of their potential, successful socialization and, moreover, their value-notional, motivational, emotional and volitional features. It is not infrequent that profound psychological culture of gifted rural seniors is the only internal key resource for the young to realize their personal competence and to demonstrate their unique skills.

Nevertheless, there is a contradiction between a social need to reveal and develop gifted senior school pupils, to research conditions of their personal and professional self-realization, to develop a harmonious personality that is capable of innovative activities and welfare for the sake of the society on the one hand, and on the other hand, an insufficient number of scientific research devoted to peculiarities and development of psychological culture as a means for cultivating giftedness of senior schoolchildren, residing and getting education in a rural area.

The subject matter of the given research presupposes revealing and analyzing developmental peculiarities and differences in psychological culture of senior pupils at a rural school and considered "talented" and "senior pupils with average and low performance". The problem in question is connected with numerous scientific ideas, reflected in the works of Bondarchyk and Litvak [11], Vysotskaya [21], Ivleva and Inozemtsev [5], Potapova [14], Savenkov [16, 17], Olszweski-Kubilius [13, p. 28], Shechtman and Silektor [19, pp. 61–72], Ziegler and Stoeger [24, pp. 131–152] and other scientists, focusing on specific components of psychological culture, social and psychological competence of gifted personalities.

2   Literature Review and Basic Assumptions

The processes of development in modern Russia set a task to educators to train competent specialists who will work in the system of Europe-wide cooperation, intensive exchange of information and human labor products. The role of education at the current stage of Russia's development is determined by challenges faced by the state that needs gifted and socially adjusted people that will be capable of determining the country's economic and political progress. Due to the fact that talented people develop civilization, create new technologies and trends, affect the speed of economic and cultural growth, studies of giftedness, identification and development of talented children occupy a leading position in the major official acts of the state. Modern surveys on giftedness, conducted by Vygotsky [20], Ivleva and Inozemtsev [5], Potapova [14] Bogoyavlenskaya and Bogoyavlenskaya [2], Savenkov [17] and others, reveal philosophical, pedagogical and psychological essence of this phenomenon. Thus, Bogoyavlenskaya, who points out that in modern conditions it is a gifted personality who is competitive and possesses a greater potential than others, is confident that the task to give a theoretical ground of giftedness is so urgent [2]. In the Russian studies on the nature of giftedness, based on the analysis of personal creative development, can be traced in the Concept of creative giftedness, elaborated by Matyushkin [12]. Leytes' long-term surveys, have allowed singling out natural laws and mechanisms leading to giftedness development at an early age [10]. Zaporozhets has put forward an idea of amplification which means enhancement, intensification and development of the potential that is specific for pre-school age [12]. Teplov and Krutetsky have greatly contributed to the studies on psychology of music and math inclinations and their importance in talent development [10, 17]. In 1998 a numerous group of psychologists and educators under the guidance of Bogoyavlenskaya, Doctor of Psychology, a member of the Russian Academy of Natural Sciences, made out and published "Operational Concept of Giftedness" [2]. The authors consider the problems of gifted children, peculiarities of their upbringing, education and cooperation.

Galton, being a proponent of the genotypic approach, was one of the first to study the nature of giftedness [12]. Piaget expressed an opposite opinion underlining the significance of intellect for giftedness development [12]. Wertheimer, Jackson, Guilford [4], Druzhinin and others studied interconnection between intellect and creativity, divergent and convergent abilities [17]. Guilford has worked out a graphic concept of giftedness, based on structural intellectual components [4]. Torrance [12] and Williams' works present an essential stage in identifying factors of personal creative abilities [21]. Winner and Landau analyze the myths and reality of gifted children [8]. A number of studies extend modern ideas of personal peculiarities and problems connected with giftedness development among people of various age groups [1, 10, 15, 23]. Thus, Bezrukova and a group of authors are examining conditions, problems and approaches to development of potentially gifted rural schoolchildren [1]; Litvak and Bondarchyk are considering the effectiveness of social adjustment of talented children, residing in various social and cultural conditions [11]; Savenkov is engaged in solving the problem of children giftedness development as an outcome of successful social adaptation, enhancement of the level of social competence and research culture of talented children [16]; psychological peculiarities of social competence of gifted senior pupils are being investigated by Kolmogorova and Parfenova [7]. Larionova is empirically studying formation of giftedness major components (intellect, creativity and spirituality) in senior pupils and students of various countries; interconnection of these components; development peculiarities of giftedness basic components, depending on the gender [8, 9].

At the same time literature review demonstrates that a number of empirical researches, devoted to the problem of psychological culture of gifted senior pupils, are not sufficient, although it is important to stress the need for new surveys of psychological culture of gifted rural senior pupils capable of great achievements and meaningful transformations of the society.

Psychological culture is the basic personal characteristic that promotes harmonious development and is a ground for social and psychological self-realization (Vygotsky, Maslow and others) [12]. The issue of psychological culture is expressly or by implication touched upon in a number of scientific research papers. Such German classical philosophers as Hegel, Herder and Kant discuss the role of culture in personal life and development in their works [12]. Many foreign authors have created a background for scientific ideas concerning personal psychological culture, for instance, Mead, a culture anthropologist [12]; Cole, the author of the book "Cultural-Historical Psychology: A Once and Future Discipline" [12]; Erikson, the author of the theory of personal psychosocial development in the context of culture [12]; Maslow, a representative of the humanistic psychology characterizing the image of modern culture [12]; Freud, a theoretician of non-classical understanding of culture as sublimation of nature, and others [12].

The notion of "psychological culture" is an integral part of categorical arsenal of Russian pedagogy and psychology. Numerous researchers stress the necessity and significance of the given phenomenon for modern educational practice. For instance, Vygotsky has proved that child's mental development is a process of his cultural growth [20]. According to Bodalyov [6], basic assumptions of psychological culture are determined by three elements: the ability to properly analyze other people and their psychology; the skill of responding in an emotionally adequate way to their behavior and condition; the ability to choose the right way of communication, reacting to the individual peculiarities of every person, and one has to cooperate with. Semikin [18], Kolmogorova and Kholodkova study a personality as a representative of high psychological culture, capable of self-organization, self-regulation and life activity [6]. Zimnyaya is viewing psychological culture as the most essential component of human culture that includes intelligence (education and civility) and basic features of personal development: all the characteristics that a person cultivates or attains under the influence of the society or other people [6]. Summarizing studies of Asmolov [6], Zinchenko [6], Klimov [6], Kon [6], Leontev [6] as well as their own, Demina and Ralnikova [3] consider the psychological culture as a part of personal culture, a complex functional system of interconnected personal components, providing awareness about the laws of psychic reflection of the surrounding environment, creation of one's own psychic reality, techniques to control internal and external activities on the basis of the functioning "world image" [3].

According to various theoretical approaches, the structure of psychological culture is represented in different ways: Kolominsky [12, 17, 23] views it from the point of professional, theoretical and worldly levels; L.V. Kulikova examines its presentation, regulating and forming components; Demina and Ralnikova [3] consider its cognitive, value-notional, reflexive and estimating, creative and interactive factors [17]. The components of psychological culture, singled out by different scholars, act as the basis for educational, preventive, correctional psychological and pedagogical practice and diagnostic examination.

The phenomenon of psychological culture also has a significant importance for harmonious personal development at a school age. Senior school years include a period of the most conscious and turbulent development of meaningful life stage. In this respect, the gifted rural children should be paid a great deal of attention. Yavorsky underlines that children, born both in cities and small communities, are initially allocated equally in geographical and social terms [23]. Children giftedness can frequently fail to be manifested because of the low quality of education in rural areas in comparison with their urban counterparts; low educational and cultural level of a rural family; remote location of rural schools from various methodological and education centers, artistic societies and institutions of additional education and etc. Nevertheless, studies of Bezrukova, Barkanova and others demonstrate that in a megalopolis, despite greater opportunities to get a good education, the percentage of talented children does not exceed the number of gifted ones in the province. The authors give empirical data proving that the quantity of potentially gifted schoolchildren unidentified in rural areas is considerably higher than the number of potentially talented children overlooked in the cities. The reason may lie in the fact that rural schools do not conduct appropriate diagnostic examinations, or else they are not professionally carried out; in addition, the empirical data received are interpreted superficially and do not allow teachers to adequately evaluate a child's abilities and reveal potential or underlying giftedness [1].

Thus, in the given survey giftedness is understood as a psychic feature which is systemic and evolving within life time and determines a possibility of high personal achievements (unusual and exceptional) in one or more types of activities in comparison with other people. A gifted personality stands out due to a bright, visible, sometimes outstanding progress (or a person possesses background for such achievements) in one or several activities. Psychological culture is viewed as a component of basic culture which is a personal system feature (Kolmogorova), enabling effective self-determination in the society and self-realization in life; providing self-development and successful social adjustment and life satisfaction. Kolmogorova singles out the following structural components of personal psychological culture: cognitive (psychological literacy i.e. elementary knowledge and skills); competency-based (psychological competence); value-notional; reflexive and estimating; cultural and creative [6]. Therefore, Kolmogorova's quinary structure of personal psychological culture is the basis for the given empirical survey which is aimed at revealing peculiarities and differences in development of psychological culture (cognitive and value-notional components) of gifted senior pupils and senior high school students who are not identified as talented ones (with average or low performance), trained in rural schools.

3 Materials and Methods

While conducting a research, there have been raised a number of questions: what is the level of psychological culture and what are the peculiarities of cognitive and value-notional components of psychological culture of gifted senior pupils in rural schools? Does the level of psychological culture of gifted rural senior pupils promote social realization of their high potential? In order to solve the given questions, there has been an empirical research (2016–2018), directed at identifying peculiarities and differences in psychological culture of gifted senior pupils and their counterparts with average and low performance studying in rural schools. The target of the research included personal psychological culture; the subject of the research embraced peculiarities of personal psychological culture of gifted rural senior pupils. The hypothesis included two statements: (1) components of psychological culture of gifted rural seniors are considerably developed; (2) development of psychological culture components in gifted senior pupils and their counterparts with average and low performance has specific peculiarities.

The stages of empirical analysis have been planned in accordance with the following tasks: (1) to identify gifted senior pupils out of respondent total sample; (2) to research peculiarities and component development of psychological culture in the gifted senior pupils and their counterparts with average and low performance; (3) to single out differences in psychological culture component development in gifted senior pupils and their counterparts with average and low performance, using methods of math statistics; (4) to draw a conclusion if a level of psychological culture of gifted seniors promotes social realization of their high potential.

The authors used the following empirical techniques: observation, expert evaluation method, the test and questionnaire method. The given techniques are employed to solve two types of tasks:

1. **To identify the respondent's giftedness,** the researchers have made use of Renzulli's pattern of giftedness [15, 17] which includes creative, intellectual and motivational components. In reference to techniques, the authors have employed the expert evaluation method to identify motivational and creative components; Raven's "Progressive Matrices" test to find out an intellectual component; Williams' creative tests to detect a creative component [22].
2. **To research the component of psychological culture development level of senior pupils** (cognitive and value-notional), the authors have used "Psychological Literacy for Senior Pupils Questionnaire" by Kolmogorova [6]; "Readiness for Self-Development" test by Ratanova and Shlyakhta; "Life-Purpose Orientation" test by Leontev [6]. A statistical analysis of the results has been carried out with the help of Mann-Whitney U-test. Respondent total sample embraced 164 pupils of the 10th and 11th grades, including 87 girls and 77 boys aged 16–18 years old. The study was carried out at the Secondary comprehensive school, a municipal budgetary educational institution of Altaiskoe village with population of 14 000 people, of the Altai Krai.

The empirical research has been conducted in several stages.

Stage I: **Identification of giftedness, pedagogical diagnostics.**

Four educators whose schoolwork experience comprises 5 and 15 years have carried out an expert evaluation of senior pupils' abilities. The experts (the form tutor, teachers of Russian, Math and History) have expressed their opinion about seniors, taking into account their long-term work experience. Pupils' intellectual, creative, motivational, emotional and volitional qualities have been estimated using a specific scale. After this stage the experts have singled out 64 (39%) out of 164 (100%) senior pupils possessing various traits of giftedness, according to the techniques offered by Savenkov, Burmenskaya, Shumakov and others [7, 17]. All the teachers are emphasizing a high level of different abilities and motivation in these respondents. Some of the pupils selected by the experts have already achieved great results in educational and creative spheres (academic and sport competitions, contests and so on).

Stage II: **Identification of giftedness, psychological diagnostics.**

The experts have diagnosed 64 senior pupils, selected at the previous stage. According to Raven's "Progressive Matrices" test (sample for ages 14–65), the teachers have got information about the intelligence level of the respondents (advanced, upper intermediate, intermediate, elementary, starter) and evaluated their ability for the planned, classified, systematic intellectual activity (logic thinking). The same senior pupils have been examined with the help of Williams' creative tests to detect an artistic component of giftedness (a non-verbal test type) [22]. The survey has been conducted in a group. F. Williams' tests are intended to give an effective, practical and efficient evaluation of cognitive and divergent factors (fluent, flexible and original thinking, critical thinking and ability of associational vocabulary composition). As a result, the experts have received 5 characteristics, expressed in rough points, and using the test key identified the level of respondents' creative thinking as above the standard, average, below the standard, below the average, and above the average. Based on these three methods and analysis of all the respondents' empirical characteristics, the experts have singled out two groups to study their peculiar features of psychological culture at the next stage of the research: "Group of Gifted Seniors" – 16 people (hereafter referred to as GGS) and "Group of Seniors with Average and Low Abilities" or "Group of Average Seniors" – 19 people (hereafter referred to as GAS) whose giftedness has not been identified.

GGS includes 11 girls and 5 boys, some of whom are prize winners in the regional and All-Russia Contests in Literature, Social Studies, Geography, Ecology, History, Law and Russian; winners in municipal reading contests; school newspaper reporters; regular performers at school and municipal artistic events; winners of children's creative and research competitions; leaders of youth and volunteering movement; winners of choreographic contests and various festivals; holders of junior category in judo and hand-to-hand combat; multiple winners of regional sport competitions in the Siberian Federal District; a member of the Youth Duma of Altai Region and others.

GAS includes 12 girls and 7 boys who are characterized by a lower level of cognitive activity and lesser involvement in artistic and social events. Nevertheless, more than 50% of them attend additional classes in a children's art school, a youth center and are engaged in various school sections and clubs.

Stage III: **Diagnostics of psychological culture components in respondent groups.**

(1)   Identification of the psychological competence level has been carried out with the help of "Psychological Literacy for Senior Pupils Questionnaire" by Kolmogorova [6]. The respondents were handed out work sheets with questions and tasks divided into 4 parts; in every section there is a special task for the respondents, for example, to choose one of 4 variants, give a written answer to the question, etc. As a result, the experts evaluate the level of psychological competence development of senior pupils.

(2)   Identification of readiness for self-development has been performed with the "Readiness for Self-Development" test by Ratanova and Shlyakhta [6]. The senior pupils have been offered 14 statements that should be ranked as true/false from the respondent's point of view. The answers have been processed with the key, and then there has been drawn a chart showing "the respondent's readiness to learn about oneself" and "readiness to change and improve oneself" and, in addition, the respondent's prevailing level of self-evolution, for example, "A" – "I can improve myself but do not want to know myself"; "B" – "I want to know myself and can change myself"; "C" – "I do not want to know myself and do not want to change myself"; "D" – "I want to know myself but cannot change myself".

(3)   Identification of personal life-purpose orientation has been performed with the "Life-Purpose Orientation" test by Leontev [6]. During the test the respondents have been handed out work sheets with 20 pairs of opposite statements reflecting the personal idea of life comprehension factors. A respondent has been choosing one of the statements which is the most consistent with the reality from his point of view. Data processing includes summing up numbers in all scales and converting a total score to standard ideas, afterwards the results are matched with the chart key which depicts average and standard deviations from a common personal life-purpose orientation rate and all five subscales for men and women separately. According to the key, the "Life-Purpose Orientation" test analyzes 6 scales: "Life Purposes"; "Life Process"; "Life Progress"; "I am Control Locus"; "Life is Control Locus"; the total score – "Life Comprehension".

The results have been compared in accordance with the research objectives and the hypothesis.

4 Results and Discussion

The data analysis of cognitive and value-notional components of psychological culture of rural senior pupils has allowed the authors to find out the following:

Concerning a cognitive component:

(1) 100% of gifted senior pupils (GGS) have demonstrated profound psychological competence that can be proved by the effective score of 70-90 points in psychological literacy. Gifted respondents are characterized by a positive vision of cultural ideas and comprehension of human peculiarities. They also possess deep knowledge of psychological skills (social communication, psychic activity, behavior, emotions and so on), self-regulation and empathy. Gifted senior pupils have correctly answered the questions concerning personal psychological characteristics; have been able to solve psychological tests, to understand mechanisms of personal behavior and development, to explain metaphors and phraseological units characterizing a person and one's social behavior.

(2) The analysis of psychological competence in GAS (a group of seniors with average and low abilities) demonstrates that this rate is volatile from 48 to 77. The majority of respondents, which is 10 (53%), possess deep psychological ideas and understanding of personal psychological peculiarities. Over half of senior pupils from both GAS and GGS have displayed considerable psychological competence in terms of psychological knowledge (social communication, psychic activity, behavior, interpersonal relations and others). Over a third of GAS representatives, that is 7 (37%), possess an average level of psychological competence. They have been diagnosed with impediments in understanding norms, conditions and mechanisms of self-regulation and empathy and with frequent wrong answers about personal psychological qualities. They have made mistakes when analyzing psychological situations, explaining phraseological units and metaphors characterizing a person's behavior in the society. Two senior GAS pupils (10%) have demonstrated low psychological competence that is the evidence of that the cognitive component of their psychological culture is not formed yet. These pupils are not capable of understanding mechanisms of personal behavior and development, of explaining personal characteristics and social behavior; besides, they do not possess elementary culture notions and understanding of personal peculiarities.

In accordance with the value-notional component ("Readiness for Self-Development" Test by Ratanova and Shlyakhta [6]):

(1) A positive fact is that the majority of gifted senior pupils (12; 75%) "want to know about themselves" (Level "B" of readiness for self-development). They are motivated to develop and improve themselves, to analyze their mistakes, difficulties and opportunities for progress achievements due to personal growth. However, there are some problematic points. Two gifted respondents (12.5%) have demonstrated low readiness for self-development: Level "A" – "I can improve myself but do not want to know about myself" (1; 6.25%); Level "C" – "I do not want to know about myself and do not want to change myself" (1; 6.25%). Two gifted senior pupils (12.5%) have demonstrated Level "D" – "I want to know about myself but cannot change myself".

(2) As well as GGS, the majority of GAS respondents (11; 58%) "want to know about themselves and can change themselves" (Level "B" of readiness for self-

development). They possess profound readiness for self-improvement that indicates a considerable level of value-notional component of psychological culture in more than half GSA respondents. Nevertheless, there have been found some problems, for example, one in four (21%) has been diagnosed with Level "C" – "I do not want to know about myself and do not want to change myself"; 3 (16%) – with "A" – "I can improve myself but do not want to know about myself". One of the seniors (16%) has shown low readiness for self-development that is level "D" – "I want to know about myself but cannot change myself".

(3) Consequently, 25% of gifted senior pupils and 42% of their counterparts with average and low abilities, who have demonstrated "A", "C" and "D" level of readiness for self-development, need psychological assistance in development of value-notional component of psychological culture.

Concerning the value-notional component ("Life-Purpose Orientation" Test by Leontyev [6]):

(1) GGS respondents possess its normal rate – 12; 75% according to the total score of "Life Comprehension". Two gifted senior pupils (12.5%) have a "Life Comprehension" index that is below and two others – above the standard. GAS seniors, as well as their GGS counterparts, have a normal "Life Comprehension" rate (14; 74%); 10% (2 GAS pupils) have a "Life Comprehension" index that is above and 16% (3 GSA pupils) – below standard. Therefore, in both groups the number of respondents with average and high life comprehension characteristics does not have considerable differences (87.5% in GGS and 84% in GAS). It is a positive fact because life is meaningful and purposeful for the majority of gifted senior pupils (75% and 12.5%) and their counterparts with average and low abilities (74% and 10%). These respondents are capable of planning objectives, choosing tasks and achieving results. Moreover, they find it important to clearly correlate their goals with the future, emotions with the present and satisfaction with the past events. At the same time it has been identified that approximately the same number of gifted senior pupils (12.5%) and seniors with average and low abilities (16%) have a "Life Comprehension" rate that is now formed below standard. They need psychological assistance in development of value-notional component of psychological culture.

(2) According to the subscales of "Life-Purpose Orientation" test, the analysis has shown:

  – *In accordance with "Life Purposes" subscale,* gifted senior pupils have a prevailing average "Life-Purpose Orientation" level (12; 75%); two seniors (12.5%) have a high level and two more (12.5%) – a low level. An average "Life-Purpose Orientation" level (12; 63%) prevails among GAS respondents; 6 pupils (32%) have a low "Life-Purpose Orientation" level; only one pupil (5%) with average and low abilities possesses a high "Life-Purpose Orientation" level. Consequently, the majority of gifted senior pupils and seniors with average and low abilities, 87.5% and 68%f respectively, have demonstrated high and average "Life Purposes" levels. These respondents are determined and have goals that are directed at their future, give life a meaningfulness and a temporary perspective. Senior pupils with low "Life-Purpose Orientation", according to the subscale level (12.5% in GGS and 32% in GAS), live for the day or in the past, do not have any future goals, their plans are not realistic and are not supported by personal responsibility for their realization. The differences in the number of respondents, according to the levels, are considerable and make about 20%.

- *In accordance with "Life Process" subscale,* 13 (81%) and 1 (6%) gifted senior pupils have been diagnosed with average and high "Life-Purpose Orientation" levels respectively. Among GAS respondents, 12 (63%) seniors have been diagnosed with average and 3 (16%) with high "Life-Purpose Orientation" levels. Therefore, high and average "Life Process" scale levels have been identified in the majority of gifted senior pupils (87%) and their peers with average and low abilities (79%). These schoolchildren perceive their life as a process full of excitement, emotions and meaning. Nevertheless, their negative feature is that they are fond of life only today, in other words, "here and now". In reference to a low subscale level, it has been identified in 2 (13%) gifted senior pupils and 4 (21% which makes almost ¼ of the total number) GAS seniors. The pupils who have a low scale index are not satisfied with their life at present, are focused only on reminiscences about "better past", or are exclusively focus on the future. The differences in the number of respondents, according to the levels, are not considerable and make 8%.

- *In accordance with "Life Progress" subscale,* 11 (69%) gifted senior pupils have a prevailing average level, while 3 (19%) of them – a high level. Almost all GAS respondents (18; 95%) have demonstrated average and high "Life-Purpose Orientation" subscale levels. Consequently, a great majority of gifted senior pupils (88%) and their peers with average and low abilities (95%) have a prevailing average and high subscale levels. These respondents are satisfied with self-realization and their life progress for the moment and perceive the period they have lived through as productive and meaningful. A low scale level is registered only in 2 (12%) of gifted senior pupils and 1 (5%) GAS student that, according to the procedure, characterizes them as people who live in the past and who are not capable of finding life perspectives. This fact contradicts the adolescent nature and provokes intrapersonal conflicts. Low scores indicate a respondent's life discontent. The differences in the number of respondents, according to the levels, are not considerable and make 7%.

- *In accordance with "I am Control Locus" subscale,* 9 (56.25%) gifted senior pupils have been diagnosed with an average "Life-Purpose Orientation" level, while 6 (37.5%) students – with a high level. As for GAS respondents, 13 (68%) students possess an average level and 4 (21%) – a high scale level. So, 93.75% GGS and 89% GSA respondents perceive themselves as strong personalities that possess enough freedom to build their life in accordance with their own goals, tasks and life-purpose orientations. A minimum of 1 (6.25%) gifted student and 2 (10.5%) GAS pupils have a low scale level. These seniors do not believe in themselves and cannot control their own lives. The differences in the number of respondents, according to the levels, are not considerable and make about 5%.

- *In accordance with "Life is Control Locus" subscale,* 9 (56.25%) gifted senior pupils have been diagnosed with an average "Life-Purpose Orientation" level, while 5 (31.25%) students – with a high level. As for GAS respondents, 12 (63%) students possess an average "Life-Purpose Orientation" level and 4 (21%) – a high scale level. Consequently, 87.5% of gifted respondents and 84% of senior pupils with average and low abilities are willing to manage their lives and are convinced that a person is able to control events, take decisions and make them real. A minimum of 2 (12.5%) gifted and 3 (16%) GAS respondents have a low subscale level. It can be inferred that these pupils are strongly convinced that a person's life is not subject to conscientious control and it is impossible to plan one's future because "everything has been decided for me". The differences in the number of respondents, according to the levels, are not considerable and make about 4%.

5   Conclusion

Analysis of the empirical data shows that *Assumption I that psychological culture components of rural senior pupils are sufficiently developed has been proved.* Thus, 100% of gifted rural senior pupils have profound psychological competence; possess psychological knowledge about the standards of communication, psychic activity and self-regulation, behavior, emotions, empathy, etc. They are competent in questions concerning personal psychology, psychological issues; capable of understanding mechanisms of personal behavior and development; can explain metaphors and phraseological units characterizing a person, personal behavior and relations in the society. 75% of gifted rural senior pupils are motivated to develop and improve themselves, analyze their mistakes, problems and opportunities to achieve success due to personal growth. 87.5% of gifted seniors are capable of setting future goals that can make their life meaningful, focused and promising. 75% of pupils are determined while 87% of them consider their life an exciting process full of emotions and meaning. 88% of gifted pupils are satisfied with their life and self-realization and evaluate this period as a productive and meaningful time. About 94% of respondents see themselves as strong personalities who possess sufficient freedom to build their life in accordance with their own goals, tasks and ideas. 87.5% of gifted senior pupils are ready to manage their life and are convinced that a person is able to take independent decisions and make them real.

Assumption II concerning a number of characteristics that gifted rural senior pupils and their peers with average and low abilities have differences in development of components of psychological culture has also been proved. The research has been conducted with the help of Mann-Whitney U-test. According to the characteristics of psychological competence, the differences between the total scores are considerable (when Uemp. = 82; significance point is $\wp \leq 0.01$; Ucr. (≤ 0.01) = 101; Ucr. (≤ 0.05) = 25): the level of development of psychological competence of senior pupils who are not identified as gifted ones is lower than that of their talented peers. According to "Readiness for Self-Development" Test, the analysis has detected no considerable statistical differences in characteristics of either gifted senior pupils, or students with average and low abilities. Both groups of respondents are or not equally willing to know about themselves, change and develop themselves. Referring to life-purpose orientation characteristics of "Life Purposes" subscale, the research has shown differences in respondents' development but they require further tests and analysis as Uemp. = 107 when significance point of $\wp \leq 0.06$ is not high enough. Probably, the result is such because there are more people who are determined and able to set conscious goals among the gifted rural senior pupils than among their counterparts with average and low abilities, as there prevail students who live for the day or in the past, do not have future goals, build plans that at present do not have a real support and are not supported by personal responsibility. In this respect, this fact needs a further study.

The detected characteristics of psychological culture of gifted rural senior pupils make it possible to consider them as real internal resources for social self-realization of a talented personality. Nevertheless, all senior pupils need psychological follow-up to develop value-notional component of psychological culture that can be directed at understanding importance of personal adolescent self-development in terms of age-related, individual, life and professional plans, aspirations, scripts and goals.

Further research perspectives are also connected with development of practical methods that will contribute to the formation of psychological culture of gifted children, pupils and students in modern educational space.

References

1. Bezrukova, N.P., Barkanova, O.V., Bezrukov, A.A., Selesova, E.V., Tazmina, A.V.: Vyayvlenie i razvitie potentsialno odarennykh uchashchikhsya selskikh shkol: problemy i podkhody k ikh resheniyu [Identification and development of potentially gifted country schoolchildren: problems and their solution]. Sovremennye naukoyemkie tekhnologii [Modern high technologies] **10**, 84–89 (2017). (in Russian)
2. Bogoyavlenskaya, D.B., Bogoyavlenskaya, M.E.: Psikhologiya odarennosti: ponyatie, vidy, problemy [Psychology of talent: notion, types, problems]. Issue 1. Psychological Institute of RAO, Moscow Institute of open education, Moscow (2005). (in Russian)
3. Demin, L.D., Ralnikova, I.A., Luzhbin, O.N.: Psixologicheskaya kul`tura uchitelya-professionala [On the psychological culture of the teacher-professional]. Pedagogicheskoie obrazovanie v sovremennom classiheskom Universitete Rossii, 34–65 (2003). (in Russian)
4. Guitford, J.P.: The Analysis of Intelligence. McGraw-Hill, New York (1971)
5. Ivleva, M.L., Inozemtsev, V.A.: Formirovaniye filosofskikh osnovaniy psikhologicheskoy kontseptsii odarennosti v XIX-nachale XX veka [Formation of Philosophical Grounds of Psychological Concept of Giftedness in XIX-XX century], https://cyberleninka.ru/article/n/formirovanie-filosofskih-osnovaniy-psihologicheskoy-kontseptsii-odarennosti-v-xix-nachale-xx-veka. Accessed 3 Aug 2018. (in Russian)
6. Kolmogorova, L.S.: Diagnostika psikhologicheskogo zdorovia i psikhologicheskoy kultury shkolnikov [Diagnostics of Psychological Health and Psychological Culture of Pupils]. ASPU, Barnaul (2014). (in Russian)
7. Kolmogorova, L.S., Parfenova, G.L.: Psikhologicheskiye osobennosti sotsialnoy kompetentnosti odarennykh starsheklassnikov [Psychological Peculiarities of Social Competence of Gifted Pupils]. In: Proceedings of II Conference on Psychology of Development, pp. 14–16. MGPPU, Moscow (2009). (in Russian)

8. Landau, E.: Odarennost trebuyet muzhestva [Giftedness Requires Bravery]. ACADEMIA, Moscow (2002). (in Russian)
9. Larionova, L.I.: Issledovaniye struktury intellektualnoy odarennosti [Research of the Structure of Intellectual Giftedness]. In: Proceedings of II Conference on Psychology of Development, pp. 16–18. MGPPU, Moscow (2009). (in Russian)
10. Leytes, N.S.: Vozrastnaya odarennost i individualnyye razlichiya: izbrannyye psikhologich-eskiye trudy [Age-Related Giftedness and Individual Spesifics: selecred psychological works]. MODEK/MPSI, Voronezh/ Moscow (2003). (in Russian)
11. Litvak, R.A., Bondarchyk, T.V.: Zakonomernosti sotsializatsii odarennykh detey v sovremennykh sotsiokulturnykh usloviyakh [Mechanisms of Socialization of Gifted Children in Modern Social and Cultural Conditions], https://cyberleninka.ru/article/n/zakonomernosti-sotsializatsii-odarennyh-detey-v-sovremennyh-sotsiokulturnyh-usloviyah. Accessed 16 July 2018. (in Russian)
12. Martsinkovskaya, T.D.: Istoriya psixologii [History of psychology]. Academy, Moscow (2004). (in Russian)
13. Meshkova, N.V.: Research of academic giftedness in foreign studies: socio-psychological aspect. J. Mod. Foreign Psychol. 4(1), 26–44 (2015)
14. Potapova, V.V.: Issledovaniye kulturnoy dissinkhronii psikhicheskogo razvitiya intellektu-alno odarennykh podrostkov [Research of Cultural Dyssinchrony of Psychological Development of Intellectually Gifted Teenagers], https://cyberleninka.ru/article/n/issledovanie-kulturnoy-dissinhronii-psihicheskogo-razvitiya-intellektualno-odarennyh-podrostkov. Accessed 18 July 2018. (in Russian)
15. Renzulli, J.S., Owen, S.V.: The revolving door identification model: If it ain't busted don't fix it, if you don't understand it, don't fix it. Roeper Rev. 6, 39–40 (1983)
16. Savenkov, A.I., Karpova, S.I.: Problema prognozirovaniya uchebnoy i zhiznennoy uspesh-nosti v psikhologii XX v. [The Problem of Prediction of Education and Life Success in the Psychology of the XX century], https://cyberleninka.ru/article/n/problema-prognozirovaniya-uchebnoy-i-zhiznennoy-uspeshnosti-v-psihologii-hh-v. Accessed 02 Aug 2018. (in Russian)
17. Savenkov, A.I.: Psikhologiya detskoy odarennosti [Psychology of Children's Giftedness]. Genesis, Moscow (2010). (in Russian)
18. Semikin, V.V.: Psixologicheskaya kul`tura i obrazovanie [Psychological culture and education], Izvestiya RGPU im. A.I. Herzen 3 (2002), https://cyberleninka.ru/article/n/psihologicheskaya-kultura-i-obrazovanie. Accessed 15 Aug 2018. (in Russian)
19. Shechtman, Z., Silektor, A.: Social competencies and difficulties of gifted children compared to nongifted peers. Roeper Rev. 34, 63–72 (2012)
20. Vygotsky, L.S.: Psikhologiya razvitiya kak fenomen kultury [Developmental Psychology as a Cultural Phenomenon]. NPO MODEK, Voronezh (1996). (in Russian)
21. Vysotskaya, E.L.: Osvoyeniye kultury odarennymi detmi (na primerakh poeticheskogo i nauchnogo Dara) [Mastering of culture by gifted children (on the examples of poetic and scientific Gift)], https://cyberleninka.ru/article/n/osvoenie-kultury-odarennymi-detmi-na-primerah-poeticheskogo-i-nauchnogo-dara. Accessed 16 July 2018. (in Russian)
22. Williams, F.E.: Creativity Assessment Packet (CAP). D.O.K. Publishers. Inc., Buffalo (1980)
23. Yavorsky, N.I.: Sovremennyye problemy razvitiya sistemy obrazovaniya odarennykh detey [Modern Developmental Problems of the Educational System of Gifted Children]. NGU Bull. Ser. Pedag. 10(2), 2–14 (2009). (in Russian)
24. Ziegler, A., Fidelman, M., Reutlinger, M., Vialle, W., Stoeger, H.: Implicit personality theories on the modifiability and stability of the action repertoire as a meaningful framework for individual motivation: A cross-cultural study. High Ability Stud. 21, 147–164 (2010)

Intellectics Fundamentals in a Professional Activity of a Future Bachelor

Galina I. Egorova[1]([✉]) [iD], Natalya I. Loseva[1] [iD], Olga A. Ivanova[1] [iD],
Bibinur M. Chubarova[1] [iD], and Olga M. Demidova[2] [iD]

[1] Federal State Budget Educational Institution of Higher Education, Industrial
University of Tyumen, 626150 Tobolsk, Russian Federation
egorovagi@list.ru, lni99@yandex.ru,
{olga62ivanova, bchabarova}@mail.ru
[2] Tomsk Polytechnic University, 634050 Tomsk, Russian Federation
berezovskaya-olechka@yandex.ru

Abstract. The article describes a new course on "Intellectics fundamentals in a professional activity of future Bachelors" -into the curricula of Russian Universities. The following issues are solved: the reasons for a new course introduction; its contents; possibilities and forms of its introduction into the current curricula; the place of the course in the system of the professional student training.

The purpose of the study is enhancing bachelors' intellectual culture as basis of professional university training. The study results are:

(1) a new course has been developed and tested; its main idea is developing future bachelor culture and readiness to solve creative tasks;
(2) course modules have been developed and tested;
(3) a concept "intellectics" of bachelors professional activity has been added into the methodologies of bachelors training;
(4) the necessity to include informative modules, technologies, techniques of intellectual culture development into bachelors' training has been proved;
(5) it has been proved that the system of intellectual culture development is necessary and important for immersing students into their future profession.

Research methods are theoretical, empirical, and pilot testing.

Key conclusions: the concepts "intellectics" of a professional activity, "intellectual" culture have been disclosed, their social role has been shown; the course has been designed and tested. These modules are a part of basic and variable components of a fundamental professional educational program (FPEP).

The course can be applied for professional competences development with the main factor – enhancement of bachelors' professional education within the university and new employers' requirements.

Keywords: "intellectics" · Intellectual culture ·
Intellectual culture development · Course

Z. Anikina (Ed.): GGSSH 2019, AISC 907, pp. 45–52, 2019.
https://doi.org/10.1007/978-3-030-11473-2_5

1 Introduction

There is no doubt that Russia's development and stability depends completely on intellectual specialist's training, which will sustain the knowledge-driven industry in the XXI century. A strong state should aim its intellectual potential both at the development of a high-tech manufacturing and at the development of a harmonious, highly intellectual bachelors and specialists' personality [1].

Russian reforms show the importance and role of intellectual innovations, new technologies in industry. Their number has increased 20 times recently. Literature analysis on intellectual innovations shows their necessity and importance [2]. However, despite high requirements of the society to highly intellectual bachelor's and specialist's personality, current training in tertiary education system is not sufficient. It should be noted that training of an intellectual specialist is one of the priorities for the next 10–15 years. It is evident that its solution is impossible without significant focus from the society and the state, as well as corresponding state policy in the sphere of education and science, the use of the results of intellectual potential and incorporating these results into research and technology [6].

The issue arises under such conditions to support the intellectual development of a future bachelor and a specialist not only within the educational institution but on the state scale. On the nation wide scale there is a necessity to create the infrastructure which includes such components as: intellectual culture and property institute; institutes of advanced training on intellectual culture; regional centers for the development, regulating intellectual culture of the scientific and cultural sphere; development of inter discipline courses, integrated courses; developing teaching materials on the development of students intellectual culture; developing different courses for universities, schools; issue of teaching, scientific and methodological literature on development and regulating intellectual culture, intellectual property; utilization of different innovative technologies [7, 8].

The latter is of special importance as it has been already widely used in a comprehensive secondary school, but the university introduces new teaching technologies ineffectively. The problem of such inefficiency is connected with limited budget and non-budget financing, as well as low intellectual culture of employees [8].

2 Methods and Materials

The problem of the intellectual culture development requires long-term solution. However it is evident that if the students acquire intellectual culture that will solve many problems in social, economical and technical aspects [9].

Every future specialist studying in the university should have necessary knowledge and practical skills, abilities to improve his/her own intellectual culture, be able to work with the objects of the intellectual property, know ways of their effective functioning. Our pilot testing, university students questioning analysis allow to conclude that most of the university graduates do not have any idea about these categories and their functioning. Nevertheless, it is not their fault because they were not taught to know it [10].

Some basic knowledge about intellectual culture the students obtain in such courses as "Intellectual property fundamentals" and other similar in context, studied in some technical and humanitarian universities. We think that a course "Intellectics fundamentals in a professional activity of a future bachelor" should be gradually introduced on a large scale into the training courses; as future graduates will work in the sphere of intellectual work in the market economy. That is why every university graduate should have knowledge of intellectual culture and ways of its development and management. All these are important for the future specialist.

A person possessing rich intellectual culture estimates and protects the results of his/her creativity and respects others' intellectual achievements [11]. These theoretical positions and formulated conceptual approaches lie in the bases of university curriculum, which is the framework for this course. We deliberately emphasize the level of the university. According to existing regulatory documents a faculty member should go to some kind of advanced training in a different university once in five years, often it is the Institute of Further Training, Advanced Training Courses and so on. Studying theoretical and scientific material a faculty member could study issues of intellectics as well. The theoretical training obtained will allow utilizing main forms and methods of students' intellectual culture development.

The forms of students' intellectual culture development include: at the Department level—a theoretical seminar, at the Faculty level—Research and Practice Seminar, at the University level—the University Council, a School of A Young Lecturer and a Seminar for Post-graduate students.

The principally common content of the course at all levels is that the issues of students' intellectual culture development are studied both in theoretical and practical aspects. At the Department level the issues of intellectual culture are studied according to the traditions and experience of the Department. It takes into account scientific interests of the faculty. Main attention is given to the Faculty level of program utilization, where a necessary intellectual-reflexive environment is created, which forms intellectual culture. Forms and content of the work at the Faculty level do not duplicate those at the Department level, but use, systemize and summarize the work of Departments and can be recommended to others. In terms of organization this type of activity is relevant as it provides greater independence to the departments in developing curricula and syllabi, designing research and teaching materials for the students and the faculty.

At the University level the general atmosphere is determined by Departments and Laboratories of the university, as they have the professionals who are able to design, organize and implement scientific ideas. A School of a Young Lecturer, a Seminar for post-graduate students can be used in studying issues of the intellectual culture, according to the special syllabi, suggesting different aspects of reflection (intellectual, personal, communicative and so on). Variants of enhancing the level of intellectual culture include: creative cooperative forms of experience exchange based on common professional interests, intellectual personal education; training sections; personal consultations. Inter-university teaching methods centers, interdisciplinary integrated syllabi and courses as well as Research and Education centers can be used in the work.

Now we would like to consider the contents of the course. It is important to distinguish the concepts of "intellectual culture" and "intellectual property", "intellectual creativity".

Intellectual culture as an integral factor is composed of interaction and integrity of all its components (cognitive, active, and motivational). Possessing intellectual culture, a bachelor, a specialist enters a creative level which leads to the certain final product. The exclusive use of the result of intellectual creative activity in the form of an object reproduction determines intellectual property.

We would like to mention the status of the course "Intellectics fundamentals" in the system of specialists' professional training. This course is, first, complex, combining information from different spheres: history (origin, civilization intellectual culture development. Epochs of intellectual ideas development, outstanding people); pedagogy, psychology (intellectual characteristics and ways of their development, components of intellectual communication, intellectual aspects of dispute settlement); logics (some aspects of scientific cognition, principles of reasoning, main argumentation aspects and argument settlement); law (legal arrangement and copyright protection); economics (calculation of profits, fines for violation of the exclusive rights, royalties for creation and utilization of the intellectual objects), specialized areas of intellectual activities, allowing to estimate the value of the intellectual object, its novelty and significance. Second, the course is an application area of humanitarian and technical knowledge. Third, the course is one of special scientific programs, closely related to other applied sciences. This program is based on such sciences as: history, economics, logics, sociology, philosophy and others. It takes into account their absolute cultural and upbringing importance for students.

This course will allow to implement main ideas of the project "State Support concept for talented scientific youth and human resource development for Russian Science", which proves a necessity to "broaden teaching of the scientific disciplines in universities to develop young specialists knowledge about science, its organization in the modern society and abilities to work with scientific organizations and scientists" [3, 4]. We think that this course is useful for students, post-graduate students of humanitarian and scientific majors. This course will provide a wide range of intellectual training and in particular, possibility to study disciplines professionally important for them. Let's consider the contents of the course "Intellectics fundamentals in a professional activity of a future bachelor".

The main aim of the course is future bachelor's intellectual culture development, intellectual preparedness to solve creative tasks. Main course modules consider such concepts as "intellectual culture", "intellectual activity and its structure", "intellectual technologies", "intellectual property", "protection and regulation documents". The review of typical explanations and authors' approaches to different categories is given. The students learn different intellectual technologies, peculiarities of the organizational structure, contents, technology of the intellectual culture development.

The course includes lectures, practical classes and independent work. Lectures deal with methodological and scientific bases of the course, most common theoretical issues, providing understanding to the key issues of the intellectics. Practical classes deal with theoretical discussions, helping to understand leading ideas, conceptual approaches, that will widen scientific-technical scope, develop intellectual abilities.

The main idea of the practical classes is developing personal intellectual style of the bachelor student, which helps to join the work teams of different types.

Practice helps to develop intellectual readiness to the work through practicing and intellectual training [4, 5]. Independent work means development of a personal creative product, developing an author's concept of the personal intellectual culture.

Attending lectures, practical classes a future bachelor obtains core knowledge, abilities on patenting, measures to protect intellectual property.

In the end of the course students prepare creative work in the field of intellectualization of bachelor's activity. Methodological analysis; innovations as an intellectual product in the society and their significance; history of science, engineering, culture via integration; computerization as a way of society socialization; ways and forms of intellectual systems development; bachelor's intellectual culture. Main aspect of development; intellectual systems in society and regional development; intellectual property and its kinds: authorial and allied rights, industrial property; intellectual property economics. Economic benefits; intellectual property legislation in the Russian Federation. Patent system in the Russian Federation.

Creative work is based on historical analysis of the studies and analysis of the current research in this field.

The course includes several modules. Module "Social and scientific background of engineering intellectics origin" includes the following: intellectuality theory in the society, economics, engineering; society intellectuality as an imperative of human survival in the XXI century; "Intellectics" as a part of philosophical knowledge; engineering intellectics origin, history of origin; main parts of intellectics (gnosiology, heuristics, psychology, science methodology, semiotics, system theory, logics, economics); functional value and principles of construction; intellectics interaction with other sciences; engineering education principles.

Module "Basic concepts of engineering intellectics" deals with further issues: structural-functional properties of the intelligence; interdisciplinary conceptual – terminological field in intelligence characteristics; specific variety and different authors approaches to the concept "mechanical intelligence"; leading approaches and methods; personal intellectual universals; intellectual competence, intellectual style; intellectual systems.

Module "Intellectual culture in the intellectics system" deals with the concepts of "culture", "intellectual culture", "intellectual freedom"; structural components, stages, existing approaches, main purpose of the intellectual culture; ways and means of intellectual culture development; intellectual culture and engineer's creativity.

Module "Intellectual activity" concerns system characteristics and peculiarities of intellectual activity in the cognitive activity process; intellectual activity structure: information perception and understanding; information revision and understanding; accumulating, systemization, information analysis, finding regularities, common tendencies of awareness and thinking; practical level of intellectual activity (approbation, experiment, receiving of new information); acts of thinking (analysis, synthesis, induction, deduction, specification, abstracting, insight); behavior rules in intellectual collective activity; activity levels; reflexivity, creativity in intellectual activity structure; intellectual barriers and overcoming them; means of intellectual support.

Module "Technologies of intellectual culture development" deals with theoretical and practical issues of modern technologies: information technologies (the Internet, fax, case-technologies); interactive technologies (sparring partnership, business game, projects, etc.); technical thinking development technologies (TRIZ technologies).

Module "Intellectual systems" includes the following information: intellectual systems as a new organization form of engineering and people, entering anthropo-sphere epoch; intellectual systems within a synergetic approach (self-organization, self-programming, self-regulation); level analysis (concepts level, activity level, manage-ment level); intellectual systems in the structure of engineering knowledge and industrial externship.

Module "Intellectual property and legislation" deals with the following: authorial and allied rights; industrial property; copy-right and copy-right related materials, industrial property; intellectual property protection; patent and other copyright pro-tection; origin and development of the authorial and patent rules abroad; international agreements and international organizations in the sphere of intellectual property; pro-tection documents classification; International Patent Classification (IPC), its structure and composition; National Patent Classifications, their relation to IPC; intellectual product protection formatting; intellectual property law violation.

Module "Legislative aspects of intellectual property protection" includes the fol-lowing: patent system of the Russian Federation; patent services of enterprises; institute of patent agents; document package of a claim for an invention; a utility model; design invention; the trade mark; patenting of Russian inventions; international claim; con-ventional priority.

3 Results

The course is designed for 36 academic hours. It can be easily introduced into master students training, post-graduates students training. Every university and department determines the status of the new discipline for itself: compulsory or optional, obligatory or elective. The problem is in providing qualitative teaching staff. Are there enough specialists for these purposes? That is why the course should be introduced gradually, as it was done in our case. At the first stage psychological specialists, pedagogues delivered psycho-pedagogical aspects through humanitarian disciplines. At the second stage legal aspects of intellectual property were considered, where practical peculiar-ities of intellectual property management were studied. Special disciplines were taught at the third level. At this stage special discipline lecturers were involved. Full imple-mentation of the program required special training of the faculty at the career devel-opment courses in leading universities of the country in: (a) patent activity; (b) innovative technologies; (c) intellectual systems.

The theoretical part of the program incorporates practical training in terms of the work with intellectual culture objects. The practical training develops following skills: to evaluate intellectual property objects; to estimate the level of intellectual culture development; be able to solve intellectual problems; to use patent documentation of the Russian Federation and other countries; to register intellectual property objects [11].

4 Conclusion

Pilot testing of intellectual development, based on this program was done in face-to-face and distant training. Program modules were incorporated into study and scientific training of students. Peculiarities of the program implementation have been discussed during several years in seminars and Department of Chemistry and Chemical Technology meetings. In many aspects we reached our aim that resulted in significant improvement of bachelors training in major 18.03.01 "Chemical technology". Further students' achievements during last 5 years illustrate its efficiency. Students' qualitative performance increased by 25%; increase in number of students involved in research; increase in number of students publications; variety of diploma and course papers themes, recommended for implementation in industry; increase in the number of Diplomas with Honors; increase in number of graduates recommended for doing a post-graduate course – all these are results of intellectual collaboration of the teaching staff and students.

Nowadays our graduates are involved in the research of the Department. This year students, participated in the international contest, were given top places; students of dual training programs completed course papers with excellent marks on the themes, required by Tobolsk industrial site. Students' research reports are interesting as well as their activity in grants of different levels.

References

1. Vedeneev, A.: Modelirovanie elementov myshleniya [Modeling elements of thinking]. Nauka, Moscow (1988). (In Russian)
2. Vekker, L.: Psyhologicheskie process [Psychological processes], vol. 2. Myshlenie i Intellect, Leningrad (1976). (In Russian)
3. Egorova, G., Loseva, N., Belyak, E.: Otsenka sformirovannosti professional'nyh kompetentsiy bakalavrov tehnicheskih vuzov. Problemy pedagogicheskoi innovatiki v professional'nom obrazovanii [Estimating the level of bachelors' professional competences development]. In: Proceedings of 18th International Research and Science conference, pp. 43–45. Russian State Pedagogical University, Saint- Petersburg (2017). (In Russian)
4. Egorova, G., Egorov, A., Loseva, N., Belyak, E., Demidova, O.: Siberian arts and crafts as basis for development of cultural traditions and innovations of bachelors. Adv. Intell. Syst. Comput. **677**, 93–100 (2018)
5. Efimov, Yu.: Intellektualizatsiya VUZa – bazovyi istochnik rosta intellektualnogo potentsiala regiona [University intellectualization as a basic source of intellectual potential increase in regions]. In: Proceedings of the international scientific symposium, pp. 410–414. SSEA, Samara (2003). (In Russian)
6. Karpova, YuA: Innovatsii, intelekt, obrazovaniye [Innovations, intellection, education]. MSFU, Moscow (1998). (In Russian)
7. Razumov, V.I.: Kategorial'no-sistemnaya metodologiya i formirovanie novih strategii myshleniya [Category-system methodology and formation of new thought strategies]. In: Proceedings of the international scientific symposium "Chelovek: fenomen subjektivnosti". P.I, P. 78–83. Omsk, OmSU, (2014). (In Russian)

8. Ladenko, I.: Razvitie intellectualnyh innovatsiy v sovremennom obshestve. Kompleksnaya programma issledovaniy [Intellectual innovations development in the modern society. Complex research program]. SBRAS, Novosibirsk (1990)
9. Manuilov, V., Fedorov, I.: Modeli formirovaniya gotovnost k innovatsionnoi deyatelnosti [Formation models of preparedness to innovation activity]. High. Educ. Russ. **7**, 56–70 (1987). (In Russian)
10. Finn, V.: Intellektualnye sistemy i obshestvo [Intellectual systems and the society]. In: RSHU Collection of works, pp. 309–312. RSHU, Moscow (2000)
11. Fokin, Yu.: Prepodavanie i vospitanie v vysshey shkole: metodologiya, tseli, soderzhanie, tvorchestvo [Teaching and upbringing in a University: methodology, aims, contents, creativity]. Academy, Moscow (2002). (In Russian)

Nomadic School: Problem of Access to Quality Education

Evdokia Nikiforova⑩, Viktor Nogovitsyn⑩, Lena Borisova⑩,
and Anatoliy Nikolaev⁽✉⁾⑩

North-Eastern Federal University, Yakutsk 677000, Russian Federation
nevdokia@bk.ru, v_nogovitsyn@mail.com, {monep,
nickan07}@mail.ru

Abstract. The article is devoted to problems of accessibility to quality education in small nomadic schools. In this paper the problems of nomadic schools are considered from the point of view of current modernization processes, taking place in the Russian Federation in the field of general secondary and higher professional education. Herein we consider the problems of small schools, including nomadic, which teach children of small peoples of the North. The development of nomadic schools, in our opinion, should be considered comprehensively, not as a separate segment, but in the context of improving the quality of life and well-being of the Arctic territories inhabitants who live in severe climatic conditions.

In the Republic of Sakha (Yakutia), there are 11 nomadic schools and kindergartens in places of compact residence of small peoples of the Russian North. They are created to preserve the languages, original culture and traditions of small peoples. In the near future, two more schools may appear, this issue is being worked out in the Ministry of Education and Science of the Sakha Republic (Yakutia) and two municipalities.

Keywords: Small educational organizations · Indigenous peoples
Nomadic school · Places of compact residence of small peoples
Regional and ethno-cultural features · Minority languages

1 Introduction

In recent years, interest in the Arctic has increased. The Arctic is a geopolitical region. The Russian President Vladimir Putin said at the IV International Arctic forum "The Arctic – Territory of Dialogue" the following: "Our state's goal is to provide a sustainable development of the Arctic, create a modern infrastructure, develop resources, develop the industrial base, improving the life quality of the Russian North indigenous peoples, preserving their unique cultures, their traditions" [6, p. 1]. It follows that the education system has an important role to play in solving these problems.

Functioning of nomadic educational organizations is regulated by the Federal law "On Education in the Russian Federation" and the laws "On Education in the Sakha Republic (Yakutia)", "On Nomadic Schools of the Sakha Republic (Yakutia)". At present, the Federal State Educational Standards are being implemented in nomadic

© Springer Nature Switzerland AG 2019
Z. Anikina (Ed.): GGSSH 2019, AISC 907, pp. 53–58, 2019.
https://doi.org/10.1007/978-3-030-11473-2_6

schools, as we as in all educational institutions of the Republic. The pedagogical activity in nomadic schools should also meet the requirements of the FSES for general education: results of development, structure and conditions of basic educational programs implementation. In this regard, a special attention should be paid to the study of native languages.

According to the 2010 census, more than 40 thousand representatives of the Russian North small peoples live in Yakutia, including the Evenks - 21008, the Evens – 15071, the Dolgans - 1906, the Yukaghirs – 1281 and the Chukchi 670 people. Meanwhile, the birth rate is higher in the Arctic regions of Yakutia than the average in the country. According to the law of the Republic of Sakha (Yakutia) "On the status of the North indigenous peoples' languages in the Sakha Republic (Yakutia)", the languages of the North peoples are official in places of these peoples compact residence. The official languages are used for record-keeping, court proceedings, training and organization of educational process. Education in the languages of indigenous minorities is provided in 40 educational institutions of the Republic, including small nomadic schools.

2   Nomadic School Development

Analysis of the main regulatory legal acts of the Russian Federation and the Republic of Sakha (Yakutia) in the field of small schools development, including the nomadic, showed that this problem has been taken into account in the legislative aspect in recent years. For example, the law "On Education in the Russian Federation" it is noted in article 3 that "the freedom to choose education has to be according to the inclinations and needs of the person, creation of conditions for self-realization of every person, free development of their abilities, including the right to choose the form of training, form of education, the organization that carries out educational activities, the direction of education within the limits provided by the education system…" [2, p.7]. Provisions on small schools, including nomads, were introduced for the first time in the law "On Education in the Sakha Republic (Yakutia)". Meanwhile, as practice shows, due to various objective reasons, there are difficulties in implementation of these laws.

The Federal State Educational Standard for secondary (complete) general education also attempts to take into account the regional and ethno-cultural needs of the peoples of the Russian Federation. In the Sakha Republic (Yakutia) 1 nomadic school for Dolgan, 1 for Chukchi, 4 educational organizations are for Evenk, and 5 for Even. Thus, in places of the North small peoples' residence, training is conducted in 4 official languages, taking into account regional, natural, climatic and geographical features of the Republic. Organization of such schools needs the socio-economic and cultural characteristics of the territories of traditional nature management and economic activity. Organization of nomadic schools is aimed at strengthening the unique type of family, preserving and restoring the traditional economy and way of life of the indigenous peoples of the Arctic and allows children to preserve their identity, mother tongue, and join the original culture of their native people.

In 2016, the State Assembly (in Tumen) of the Republic of Sakha (Yakutia) adopted the law "On Nomadic Family", which defines a system of economic, social and legal guarantees, aimed at improving the quality of life for nomadic families. According to the law, the main functions of a nomadic family are the upbringing and handing on the family values of indigenous peoples of the North from generation to generation who lead a nomadic lifestyle, based on the historical experience of their ancestors in the field of nature management, original culture, language and traditional knowledge [4].

Problems in the study of minority languages and literature of the North indigenous peoples, dedicated to the research of such scientists of Yakutia, as: Belolyubskaya, Lekhanova, Neustroev, Robbek, Sharina Vinokurova, Zhirkova and others [1, 4, 6, 8–11].

The Ministry of Education and Science of the Sakha Republic (Yakutia), Institute of National Schools of the Republic have developed approximate regulations on nomadic educational organization, basic educational programs and schedules of educational activities. More than 100 teaching materials, fiction books that have been published in the North small peoples' languages in recent years. Funding of teaching materials is under the state program of the Republic education development. In conditions of a nomadic life it is very important that textbooks and teaching materials have an electronic format. A distance Education Centre has been set up in the Republic in order to form an e-learning infrastructure which provides distance education for students of small schools and children with disabilities in unavailable basic subjects. Unfortunately, due to the technical reasons these services are not fully available to nomadic schools.

The Federal State Educational Standard for General Education sets the specific requirements to conditions of educational process, to the materials and technical base. In this regard, the Republic is working on equipping the nomadic educational organizations with "mobile schools". A mobile school is a classroom with converting desks, interactive equipment and toilet area. Learning in comfortable conditions will certainly help to improve the education quality for nomadic school students.

There are the following types of nomadic educational organizations in Yakutia: hub secondary (basic) and underfilled general school (nomadic); nomadic school as a branch of the secondary school; a primary nomadic nursery school; network nomadic nursery school and seasonal (summer) nomadic school-camp.

The main forms of educational process organization are full-time sessions of students with teachers in reindeer herding teams when they are visited by teachers from the hub schools, usually from educational institutions located in district centers. Another form of education involves organization of face-to-face sessions in hub schools. Such forms as education of children in the nomad families by their parents are also practiced; they are also consultants and tutors. In connection with the information and communication technologies development and expansion of the Internet, distance learning can also be organized in such schools.

In the Sakha Republic (Yakutia), 71% of educational institutions are situated in rural areas. The distance between schools in the Northern regions is 300–400 km. There is usually a shortage of teachers in small schools. For example, in the academic year 2015–2016, the Ministry of Education and Science of the Sakha Republic (Yakutia) opened 58 vacancies for teachers of the Russian language and literature, in 2016–2017 – 55, and in 2017–2018 – 59. This means that in these educational institutions the Russian language and literature are taught by teachers of other subjects, which naturally reduces the quality of education.

This academic year, 26 bachelors in the field of "Pedagogical education" have graduated the faculty of Philology of the NEFU (North-Eastern Federal University). 70% of them are employed in educational institutions of the Republic, including the Arctic regions, the rest of them continue their education in the master's degree. Training of native language specialists is conducted at the Institute of Languages and Cultures of the North-East Peoples at the Department of Northern Philology, and tutors-bachelors of the basic small and nomadic school - at the Pedagogical Institute of the NEFU, as well as in previous years, the teacher training for nomadic schools at extramural department was conducted at the Yakut Pedagogical College named after S. F. Gogolev.

A Special Dissertation Board for doctoral and candidate theses on the following specialties: 13.00.01 – General Pedagogics, History of Pedagogy and Education; 13.00.02 – Methods for Teaching Native Languages (Barakhsanova, Chairwoman of the Board, Doctor of Pedagogical Sciences, Professor; and Nikiforova, Vice-Chairwoman, Doctor of Pedagogical Sciences, Professor; Nikolaeva, Doctor of Pedagogical Sciences, Professor) has been opened at the North-Eastern Federal University named after M.K. Ammosov [3]. This Board opening will provide opportunities for training highly qualified scientific personnel in the field of general pedagogy and methods for native languages teaching, including the small peoples of the North.

In order to attract teachers to educational institutions that are located in remote Northern regions, the government of the Sakha Republic (Yakutia) has been implementing the program "Provision of housing for teachers and medical workers of rural institutions of the Arctic and Northern uluses (regions)" for thirteen years. This program provides housing for the teachers who passed the competition and entered into a tripartite agreement with the Ministry of Education and Science of the Sakha Republic (Yakutia), and who have worked for 5 years in rural educational institutions located in the Arctic and Northern regions of the Republic. The State provides social payments in the amount of 50% of the cost of one-room comfortable apartment at any desirable place of permanent residence for the teachers who have fulfilled the terms of the contract. Over the past period, 590 people participated in the program, of which about 30% successfully continue to work in the Northern uluses (regions).

This program contributes to the provision of teaching staff in the Northern and Arctic regions of the Republic. Despite the measures taken to support nomadic schools by the Executive authorities of the region, scientific and public organizations, a lot of efforts are still required in terms of content and organization. First of all, it concerns the organization of native languages training; they are Evenk, Dolgan, Chukchi and Yukaghir languages. It is necessary to expand the range of native languages through the development of non-formal education.

It is necessary to continue the development of educational and methodical complexes, publication of textbooks translated into native languages that meet the requirements of the Federal State Educational Standard and multilingual education. An important condition for improving the quality of education is provision of resources, which, as we know, involves economic issues, providing teaching staff, textbooks and teaching materials, creating a safe and modern communication environment, etc.

Taking into account the complexity of the work, the salaries of teachers in nomadic schools should be increased and paid according to the branch system of remuneration, regardless of the number of students.

There is a certain experience in supporting nomadic schools in the Republic. A project "Teachers of the Arctic" was developed together with experts from UNESCO and the Institute for Education Development and Training of Education workers of the Ministry of Education and Science of the Sakha Republic (Yakutia). Within the framework of this project, an information and educational portal is functioning, the materials of which help teachers of the Arctic to improve professional competence, information culture and disseminate best practices.

Thus, the results of our research can be used in educational activities of nomadic schools, in small educational institutions, in development of educational programs, textbooks and teaching materials of a new generation, taking into account the regional and ethno-cultural features of students in places of compact residence of the North small peoples and the Arctic, in training courses and retraining of teachers and tutors for nomadic schools. This model of education can be applied in the practice of teaching other peoples who lead a nomadic lifestyle, as well as it can be implemented in training of children, travelling with their parents in long work and business trips and where access to educational organizations is impeded.

3 Conclusion

Summing up the results, we can suggest the following effective development ways for nomadic schools. For development of this unique model of education it is necessary: to develop the infrastructure, material and technical base of nomadic schools; to improve educational programs in accordance with the new requirements of the time, to introduce innovative information and communication technologies; to prepare teams of authors among scientists, teachers, practitioners in development of a new generation textbooks on native languages; to carry out purposeful work on socialization of nomadic children; to consolidate the intellectual potential of teachers of the Arctic regions of the Russian Federation and foreign colleagues in order to support and develop small educational organizations; to determine the coefficients to the standards of payment for teachers who teach in the languages of small peoples; to develop measures of social support for teachers who work in nomadic educational institutions.

References

1. Belolyubskaya, V.G.: Yazyki korennyh narodov severa: istoriya i sovremennost [The Languages of the indigenous peoples of the North: history and contemporaneity]. Higher education in Russia 5, Moscow, 110–115 (2014). (in Russian)
2. Federalnyj zakon "Zakon ob obrazovanii v Rossijskoj Federacii [Federal Law on education in the Russian Federation]. № 273-FZ dated 29 December 2012 (as amended in 2017). http://www.zakon-ob-obrazovanii.ru. Accessed 10 Nov 2018. (in Russian)

3. Dissertacionnyj Sovet [Dissertation Council] Homepage. North-Eastern Federal University, Yakutsk, Russia. https://www.s-vfu.ru/universitet/rukovodstvo-i-struktura/strukturnye-podrazdeleniya/unir/otdel-dissertatsionnykh-sovetov/dissertatsionnye-sovety/pn/sostav/. Accessed 10 Oct 2018. (in Russian)
4. Zakon Respubliki Saha (Yakutiya) "O kochevoi semje" [Law of the Sakha Republic (Yakutia) "On nomadic family"]. № 1660-Z №963-V dated 15 June 2016 (as amended in 2017). http://www.docs.cntd.ru/document/439090024. Accessed 10 Nov 2018. (in Russian)
5. Lekhanova, F.M.: Polozhenie yazykov korennyh malochislennyh narodov Severa, Sibiri i Dalnego Vostoka v Rossijskoj Federacii [The provision of the indigenous peoples of the North, Siberia and the Far East of the Russian Federation] (2018). http://www.ifapcom.ru/files/publications/lehanova.pdf. Accessed 10 Nov 2018. (in Russian)
6. Neustroev, N.D.: Specifika podgotovki tyutora-pedagoga dlya malokomplektnoj i kochevoj shkoly Severo-Vostoka Rossii v vuze [The specificity of the tutor-teacher training for small and nomadic school at the University of the North-East of Russia]. Research and innovation center, Krasnoyarsk (2015). (in Russian)
7. Putin, V.V.: Vystuplenie na IV Mezhdunarodnom arkticheskom forume "Arktika territoriya dialoga" [Speech at the IV international Arctic forum "The Arctic-the territory of a dialogue"] 30 March, Moscow [Online] (2017). http://kremlin.ru/events/president/news/54149. Accessed 10 Nov 2018. (in Russian)
8. Robbek, V.A.: Koncepciya sistemy kochevyh obrazovatelnyh uchrezhdenij Respubliki Saha (Yakutiya) [The concept of the nomadic educational institutions system in the Sakha Republic (Yakutia)]. The Institute for Humanities Research and Indigenous Studies of the North (IHRISN), Yakutsk (2004). (in Russian)
9. Sharina, S.I.: Osobennosti yazykovoj situacii u malochislennyh narodov Severa Respubliki Saha (Yakutiya) [Features of the language situation in the small peoples of the North in the Sakha Republic (Yakutia)]. Eur. Soc. Sci. J. International research Institute. 6 (22), Moscow (2012). (in Russian)
10. Vinokurova, E.I., Nikitina, R.S.: Socializaciya i vospitanie detej v tradicionnom uklade zhizni korennyh narodov Severa Respubliki Saha (Yakutiya) [Socialization and education of children in the traditional lifestyle of indigenous peoples of the North in the Sakha Republic (Yakutia)]. Eur. Soc. Sci. J. 2(18), 262–269 (2012). (in Russian)
11. Zhirkova, Z.S.: Vliyanie pedagogicheskogo potenciala ehtnokulturnyh tradicij kochevyh narodov na formirovanie lichnosti uchashchihsya [Influence of the pedagogical potential of nomadic peoples ethno-cultural traditions to the formation of students' personality]. Historical and Social Educational Ideas, 257–260 (2015). (in Russian)

Peculiarities of Leadership Potential Development in Educational Environment

Elena M. Pokrovskaya, Margarita Yu. Raitina[(⊠)], and Viktoria D. Chernetsova

Tomsk State University of Control, Systems and Radioelectronics, Tomsk 634051, Russian Federation
elena.m.pokrovskaia@tusur.ru, raitina@mail.ru,
greta@mail2000.ru

Abstract. The problem of leadership is actualized everywhere, the society needs competent and active specialists who are able to take decisions independently, ready to assume responsibility for their implementation. Specialists can clearly define the goals of their activity, predict the options for achieving it, analyze the course and results, overcome difficulties, and also know how to properly build relationships with other people, work in a team, that is, show leadership qualities. The authors consider the main concepts and approaches to leadership. The methodological basis for this is the interdisciplinary and comprehensive approaches, practical usefulness principles focusing on students' personal qualities. The article considers the influence of personal reference points on development of leadership qualities in the educational environment on the example of Tomsk State University of Control Systems and Radioelectronics (TUSUR). Practical significance of the work lies in the fact that the most important role in the stage of formation and development of youth as groups is played by the leaders - students who ensure the success of group activity by their behavior, mobilizing influence on others, initiative, efforts to achieve educational and educational tasks. The authors conclude that extracurricular activities of universities are the most important components of training specialists. It allows students to form a conscious civic position, the desire to preserve and multiply moral, cultural and universal values, develop skills of constructive behavior.

Keywords: Leadership · Education · Motivation · Leadership potential · Personal qualities

1 Introduction

In the conditions of social processes that are characteristic of the world, the issues of scientific management of society and its various components, and hence issues of leadership, are becoming increasingly important.

In modern society at the beginning of the 21st century, radical changes took place in all spheres of life, transformation in the social, economic, political and social spheres, reassessment of the values of youth, change in the content of social being and consciousness. The problem of leadership is actualized everywhere - in large and small

© Springer Nature Switzerland AG 2019
Z. Anikina (Ed.): GGSSH 2019, AISC 907, pp. 59–64, 2019.
https://doi.org/10.1007/978-3-030-11473-2_7

organizations, in business and in religion, in trade unions and charitable organizations, in companies and universities. The conditions for the existence of leadership is the existence of people groups. The society needs competent and active specialists who are able to take decisions independently, who are ready to assume responsibility for their implementation, who can clearly define the goals of their activity, predict the options for achieving it, analyze the course and results, overcome difficulties, and also know how to properly build relationships with other people, work in a team, that is, show leadership qualities, which determined the relevance of the study.

Leadership as an aggregate of certain mental qualities was described by the American sociologist Bogardus [2]; analysis of studies of the features of leaders, describing the personality of the leader and its characteristics, are contained in the works of Stogdill [13], Jennings [9]. These concepts can find application in management.

The second group of theories is related to the characterization of the leader's behavior and the leadership functions. Among the concepts of this group are behavioral, motivational theories, theories of exchange and transactional analysis: Levin [7], White [15], McGregor [8], etc. Leadership in the situation of changes becomes the main object of research psychologists, on whom the works of Dewey [3], and Cooley [6] exerted a significant influence. In the development of the value approach to leadership, a significant contribution was made by Kuchmarski [4], Feyrholm [9] and others.

Modern youth has a huge range of opportunities for social development. It is difficult to overestimate the importance of the right choice of young people; their role in the development of the country, so today the attention of the public and state authorities is turned to the problem of training qualified personnel for work in various fields. Proceeding from the foregoing, the state sets itself the task of training leaders, and therefore the disclosure of the meaning of the leadership phenomenon in the educational system is an important stage in identifying one of the main problems of modern social practice.

2 Theory and Methods

Numerous studies show that the skills and abilities of leaders are estimated by people and are usually significantly higher than the corresponding characteristics of other members of the group. The leader usually has certain psychological qualities, such as: self-confidence, a flexible and sharp mind, strong will, competence and thorough knowledge of his work, ability to understand the characteristics of human psychology, and organizational abilities [14].

However, the analysis of some real groups has shown that a person who does not possess the above qualities can become a leader, and vice versa a person can possess these qualities and not be a leader. There was a situational theory of leadership, according to which the leader becomes the person who, when a group has a situation, has the qualities, properties, abilities, experience necessary to optimally resolve this situation for a given group. In different situations, the group nominates different people as leaders [5].

A leader is an authoritative member of an organization, social group, and society as a whole, whose personal influence allows him to play an essential role in social processes.

The leader acts as the center of the concentration of the informal group's interests. Therefore, in the literature it is noted that leadership is a phenomenon entirely intra-group. The leader is considered to be the person who has the greatest authority and informal recognition in a small group. The leader is not appointed; he is nominated himself due to his personal qualities.

The term "leader", as evidenced by the Oxford Dictionary, appeared around 1300. However, other specialists, in particular Stogdill, believe that this happened not earlier than 1800. One of the first definitions of leadership was noted by Cooley (1902) that the leader is the focus of group processes [6].

According to Hellman [6], at the moment there are new signs of leadership. In the past, being a leader meant being ahead. Take first place in the rankings. Occupy the highest post. To score the greatest number of goals or to get the most votes. But over time, the concept of leadership is changing. We live in a new world and must view leadership from a new angle. Here are a few signs of a "new leadership":

1. Leadership does not exist in isolation from people. Leadership implies not only the management of systems, sales volumes, equipment and indicators. It is necessary to change the benchmark and focus on other issues. "Could I realize myself? Could I make people happier and stronger? Has my quality of life improved because of me?" A new criterion is also needed to measure success in achieving the goals. Reasons like profit or gross domestic product are no longer sufficient [6].
2. To be a leader means to control yourself. Set a goal, reach it - and you will immediately become a leader for others. Live your own life. Do not be equal to others. Live according to your own plan.
3. Leadership is inextricably linked with inner motivation. For many of us, the leader is associated with a stern army officer, shouting in a thunderous voice: "You will have to get used to it, otherwise go out!" Many owners of companies treat their employees in the same way. This is understandable, since the structure of companies was formed on the basis of the military model. It still operates in some countries and areas of industry.

Thus, this form of leadership ceases to be relevant. The command-and-control approach will be replaced by an approach that teaches and enriches. Sooner or later people begin to resist and resist external control. When this happens, job satisfaction is lost and productivity falls.

Today, the leader is called a member of the group, for which it recognizes the right to make decisions in important situations. The leader plays a central role in the organization of joint activities and regulates interpersonal relations. Leadership is an element of informal relations [9].

In the development of the value approach to leadership, a significant contribution was made by Kuchmarski [4], Feyrholm [10] and others. Kuchmarsky developed a value concept of leadership, according to which the leadership process is associated with the possibility not only for a certain person, but for all members of the group to exercise their leadership abilities [4].

Feyerholm proposed a new paradigm of leadership values, the main one on the choice of actions that determine the modeling and development of values that unite people who create a culture of trust and high motivation and morality. The scientist formulated the value orientations and principles that are basic for sociology of leadership, showed that the appointment of a leader is in the internationalization of value-loaded principles that contribute to solving common problems for the organization, and also analyzed leadership as a process reflected in the leader-follower relationships, followers, strengthening their confidence in leaders [10].

In the studies of Panarin, the types of leaders are distinguished depending on the nature of the activity (the universal leader, the situational leader), its content (leader-inspirer, leader-executive, business leader, and emotional leader) and leadership style (authoritarian leader and democratic leader) [11].

Thus, the group leader can only be the person who can lead the group to resolve certain group situations, problems, tasks, who carry the most important personality traits for this group, who share the values inherent in the group. A person who is a leader in one group does not necessarily become a leader again in another group (another group, other values, other expectations and requirements for the leader) [1].

3   Results and Discussion

At present, there is no clear understanding of the phenomenon of leadership. Each scientist, giving their definition, emphasizes only one or another aspect. The most common and universally accepted theories are the theories of personality traits, situational, situational and personal.

Thus, the leader is an authoritative member of an organization, a social group or a society as a whole, whose skills make it possible to represent the interests of society, and also lead a group with them.

The research conducted at the Tomsk University of Control Systems and Radio-electronics (TUSUR) was aimed at realizing the goal. The purpose of the study is to examine the characteristics of the influence of students' personal qualities on the process of forming leadership qualities. The subject of the research: Influence of personal reference points on development of leadership qualities in the educational environment. The study was conducted using the questionnaire method. Developed by authors, the questionnaire, consisting of 12 questions, has been filled by 229 full-time students.

According to the results, the percentages of the humanities, economic and law faculties take leading positions. The Humanities - 15%, Economic - 14%, Law- 13%. Engineering faculties occupy the lower positions.

According to the results of the study of leadership qualities among TUSUR students, in conformity with the region of arrival, note that the leadership qualities are most pronounced in students who came to study from the Republic of Kazakhstan. Students who study at the Humanities, Economics and Law faculties have more leadership qualities and actively manifest themselves at various intra-university activities and extracurricular activities. Students of the Faculty of Computer Systems showed high results which prove that there are many students with leadership qualities.

One of the reasons identified in the study was that a large number of foreign students, mostly from the Republic of Kazakhstan, who have a high potential for leadership skills, are trained there.

Being in different divisions, students can apply their organizational skills in practice, take responsibility for organizing and conducting events at various levels. Taking part in the events of the university and the city, students receive important professional skills - initiative, activity, leadership, independent decision-making, and responsibility. Also, working in groups, students learn to act in a team. During work on the projects, students need to coordinate events with the Faculty and university administration, departments, sponsors and partners, which develops the ability to communicate and build relationships with people [12].

Thus, extracurricular activities are an important tool for consolidating professional skills, forming professional competencies, and having a significant impact on the process of professional development of students in managerial specialties. The result of the complex influence of educational and extracurricular activities is the university graduate who is an independent personality, a specialist, a professional, and a responsible citizen.

4   Conclusion

The leadership initiative requires further reflection and development. This problem is of considerable interest to the world community. Leadership becomes a desirable social and personal value. There is a growing number of diverse leadership schools and leadership programs. Leadership is developing, becoming more complex and changing. The development of youth's leadership potential and the full range of measures taken to form effective leadership in Russia will contribute to a qualitative renewal of the country.

The most important role in the stage of formation and development of youth as groups is played by leaders - students who ensure the success of group activity by their behavior, mobilizing influence on others, initiative, and efforts to achieve educational tasks. The study of the leadership mechanism is important from the point of view of becoming an individual with the potential to advance to leadership positions, the ability to manage this process, and use it to improve the effectiveness of group activities.

Extracurricular activities of universities are the most important components of training specialists. It allows students to form a conscious civic position, the desire to preserve and multiply moral, cultural and universal values, develop skills of constructive behavior.

Acknowledgements. The work is performed at support of the Ministry of Education and Science of the Russian Federation, Project No. 28.8279.2017/8.9.

References

1. Babintsev, V.: Liderstvo i autsajderstvo v molodezhnoj srede regiona [Leadership and outsider in the youth environment of the region]. Sociological research **2**, 76–83 (2008). (in Russian)
2. Bogardus, E.S.: Social distance in the City. Proc. Publ. Am. Sociol. Soc. (20), 40–46. Brock University, Ontario (1926)
3. Dewey, J.: Demokratiya i obrazovanie [Democracy and education]. Pedagogy-Press, Moscow (2000). (in Russian)
4. Grin, G.: Sovremennye aspekty razvitiya sotsial'noj aktivnosti studencheskoj molodyozhi [Modern aspects of development of Social Activity of Student Youth]. Socio-Humanitarian Knowledge **3**, 154–160 (2012). (in Russian)
5. Kostenko, S.: Obrazovanie kak protsess sotsializatsii s zhizneutverzhdayushhej napravlennost'yu [Education as a process of socialization with a life-affirming orientation]. Philosophical sciences **5**, 137–148 (2007). (in Russian)
6. Kryvenysheva, E.: Rol' liderskikh programm dlya molodyozhi v podgotovke konkurentosposobnykh spetsialistov [The Role of Leadership Programs for Young People in the Preparation of Competitive Specialists]. Youth and Society **4**, 82–86 (2006). (in Russian)
7. Levin, K.: Dinamicheskaya psihologiya: Izbrannye trudy [Dynamic psychology: Selected works]. Smysl, Moscow (2001). (in Russian)
8. McGregor, D.: The Professional Manager. McGraw-Hill, New York (1967)
9. Morgaeva, N.: Aktivnost' i ee rol' v strukturnoj organizatsii sub"ektnosti lichnosti [Activity and its role in the structural organization of subjectivity of personality]. Psychology of Learning **2**, 40–47 (2015). (in Russian)
10. Morozova, A.: Imidzh sovremennogo molodezhnogo lidera i osobennosti ego formirovaniya [The image of the modern youth leader and features of its formation]. Social technologies, research **4**, 59–63 (2008). (in Russian)
11. Panarin, I.: Ot aktivnosti k sotsial'noj otvetstvennosti molodyozhi [From Activity to Social Responsibility of Youth]. Uchenye zapiski RGSU **7**, 101–105 (2011). (in Russian)
12. Smirnov, V.: Vneuchebnaya rabota v vuze: nekotorye problemy i vozmozhnye puti razvitiya [Extracurricular work in the university: some problems and possible ways of development]. Bulletin of the Higher School of Almamater **11**, 17–23 (2014). (in Russian)
13. Stogdill, R.M.: Personal factors associated with leadership: a survey of the literature. J. Psychol. **25**, 35–71 (1948)
14. Werderber, K., Werderber, R.: Psikhologiya obshheniya. Tajny ehffektivnogo vzaimodejstviya [Psychology of communication. Secrets of effective interaction]. Prime-EVROZNAK, SPb (2010). (in Russian)
15. White, D., Romney, A., Freeman, L.: Research Methods in Social Network Analysis. Taylor and Francis, Somerset (1991)

Regional Aspects in a New Engineering Way of Thinking Development of Bachelors of Techniques and Technology

Galina I. Egorova[1] , Natalya I. Loseva[1] , Andrey N. Egorov[1] ,
Elena L. Belyak[1] , and Olga M. Demidova[2]([✉])

[1] Industrial University of Tyumen, Tobolsk 626150, Russian Federation
egorovagi@list.ru, {lni99, elenabelyak}@yandex.ru,
egorovandre@mail.ru
[2] Tomsk Polytechnic University, Tomsk 634050, Russian Federation
berezovskaya-olechka@yandex.ru

Abstract. In an innovative knowledge economy in Russia there must be a unified complex of bachelors' training based on integral interaction in the system "bachelor-faculty member – an engineer at the enterprise". The purpose of the study is to enhance the level of a new engineering thinking of future engineers, as the bases for professional competences formation. The results of the study: (1) the concept has been developed and tested to select individual educational trajectory, that provides the development of a new engineering thinking with the account of regionalism; (2) the importance of the concept is proved by scientific ideas, approaches, and principles; (3) it has been proved that a new engineering thinking creates flexible and dynamic systems responding fast to the regional market requirements; (4) the peculiarities of the process have been determined, by criteria-estimating component based on the integration principle, allowing to analyze mental operations, develop creativity, research scientific problems in stages thus entering a new level of thinking development; (5) conditions of methodological system are justified.

Methods of research are theoretical, empirical, pilot testing, mathematical treatment methods. Conclusion: grounds for goal-setting have been defined, the concept; syllabi and testing materials; content modules have been developed. It has been proved that it is possible to develop new engineering thinking to enhance the quality of the professional education of bachelors of Science within innovative society development.

Keywords: New engineering thinking · Conception of an individual educational trajectory · Innovation principles · Regionalism · Ecological compatibility

1 Introduction

Engineering thinking even having numerous definitions has three main forms. Taking into account classical bases of the concept, we can note that a peculiarity of the engineering thinking is in acquiring necessary professional knowledge of future

© Springer Nature Switzerland AG 2019
Z. Anikina (Ed.): GGSSH 2019, AISC 907, pp. 65–71, 2019.
https://doi.org/10.1007/978-3-030-11473-2_8

activity. Here, such parameters as foreseeing and forecasting ability for the results of professional activity are important. High intellectual potential of forecasting the results of the professional activity, high level of review and collation of the grounded facts which will determine peculiar features of the object of activity and of the process of its production.

A classical variant of engineering thinking (knowledge and skills in a professional activity, independence, creativity and so on) should include intellectuality, and innovation and ecology parameters [1]. Intellectual parameters are considered as a specific form of active thinking in the activity with the account of morphological and functional interactions of object structures from theory to practice [2]. Innovation parameters are based on studying and application of innovative methods for solving engineering problems. Ecological parameters are aimed at understanding, knowledge of ecological problems and ways of their solution. Integrity of innovation and intellectual parameters are aimed at new technological knowledge, ways and methods for development of innovative technical means and innovative technologies [3].

2   Methods and Materials

Why is a new engineering thinking of bachelors of science important in Tyumen region? It is known that about 80% of bachelors work in the region after graduating. In work and everyday life this rather numerous part of the population should use that professional knowledge on oil processing, and chemical technology obtained in a technical university. Regional chemical knowledge forms regional chemical reality in a learner's mind, creates conditions for developing new engineering thinking, which is a part of higher professional education. Students understand new engineering thinking in a complex consenter of relations through the relations in different systems "man – substance", "substance–regional material – practical activity", "profession– region", "a university – industrial enterprise – region" [4].

New thinking of bachelors of Science is developed during educational activity (in-class and extracurricular). Elements of scientific and research activity are especially important, as they are incorporated into the educational processes (research practicum, a creative project, scientific problem). In extracurricular activity a new engineering thinking is formed through active participation in grants, interuniversity, international contests, and industrial projects. Here, a learner is not only a future specialist, bachelor, and a master – they represent future innovative personality, whose high level of new engineering thinking is a base for professional development. Significant number of professional scientific disciplines in the educational process when students do not see their practical value causes dissatisfaction of the learners. The process of a new engineering thinking is aimed at the highest criteria of quality. The sphere of a professional activity as a "transfer of the society requirements" reacted to the same situation by introduction of early specialization and professionalization [5].

Developing a new engineering way of thinking is an integral, systematic continuous process where disciplines are not a set of autonomous traditional courses but integration of unified cycles of disciplines, bounded by common aim and interdisciplinary relations.

Developing a new engineering thinking is a general way of leading universities, satisfying the requirements of knowledge-driven industry and modern innovative conditions based on projecting and goal-setting. Let's disclose conceptual grounds of this statement [6].

The main conceptual grounds are determined by dialectal interaction of the components in the development process of a new engineering thinking with the parameters of integration systems of a "technical university and industrial enterprise" [7]. This enterprise as a future employer of a competitive specialist takes active part in the final result of the educational process basing on principles of regionalism, science, systematicity, integrity and succession.

It should be noted that a conservative system of technical universities is substituted with upbringing medium. In regional conditions a university becomes a school of creative personalities. It is a priority for developing a new engineering thinking. This task is relevant not only in regional conditions but also all over the Russian Federation. [8].

To solve this general problem in a university we provide learners, first, with the possibility to tackle creative regional problems, and second, give methods and means to solve these problems. We adjust educational process so that laboratory and practical classes contain large amount of creative problems, and course and diploma papers are to seek for the solution to real innovative problems, creative projects are related to industrial sites. Besides, learners have a possibility to acquire modern innovative means, practically tested, ready for mass utilization in universities, heuristic methods of engineering creativity, computer methods of research design and construction; we construct computerized banks of engineering innovative knowledge and expert systems.

This concept of a new engineering way of thinking development during professional training in a technical university determines the way of system construction based on the approaches, principles, and ideas – initial methodological landmarks [9]. Our methodological idea is that development of a new engineering thinking has many grounds of goal-setting.

The first is social requirement shown in tendencies of development of an innovative society, innovative region, and conscious necessity of development of a new engineering thinking.

The second is a learner, whose engineering thinking has individual value not only during training at the university but during future professional activity at the enterprise.

The third source is a professional requirement of a regional employer to the level of engineering thinking of a future specialist.

Goal-setting assessment, regulating documents, and FSES requirements allowed developing and fulfilling a methodological system with the account of changes in contents, activity, evaluative components. Contents component of the system is based on the basic structure of professional knowledge as bases for new engineering thinking of learners with the account of regional material (Table 1).

These modules were included into the bachelors program that allowed to solve the following problems:

Table 1. Basic structures of knowledge in developing learners' new engineering thinking.

№ Module	Developing blocks	Basic structure of learners' knowledge and skills
First	*Philosophic-cultural*	• Methodological, cultural knowledge and skills; • Understanding wide interdisciplinary context of engineering science; • Understanding the role of engineering activity in the civilization history; • Ability to apply knowledge for conceptualization of engineering models, systems and processes; • Understanding methodology of engineering activity
Second	*Innovative-technological*	• Innovative, technological, design knowledge and skills; • Ability to apply innovative methods for solving engineering problems; • Ability to use a creative approach for designing new original ideas and methods; • Ability to plan and conduct analytical, imitational and experimental research; ability to estimate data critically and draw a conclusion; • Ability to investigate new technologies in the sphere of own specialization; • Ability to use knowledge for solving new problems, applying knowledge from other disciplines; • Ability to solve innovative engineering problems with technical uncertainty and lack of information
Third	*Social–economical*	• Ability to be effective both individually and in a team; • Ability to use different methods with to interact effectively with the engineering society and society in general; • Understanding safety issues, legal aspects and responsibility for engineering activity, influence of engineering decisions on the society and environment; • Following the code of professional ethics and norms of engineering practice; • Being aware of project management and business, knowledge and understanding of risks and changing conditions; • Ability to act effectively as a team leader; the team can consist of specialists from different spheres and qualifications; • Ability to interact effectively in a national and international context
Fourth	*Cognitive-educational*	• Cognitive, metacognitive, intentional knowledge and skills; • Awareness of necessity to study independently and increase the level of thinking during the whole life; • Ability to plan, manage activity, including innovative
Fifth	*Regional-ecological*	• Territorial-industrial complex (TIC) with the closed structure of raw materials and waste in the TIC; • Branch diversification according to the priorities in the development of extraction branches and raw hydrocarbons deep processing; • Regional enterprises of Rosneft, Oil-Gas processing plant in Antipinsk, "SIBUR Tobolsk" and others.; • Waste-free regional zone; • Development of a cluster of globally distributed of oil-gas-chemical plants; • Regional ecological knowledge and skills; • Ecological compatibility and safety of processes: from deep processing to creating consumer polymer materials

- To form learners' professional knowledge and skills basing on deep, comprehensive ties and relations: functional, genetic cause-and-effect;
- Provide constant development of innovative thinking while studying concepts and facts – from evident to the deep, from abstract to specific, from the entire to its parts, from particulars to generals;
- To develop corresponding nomenclature for studying facts, concepts – system of description, explanation, transformation – and consequently develop cognitive abilities and skills;
- To provide increase of professional knowledge, providing experience of creating own engineering worldview of a modern civilization.

Pragmatic component of new engineering thinking developed with the account of elective curricula and courses, which were studied gradually.

At the first stage (1–2 courses) the learning process included elements of intellectual training, to train cognitive operations (analysis, synthesis, comparison, cause-effect relations and so on). At the second stage (2–3 courses) the emphasis was given to solve unconventional inventor's problems with a lecturer's help, including elements of TRIZ technology. At the third stage (3–4 courses) learners completed projects independently on the studied modules, took part in the contest of innovative ideas, technologies, suggested own authorial projects on key technologies.

At senior courses certain research problems were solved. These are: determining key trends of long-term development of oil-gas processing waste-free industry with the account of its modern structure; arrangement and analysis of the most popular trends in wastes recycling; forecasting foresight of oil-gas processing industry with the account of rational wastes disposal; determining key innovative trends and priorities of fuel-and-energy company branches; conducting comparative analysis of key innovations in a long-term run for different companies; denoting global trends of oil-gas processing development with the account of green chemistry parameters; developing a route map of oil-gas processing up to 2050 year [4].

The course includes innovative achievements in chemistry, petroleum chemistry, chemical engineering, organic chemistry, shows innovative approaches in synthesis, production technologies of new organic and mineral matters, polymers, catalytic agents, and other important substances and materials for fuel-and-energy company.

Special attention is given to production of new polymer materials and plastics. The course "Innovations at oil-gas processing enterprises" includes several scientific and research, research and development papers for a particular enterprise. The learners analyzed the following trends: global achievements in petroleum chemistry, compared engineering, raw material production factors with world analogues (benchmarking), conducted patent surveys; analyzed existing technologies and released products; investigated new technologies and goods development, able to increase competitive advantage of the company on the global market.

3   Results

During implementation of our system mini-studies were in demand, subject synopsis on regional issues: innovative ideas and main characteristic of the enterprise OJSC "SIBUR PETF"; innovative ideas and main characteristic of the enterprise LLC "SIBUR TOBOLSK"; innovative ideas and main characteristic of the enterprise LLC "Tomskneftehim"; innovative principles of constructing chemical, petrochemical production; innovative ideas and principles of by-products disposal, spent catalyst.

Additionally, we included problem questions; to answer them it was necessary to analyze patents, monographs, scientific articles on the given problem [10]. Sample questions are: "Can we imagine modern life without functionally important chemical substances and materials on their bases, such as: fuel, polymer materials and plastics, rubber and synthetic fabrics, paints and lacquers, metals and alloys, solvent and extracting agents, freon and sprays, materials for communication devices, radio, television, transport, techniques? Explain socio-cultural role of these substances and materials on their bases". "Have you ever thought what innovative ways of polypropylene and polyethylene production exist? What the ratio is of casts and wastes in consumer good production? What are wastes? What is their chemical base and how can they be used? What innovative technologies of wastes disposal can you name?"

This methodological system was estimated by qualitative and quantitative criteria, evaluating dynamics of new engineering thinking development: according to the quantity criteria learners of the experimental groups demonstrated higher rating, higher score, and registered by portfolio storage system. Effectiveness of the research, educational rating proved high level of learners' new engineering thinking. Experimental groups showed increase in semester scores and continuous assessments. A system of tasks of different level on disciplines of basic and variable part of the general professional educational program has been developed. They evidence the formation level of cognitive operations.

4   Conclusion

Efficiency of information processing process was estimated according to the qualitative criteria: correctness parameters and speed of finding a single correct answer according to the given situation; level of development of verbal and non-verbal abilities (operational efficiency, attention, scientific and engineering view); creativity level (ability to produce various original ideas in non-standard conditions of activity).

The experimental group shows higher degree of the formation level according to the qualitative and quantitative criteria parameters, then in the control group.

According to the qualitative criteria the increase is: experimental group – 32%, control group – 10%. According to the qualitative criteria the increase is: experimental group – 27%, control group – 8%, that proves high efficiency of the methodological system for development of undergraduates' cognitive abilities.

References

1. Vedeneev, A.: Modelirovanie elementov myshleniya [Modeling elements of thinking]. Nauka, Moscow (1988). (In Russian)
2. Vekker, L.: Psyhologicheskie process [Psychological processes], vol. 2. Myshlenie i Intellect, Leningrad (1976). (In Russian)
3. Egorova, G., Loseva, N., Belyak, E.: Otsenka sformirovannosti professional'nyh kompetentsiy bakalavrov tehnicheskih vuzov. Problemy pedagogicheskoi innovatiki v professional'nom obrazovanii [Estimating the level of bachelors' professional competences development]. In: Proceedings of 18th International Research and Science conference, pp. 43–45. Russian State Pedagogical University, Saint- Petersburg (2017). (In Russian)
4. Egorova, G., Egorov, A., Loseva, N., Belyak, E., Demidova, O.: Siberian arts and crafts as basis for development of cultural traditions and innovations of bachelors. Adv. Intell. Syst. Comput. **677**, 93–100 (2018)
5. Karpova, YuA: Innovatsii, intelekt, obrazovaniye [Innovations, intellection, education]. MSFU, Moscow (1998)
6. Razumov, V.I.: Kategorial'no-sistemnaya metodologiya i formirovaniye novih strategii myshleniya [Category-system methodology and formation of new thought strategies]. In: Proceedings of the international scientific symposium "Chelovek: fenomen subjektivnosti", pp. 78–83. OmSU, Omsk (2014). (In Russian)
7. Chuchalin, A.: Kachestvo inzhenernogo obrazovaniya: mirovye tendentsii v terminah kompetentsii [Quality of engineering education: world trends in the terms of competences]. Higher education in Russia **8**, 9–17 (2006). (In Russian)
8. Ladenko, I.: Razvitie intellectualnyh innovatsiy v sovremennom obshestve. Kompleksnaya programma issledovaniy [Intellectual innovations development in the modern society. Complex research program]. SBRAS, Novosibirsk (1990)
9. Manuilov, V., Fedorov, I.: Modeli formirovanya gotovnosti k innovatsionnoi deyatel'nosti [Formation models of preparedness to innovation activity]. Higher Education in Russia **7**, 56–70 (1987). (In Russian)
10. Shapovalov, E.: Obshestvo i inzhener: filosofskie I sotsiologicheskie problem inzhenernoi deyatelnosti [Society and an engineer: philosophic and sociologic problems of engineering activity]. LSU, Leningrad (1984). (In Russian)
11. Efimov, Yu.: Intellektualizatsiya VUZa – bazovyi istochnik rosta intellektualnogo potentsiala regiona [University intellectualization as a basic source of intellectual potential increase in regions]. In: Proceedings of the international scientific symposium, pp. 410–414. SSEA, Samara (2003). (In Russian)

Conditions Affecting the Professional and Project Training of Engineering Personnel

Margarita Yu. Raitina$^{(\boxtimes)}$ ⓘ, Larisa V. Smolnikova ⓘ,
Olga V. Gorskikh ⓘ, and Tatyana I. Suslova ⓘ

Tomsk State University of Control Systems and Radioelectronics, Tomsk
634051, Russian Federation
{raitina, gormnoj2004}@mail.ru, smol.lora@gmail.com,
tania.suslowa2010@yandex.ru

Abstract. The article presents the results of the study aimed at identifying factors that affect the professional and project training of engineers in TUSUR. The authors focus on description of key factors that ensure a productive development of modern technologies for engineering, organize educational and cognitive activities for students. The level of education and training of highly qualified engineers are determined by the progress of students which depends on a number of motivational conditions: methodological support of educational process, informing students about the university life and background, relations with university teachers and administration, degree of satisfaction with the training in the university. The data analysis made it possible to establish an important factor in the competitiveness of TUSUR graduates at the labor market is graduates' entrepreneurial activity, and conformity of acquired knowledge with the real needs of the economy. This is directly related to organization of the educational process, built according to the logic of positioning the university as an entrepreneurial institution in connection with the need to develop and apply innovative pedagogical technologies, methods for professional and project training of engineering personnel. Whereas the use of new design methods and techniques by the faculty in future specialists training in line with the declared ideology of the university makes it possible to effectively attract undergraduate students to further study in masters and postgraduate studies.

Keywords: Professional and project preparation ·
Factors of professional training · Project training ·
Entrepreneurial high school · Engineering personnel

1 Introduction

In the era of innovative development of the educational system and increasing competitiveness of university graduates in modern conditions, the issue of identifying factors affecting the professional and project training of engineering personnel is especially relevant [7].

At present, there is a large number of works devoted to the study of personnel professional and project training in the psychological and pedagogical literature. Vocational project preparation is an educational technology aimed at acquiring new

© Springer Nature Switzerland AG 2019
Z. Anikina (Ed.): GGSSH 2019, AISC 907, pp. 72–79, 2019.
https://doi.org/10.1007/978-3-030-11473-2_9

knowledge by students based on real life practice, formation of individual skills and skills through a specially created situation of problem-oriented search. In this sense, professional design of training is a kind of educational activities organization, motivating the activity of students to achieve the goals of learning. The peculiarity of educational activity lies in the fact that in the process of its implementation a person not only assimilates professional knowledge, but also forms personality. The educational activity of studying universities is characterized by strengthening of the role of professional motives for self-education. Thus, learning activity is viewed as a complex dynamic system, determined by the levels of relations (student-student, student-teacher, student-administration, and student-company), behavioral models within the collective, cognitive energy and psychologically comfortable conditions.

Vocational training means mastering the theoretical knowledge, professional skills and skills in the specialized field of training, and preparedness of the educational process subjects to act independently in a situation of uncertainty, going beyond the scope of the subject field; find a solution to a particular problem-oriented problem, involving interdisciplinary, integrated knowledge for this purpose, as well as predict the results and possible consequences of these decisions as in real life.

To solve this problem, the Tomsk State University of Control Systems and Radioelectronics (TUSUR) conducted a study aimed at identifying the factors affecting the professional and project training of engineers in TUSUR in October-December 2017. The study involved 959 students of the 3rd and 4th years of all faculties, including 546 girls (57%) and 413 boys (43%). Out of the total number of respondents, 65% (628 students) get the state-funded education and 35% (331 students) pay for their education.

2 Methods and Results

The questionnaire was intended to identify the level of student achievement during the session (2016/2017 academic year), students' attitude to the methodological support during educational process and to raise awareness about life in the University, peculiarities of psychological climate, the existing relations with the University faculty and administration, degree of satisfaction with studying in TUSUR and important factors when choosing a University for graduate/postgraduate study.

The survey allowed carrying out a quantitative analysis of independent work hours for students, identifying their attitude to the idea of TUSUR development as an entrepreneurial University and a degree of confidence in student employment which will fully meet their expectations. In addition, the students assessed the teaching of non-core subjects: Mathematics and Physics, Philosophy, History, Foreign language and general knowledge for successful employment.

Analysis of the last session results of TUSUR senior students showed that 60.1% of students closed the session with "excellent" and "good", 28.5% –"satisfactory" and 11.4% had a poor result, i.e. "didn't pass". Thus, the progress level of TUSUR undergraduate students at the last session of 2016–2017 academic year is quite high – 88.6%.

As for the methodological support of educational process, the ratio of students is as follows: 65.6% of students are satisfied with the methodological support and 11.5% are not satisfied. Thus, the methodological support of educational process in TUSUR is at a sufficient level, as proved by the data presented in the Fig. 1.

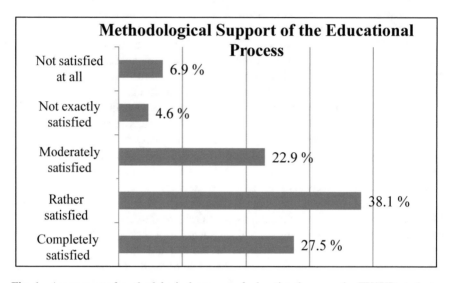

Fig. 1. Assessment of methodological support of educational process by TUSUR students.

Especially relevant is the issue of psychological and emotional health of young people, as the state of balance between a person and the outside world, balance of various mental properties and processes, an excessive influence of emotional factors can cause a nervous and mental stress. From the authors' point of view, the success of educational and professional activities determines the optimal emotional arousal as a condition of readiness for effective activity and its implementation favorable for the health of an individual [6, 11]. In pursuit of this goal, the questionnaire included a block of questions relating to a psychological climate at the University, the relationships with teachers and administration; the process of analysis revealed the assessment of these factors by senior students:

- Psychological climate in the University suits 70.9% of students and only 2.9% of students are not satisfied (see Fig. 2);
- 76.8% of students are satisfied with the faculty and only 3.8% are not satisfied;
- Relations with the administration are arranged by 75.8% of students and only 4.3% are not satisfied.

For a professional design of engineering personnel training, the quality of educational process is the most important factor [4], therefore the questionnaire survey of students assumed identification of the attitude and level of interest in obtaining education in TUSUR, and level of teaching blocks of disciplines (general subjects, Mathematics, Physics, Philosophy, History, and Foreign language):

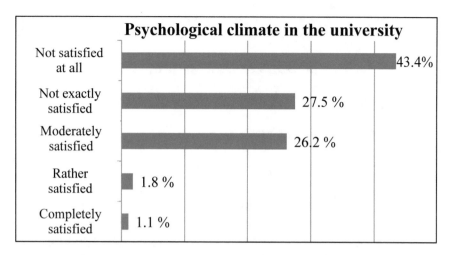

Fig. 2. Students' assessment of the TUSUR psychological climate.

- 70.1% of students study at TUSUR and only 1.9% rated it as "not happy" (see Fig. 3), and it should be noted that 79.9% of students show interest in getting education at TUSUR (see Fig. 4);

- 84% of students are satisfied with the level of teaching Mathematics and Physics, and only 3.6% are rated as "not happy" (Table 1);

- 82.5% of students are satisfied with the level of teaching Philosophy and only 5% do not like it (Table 1);

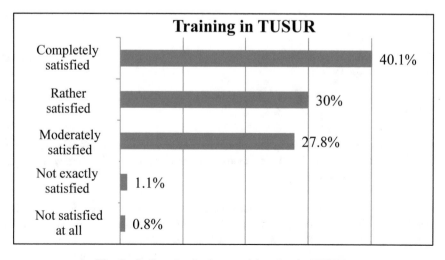

Fig. 3. Students' attitude toward learning in TUSUR.

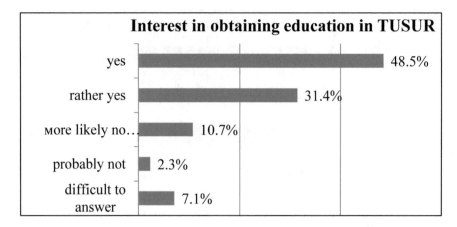

Fig. 4. Student's assessment of the interest in getting education in TUSUR.

Table 1. TUSUR students' assessment of the teaching level (%).

	Completely satisfied	Rather satisfied	Moderately satisfied	Not exactly satisfied	Not satisfied at all
Math and physics	50.7	33.3	12.5	1.5	2.1
Philosophy	44.2	38.3	12.5	2.8	2.2
Profile disciplines	45.9	39.8	11.2	2.2	0.9
History	32.3	30.1	26.3	4.7	6.6
Foreign language	48.9	29.3	14.3	5.6	1.8

- Level of teaching general subjects satisfies 85.7% of students and only 3.1% is not satisfied (Table 1);

- 62.4% of students are satisfied with the level of teaching History and 11.3% are not satisfied (Table 1);

- Level of teaching a foreign language suits 78.2% of students, 7.4% are not satisfied (Table 1).

The study was aimed at revealing the intention of senior students to get education after graduation and the most important factors when choosing a University for master's/post-graduate studies:

- Among the most significant factors that positively influence the decision to continue studying at the University (master's/postgraduate), TUSUR students distinguish the following: 75.1% – "availability of the specialty of interest", 71.6% – "guarantee of future employment", 44.3% – "opportunity to participate in promising research projects" (see Fig. 5).

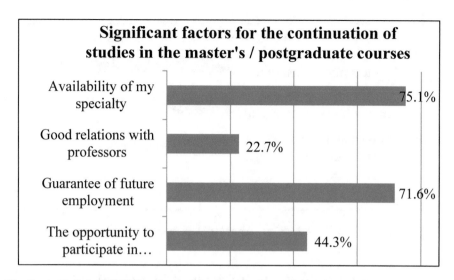

Fig. 5. Analysis by TUSUR students of the most significant factors in choosing a university for continuing education in a master's/postgraduate program.

In connection with the increased competitiveness of university graduates in modern conditions, the issue of the entrepreneurial activity for students and their further employment are particularly relevant [5, 9]. It is for this purpose the questionnaire includes a set of questions relating to the attitude of students to the idea of university developing as an entrepreneurial institution, compliance of actual educational process with the expectations of students and sufficiency of knowledge acquired at the university for their successful employment:

- 42% of undergraduate students are impressed by the direction of TUSUR development as an entrepreneurial university; consider this direction necessary and promising;

- 70.3% of students are confident in their employment (see Fig. 6);

- 42.6% consider the knowledge acquired at the university to be sufficient for a successful employment (see Fig. 7);

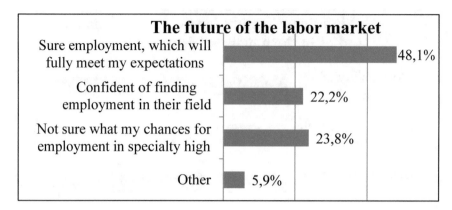

Fig. 6. TUSUR students' assessment of their future at the labor market.

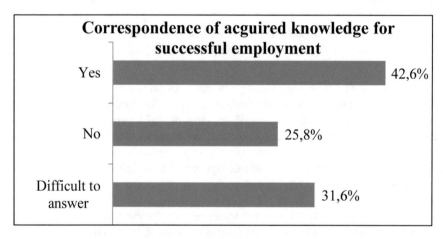

Fig. 7. Studying TUSUR by students of the knowledge acquired at the university for successful employment.

3 Conclusion

Thus, the majority of senior students is confident in their employment, and they believe that actual educational process is fully consistent with their expectations and the knowledge acquired at the University is sufficient for a successful employment [8].

After the study, it was possible to identify the factors affecting professional and project training of engineers at the Tomsk State University of Control Systems and Radioelectronics through identification of student motivation to receive education, the attitude to the learning process, relationships with the University faculty and administration, and thorough evaluation of socially and psychologically comfortable educational environment. In addition, the study revealed the attitude of students to the idea of the University becoming entrepreneurial and identified potential students for the master degree programs, and for postgraduate education [10].

References

1. Gorskikh, O.V.: Parametry analiza i ocenki predstaviteljami vuzov i shkol kachestva vzaimodejstvija [Parameters of analysis and evaluation by representatives of universities and schools of the quality of interaction]. Vestn. TGU **358**, 117–120 (2012). (in Russian)
2. Gorskikh, O.V., Tanzeva, S.G.: Sostoyaniye vzaimodeystviya vuzov i shkol v regione [The state of interaction of universities and schools in the region]. In: Prozumentova, G. (ed.) Interaction of Universities and Schools for the Formation of an Open Educational Space: Potential, Problems, Management Problems, pp. 65–75. TML-Press, Tomsk (2013). (in Russian)
3. Gorskikh, O.V., Pokrovskaya, E.M.: Kul'turno-obrazovatel'naja sreda universiteta i ee funkcii kak resursj effektivnogo mezhkul'turnogo vzaimodejstvija [Cultural and educational environment of the University and its functions as a resource for effective intercultural interaction]. Pedagog. J. **5**, 60–72 (2014). (in Russian)
4. Il'in, G.L.: Kak obespechit' kachestvo obuchenija v informacionnom obshhestve? [How to ensure the quality of education in the information society?]. Public Educ. **5**, 131–135 (2011). (in Russian)
5. Kolin, K.K.: Chelovecheskij potencial i inovacionnaja jekonomika [Human potential and innovative economy]. Bull. Russ. Acad. Nat. Sci. **3**(2), 16–22 (2003). (in Russian)
6. Odarushchenko, O.I.: Osobennosti psikhologicheskogo zdorovia studentov [Features of psychological health of students]. Russ. Sci. J. **4**(29), 266–276 (2012). (in Russian)
7. Raitina, MYu., Pokrovskaya, E.M., Gorskikh, O.V.: Mezhdisciplinarnye podhody v obespechenii obrazovatel'nogo processa v vuze: osnovnye tendencii, celi, zadachi [Inter-disciplinary approaches to the educational process in higher education: the main trends, goals, objectives]. Pedagog. Rev. **2**, 7–11 (2014). (in Russian)
8. Selezneva, H.A.: Kachestvo vysshego obrazovanija kak obek tissledovanija [Quality of higher education as an object of study]. Issledovatel'skijcentr problem kachestva podgotovki specialistov, Moscow (2003). (in Russian)
9. Smol'nikova, L.V., Pokrovskaya, E.M., Larionova, A.V.: Sistemnaja integracija studentov v innovacionno-predprinimatel'skuju dejatel'nost' [System integration of students in innovation and entrepreneurship]. TUSUR, Tomsk (2012). (in Russian)
10. Suslova, T., Raitina, M.: Problems of education in the context of technoscience: tradition and innovation. In: Filchenko, A., Anikina, Z. (eds.) Linguistic and Cultural Studies: Traditions and Innovations (LKTI 2017). Advances in Intelligent Systems and Computing, vol. 677, pp. 75–80. Springer, Cham (2017)
11. Thibaut, J.W., Kelley, H.H.: The Social Psychology of Groups. Wiley, New York (1975)

Language Teaching and Learning

An English Language Textbook: A Friend or a Foe

Margarita A. Ariyan[1] and Lyubov V. Pavlova[2(✉)]

[1] Nyzhniy Novgorod State Linguistic University, Nyzhniy Novgorod 603155, Russian Federation
fencot603@yandex.ru
[2] Nosov Magnitogorsk State University, Magnitogorsk 455000, Russian Federation
pavlovaluba405@mail.ru

Abstract. The article considers the basic issues of the foreign language textbook: characteristics and functions, areas of improvement, effective use and role in the students' social development. Being a model of the actual educational system, an English language textbook acts as the syllabus for the course; gives students confidence and a sense of achievement; performs the main principles of language education. However, there are only a few possibilities to create favorable conditions for students' comprehensive development, to make foreign language lessons motivating, learner-centered and with regard to the students' different learning styles and attitudes, remaining within the limits of the textbook. Investigation shows that the following areas of English language textbooks, used in Russia, need improvement: language content, subject matter, methods, balance of skills, cultural and developing content, conformity with the educational goals and teaching/learning conditions; authenticity of texts and language material; variety of activities and exercises, availability of socially oriented situations, etc. Modern theory of language education focuses on learners' comprehensive development (social, cultural, emotional, cognitive), which ensures their interaction with other people and successful socialization. Hence, the content of a foreign language textbook should provoke a variety of emotions, allow students to assimilate universal humanistic values, social norms and elements of foreign culture. Universal learning skills development is the other important task of a foreign language textbook. So, it is essential to bring the content and the methods of the textbook in line with the needs and interests of the students in the particular teaching and learning environment.

Keywords: English language textbook · Areas of improvement · Social development

1 Introduction

According to Russian Federal State Educational Standard (FSES) language education should be person-centered in terms of educational content and form communicative, cognitive and social competences. It emphasizes the importance of selecting teaching

© Springer Nature Switzerland AG 2019
Z. Anikina (Ed.): GGSSH 2019, AISC 907, pp. 83–90, 2019.
https://doi.org/10.1007/978-3-030-11473-2_10

material and designing textbooks on the basis of developmental principles which will allow to foster creativity and autonomy of the language learner.

In the article we will look at the challenges faced by novice teachers who have to choose a foreign language textbook from the list of the teaching materials recommended by the Ministry of Education, and adapt it to the needs of their students.

The term "textbook" is used here to mean a book which contains a number of items graded into a certain sequence which are to be followed systematically as the basis for a language course. A foreign language textbook contains samples of normative oral and written speech, language and cultural material, selected and organized according to its functional role in different forms of communication and speech activity, considering students' positive experience in their native language and preventing language interference. The debate about the role of a textbook in the teaching and learning process appears to be growing in intensity. There are compelling arguments in favor of using a foreign language textbook which include the following. A textbook:

– is a model of the actual educational system and puts into practice its goals and tasks;
– acts as the syllabus for the course which is being taught;
– gives students confidence and a sense of achievement;
– gives students possibilities for revision, when necessary;
– reduces preparation time, offers teachers a rich bank of material and ideas for teaching;
– acts as ongoing training for novice teachers [1–4].

There are also very sound arguments on the other side of the debate. Advocates of a textbook-free approach claim that a textbook:

– creates a "one size fits all" situation, when interests and needs of different individuals and groups of students are ignored. The language and topics may not be appropriate to a particular student or group of students;
– can be predictable and monotonous;
– gives a false sense of security both to the teacher and students;
– hinders teachers' creativity. As the time goes there is a tendency for teachers working with textbooks for all classes to become more conservative or less open to change.

2   Characteristics and Functions of a Textbook

However, in Russia textbooks are taken for granted. The reality for Russian foreign language teachers for the foreseeable future is that textbooks are here and here to stay even though the use of Information and Communications technology in the classroom, on-line school books become increasingly important [5]. Most institutions require teachers to use paper-based textbooks with their classes whether they like it or not.

The essential characteristics of a foreign language textbook are described in the research conducted by Ariyan [6], Bim [2], Gorlova [7], Passov [8], Pavlova [9], Vitlin [10], Yakushev [11]. They point out that a textbook: (a) models the actual educational system, (b) allows to put into practice its goals and tasks, (c) covers all educational components: topics, communicative situations, language and speech material, exercises, socio-cultural issues, (d) performs the main principles of language education: provides motives for studying, allows to take into account the students' native language, fulfills student centered approach, represents the idea of communicative learning, and provides conditions for socio-cultural competences development.

According to Bim [2], the outstanding Russian researcher of foreign language teaching methodology, a textbook performs the following important functions:

- communicative, providing for the students' ability to communicate in the target language;
- informative, as a textbook describes a new language system and gives information about the country of the target language;
- managerial, i.e. a textbook manages the teacher's and students' activity;
- educative and developing, i.e. a textbook is aimed at developing a student's personality;
- motivating, as a textbook maintains a student's interest in the language due to its contents, structure, topics and problems for discussion, etc.;
- controlling, i.e. a textbook ha special instruments for assessing students' performance.

These principles and functions are commonly acknowledged in Russia and constitute the methodological basis of foreign language textbook theory.

3 Effective Use of a Textbook

A questionnaire survey held among 78 foreign language teachers in Magnitogorsk and 64 in Nizhny Novgorod regions of Russia showed that most teachers are in favor of using textbooks, rather than a sparse collection of photocopies. However, they point out several problems in the form of questions: How can we create favorable conditions for students' comprehensive development (social, cultural, emotional, cognitive, etc.) remaining within the limits of the textbook? How can we comply with the demands of the Federal State Educational Standard to make our lessons motivating, learner-centered and suitable for the different learning styles and attitudes of our students? How can we achieve the goals, defined by the FSES, if the textbook we are working with does not contain the necessary set of exercises and lacks informative, interesting and culturally appropriate texts?

From the experience of observing teachers we can state that there are those who can use textbooks as effectively as possible and those who cannot. In the initial stages of a teacher's professional development textbooks play a valuable role, but new teachers should also learn not to rely exclusively on what a textbook can offer. Unfortunately, few teacher training courses recognize using a textbook as a skill in its own right. As a result, teachers sometimes just "go through" the units of the book, following the authors' instructions thoroughly, never deviating from them. Their students follow in turn and perceive all the vocabulary, do all the activities, learn all the texts given in the book whether they suit them or not. Such simplistic approach can hardly create vast learning opportunities for students.

Every teaching/learning situation is unique and depends on a wide range of factors, such as the personalities involved, the syllabus, the teacher's experience and enthusiasm, the students' interests and motivation, the quality and availability of resources. It is essential that a textbook should be seen as a means of forming intercultural communicative competence. According to Acklam [1], it is the teacher's responsibility to dream up lessons from scratch, which means to define the objectives complying with the FSES and the syllabus as well as to think up all the necessary activities and exercises required for their achievement. As for the instructional material offered by the textbook, however good it is, an innovative teacher feels the need to add his/her own ideas, which stem from previous experience and help to make the material more suitable for the students.

It is commonly recognized that there are four alternatives that allow to bring the content and the structure of the book in line with the needs and interests of the students: leaving out some parts of the material, adding material (published or your own), replacing material with something more suitable, changing the published material to make it more suitable for the students [12].

4   Areas of Improvement

Investigation shows that the following areas of English language textbooks used in Russia need improvement:

- *Language content.* The authors of English textbooks whose target groups are mainly French, Spanish or German students put much emphasis on some grammar items that Russian students do not need to learn because there is the same a phenomenon in the Russian language. They just transfer the skill from their native language to the foreign language they are learning, e.g. degrees of comparison of adjectives or place of an adjective in front of a noun. At the same time, grammar items which present real difficulties because do not exist in the mother tongue are underestimated. For Russians they are: word order, modal verbs, articles, prepositions, Progressive and Perfect tense forms, Conditionals, Complex Object and many others. Russian students find them difficult and need many realistic contexts to understand the use of the grammar phenomenon. They also need many more exercises and communicative tasks to master them.
- *Subject matter.* The topics and tasks, modern textbooks are structured around, may not suit students at the time. Creative teachers use the first lesson to talk about the students' interests through questionnaire and survey tasks which may include different questions: what is new in their lives and in the world, what they like doing in their free time, what will change in our life in the next twenty years, what will affect their lives in the future, etc. Teenagers like to speak about their future profession, about love and friendship, dedication to duty and the native country. From the authentic texts they want to learn more about their foreign peers, about the country and the people whose language they are learning. The outdated topics or topics that are not authentic enough never initiate any interest.

- *Methods.* The exercises offered by foreign language textbooks may be too mechanical, lacking in meaning. They do not facilitate the students' creativity, do not arouse their interest. Live communicative activities which are meaningful, create information or opinion gaps, and involve students into problem solving are few as well as the activities which foster reading and listening for understanding rather than decoding of words.
- *Balance of skills.* Foreign language textbooks are focused on skills in oral speech (speaking and listening comprehension), but do not put enough emphasis on skills in reading and writing and almost neglect universal learning skills and learning strategies.
- *Cultural and developing content.* Some textbooks tend to project an unfriendly image of a foreign peer through inappropriate visuals or comments. In some textbooks the content is not focused on students' comprehensive personal development.

Among the areas of improvement of a textbook the participants of the survey also mentioned: conformity with the educational goals and teaching/learning conditions; authenticity of texts and language material; variety of activities and exercises; focus on A, B or C language levels of difficulty; availability of grammar and culture reference guides and an electronic appendix.

Given these areas, we will look at the problems pointed out in the questionnaire survey mentioned above.

5 Learners' Development

Modern theory of language education is focused on learners' personality, i.e. on their comprehensive development, which ensures their successful interaction with other people and active cognition of the world. Language education is viewed as an instrument of cognition of world culture and an effective means of socialization [13]. Humanitarian, socially-developing nature of foreign language education serves as an incentive for learners' cognitive, cultural, esthetic, social and emotional development. In this regard, higher demands are set to the content of a foreign language textbook which should allow students to assimilate universal humanistic values, social norms and elements of foreign culture [9]. The problem situations described in a text can serve as a basis for discussion reflecting a dialog of cultures. Contrasting any elements of foreign language culture exposed by the textbook with students' native culture often offers a rich source of ideas for discussion.

Textbooks often have some developing, personalization activities built in, which provoke learner emotional reaction and response, but there is always room for more, so adaptation is welcome. The learners' creativity and autonomy is developed within classroom atmosphere where learners' response to the text is expected, their personal opinion about the problems raised in the text is encouraged and valued.

It is important that the teaching process is emotionally saturated. The term "emotion" is interpreted by the psychology as the process which in the form of strong feelings reflects the importance of internal and external situations in a person's life activity. A key principle of humanistic foreign language teaching is that instruction should focus on both thinking and feeling. In other words, it should create conditions for learners' cognitive, emotional and social-emotional development. Modern foreign language teaching methodology emphasizes the importance of emotionally appropriate learning environment in the classroom. To a great extent it can be achieved through the selection of intonation patterns and vocabulary, as well as overall positive emotional attitude of the teacher's utterances. However, as the investigation shows, the quality of the instructional material also goes a long way. Texts should provoke a variety of emotions and generate the values, attitudes and behaviors necessary to developing future citizens who will be sensitive and respectful of other people, i.e. well developed both socially and emotionally [6].

Social and emotional development is a process of socially valued qualities development which allows the person to master different ways of emotional response to a wide range of objects and people, ability to realize and control own emotions in various life situations; ability to establish and maintain emotional contact with other people. Effective language education in general and foreign language learning in particular involve students' inexhaustible sense of achievement, excitement and satisfaction over successful activity and communication. Students come to understand what emotions and feelings are, learn how they occur, recognize their own emotions and those of other people, and also develop realization of how to manage them [6].

Accordingly, special attention should be paid to the communicative tasks aiming at learners' social-emotional development in a foreign language class. This is also an important area of a textbook adaptation.

6 Universal Learning Skills

To comply with the demands of the FSES teachers and learners should never lose sight of universal learning skills formation. These skills are multi-functional: they allow students to define aims correctly, find appropriate means for their achievement, control progress of the activity, and evaluate its result. In other words, they provide a person with suitable tools for effective learning, create appropriate environment for comprehensive development and self-realization on the basis of readiness to continuous learning. In the process of foreign language learning, student's universal skills not only secure his/her ability to understand other mentality, other social values, but also develop an ability to accumulate foreign cultural social experience, master the ways of cognizing the surrounding world, and develop linguistic creative thinking [14].

Within an extreme diversity of universal learning skills mentioned in Russian educational literature we emphasize the following: moral and ethical evaluation of the content of the learning material; goal-setting, planning, previewing, predicting, correction, evaluation; search for information, including with the use of a computer; structuring of information; correct selection of effective ways of problem solving depending on the specific factors involved; reflection on the ways and specific conditions of the activity; problem statement and formulation; analysis, synthesis, comparison, categorization, finding the causal link, proposing hypothesis and its justification. The list of skills, though not exhaustive, shows that possessing them turns a person into an effective autonomous learner in any subject area. Moreover, students' universal learning skills condition their social adaptation, academic mobility and insure their success in different spheres of life.

When planning a foreign language lesson, it is essential to consider the textbook critically, i.e. to understand the learning skills that are in the focus of the authors' attention and what the teacher finds lacking. In the latter case adding some specially invented exercises seems inevitable.

Owing to special communicative nature of the foreign language as an academic discipline the following skills play a particular role: ability to take into account the speech partner's position and opinion; ability to listen and enter into dialogue, participate in brainstorming sessions, build partnership, enter into productive interaction and cooperation, etc. These skills belong to social competence which constitutes an important part of communicative competence.

Discussions generated by textbook activities often sound unnatural even bizarre because students do not realize why they should speak about (have no incentive to speak) and what they are supposed to say (have no communicative purpose). Investigation shows that most textbook authors do not want to create socially oriented situations which will help learners become better in expressing what they want to communicate taking into account the social status and cultural identity of the partner. Textbook adaptation involves building of a variety of problem situations challenging linguistically and stimulating in their content.

7 Conclusion

Observations in lots of foreign language classes, reflection on our own experience and discussions with colleagues have led us to the conclusion that a textbook is neither a loyal friend nor an insidious enemy to a foreign language teacher. Moreover, we do not believe that there are "good" or "bad" textbooks.

Teachers are different in many ways: in their level of proficiency, command of the foreign language, classroom experience, teaching style, personal beliefs, etc. The teaching contexts and the learners' needs, levels and interests also differ a lot. It explains why the choice of teaching materials, including textbooks, should be vast and rich. There must be an opportunity to choose a more or less appropriate textbook which can be an aid in the hands of a teacher who knows how to approach it.

In view of this, both, pre- and in-service training should include a special course on how to bring the content and the methods of the book in line with the needs and interests of the real students in the particular teaching and learning environment.

It stands to reason that students in our heterogeneous classes need a wide choice of topics, texts, practice exercises and communicative activities which cannot be provided by a textbook even if written by very innovative and proficient authors. That is why adaptation is inevitable.

It is a real advantage when an English language textbook gives an opportunity to practice certain language material, develop grammar, vocabulary and speech skills, and at the same time contributes to learners' gaining skills of expressing their emotions, sharing opinions and raising their self- appraisal.

Textbooks should never be adapted or supplemented just for the sake of it. However if we want to help our learners become more autonomous and creative language users we have to provide them with a variety of opportunities, including a carefully prepared user-friendly textbook, which will offer motivating and culturally acceptable texts, well structured, explicit, informative instructions and inspiring activities which will give students a sense of direction, provide objectives for their learning and improve their performance.

References

1. Acklam, R., Burgess, S.: Advanced Gold Coursebook, 2nd edn. Longman, Harlow (2001)
2. Bim, I.L.: Nekotoriye iskhodniye polozheniya teorii uchebnika inostrannogo yasyka [Some basic issues of the foreign language textbook]. Inostrannye yazyki v shkole **3**, 5–12 (2002). (in Russian)
3. Harmer, J.: How to Teach English. Longman, Harlow (1998)
4. Ur, P.: A Course in Language Teaching. Cambridge University Press, Cambridge (1996)
5. Dudeney, G., Hockly, N.: How to Teach English with Technology, 3d edn. Longman, NY (2008)
6. Ariyan, M.A.: Sotsial'no-emotsional'noye razvitiye obuchayushikhsya sredstvamy inostrannogo yazyka [Students' social-emotional development by means of foreign language learning]. Yazyk i kultura **38**, 138–152 (2017). (in Russian)
7. Gorlova, N.A.: Otsenka kachestva y effektivnosty UMK po inostrannym yasykam [Appraisal of a FL textbook quality and effectiveness]. Inostrannye yazyki v shkole **8**, 19–24 (2005). (in Russian)
8. Passov, E.I.: Uchebnik kak fenomen sfery inoyazychnogo obrazovaniya [Textbook as a phenomenon of foreign language education]. Inostrannye yazyki v shkole **4**, 39–45 (2004). (in Russian)
9. Pavlova, L.V.: Gumanitarno-razvivayushcheye obucheniye inostrannym yazykam v vysshey shkole: monografiya [Humanitarian and developing foreign language teaching in at university]. Nauka, Moscow, FLINTA (2015). (in Russian)
10. Vitlin, ZhL: Teoreticheskiye i metodicheskiye osnovy uchebnikov pervogo inostrannogo yasyka dlya vuzov [Theoretical and methodical fundamentals of the first FL textbooks for institutions of higher education]. Inostrannye yazyki v shkole **3**, 45–51 (2007). (in Russian)
11. Yakushev, M.V.: Nauchno obosnovanniye kriterii analiza i otsenky uchebnika inostrannogo yasyka [Theoretically grounded criteria of the analysis and appraisal of a FL textbook]. Inostrannye yazyki v shkole **1**, 19–22 (2000). (in Russian)
12. Cunningsworth, A.: Choosing Your Coursebook. Heinemann, Oxford (1995)
13. Leont'yev, A.N.: Deyatelnost, soznaniye, lichnost [Activity, mind, personality]. Kniga po trebovaniyu, Moscow (2012). (in Russian)
14. Obdalova, O.A., Gural, S.K.: Kontseptualniye osnovy rasrabotki obrasovatelnoy sredy dlya obucheniya mezhkul'turnoy kommunikatsii [Conceptual foundations for educational environment development when teaching intercultural communication]. Yazyk i kultura **4** (20), 83–96 (2012). (in Russian)

Combinatorial Linguodidactics as a New Direction in the Methodology of Foreign Language Teaching

Elena I. Arkhipova$^{(\boxtimes)}$ (iD) and Marina V. Vlavatskaya (iD)

Novosibirsk State Technical University, K. Marx 20, 630073 Novosibirsk,
Russian Federation
elena1503@inbox.ru, vlavatskaya@list.ru

Abstract. The article describes combinatorial linguodidactics – a direction in the foreign languages teaching methodology that emerged due to the demand of modern society and focused on the formation and development of collocation and communicative competence of the secondary linguistic personality. The purpose of the article is to define the advantages of this direction within modern society as well as to give it a scientific substantiation: to identify the objectives, sections, means and methods of teaching. Combinatorial linguodidactics is a scientific discipline based on the theory of cognitive, psychological, linguistic and methodological aspects.

Current conditions of society development set new requirements to linguistic education. Today, one can observe rapid changes in the educational system, the problems that primarily relate to intercultural communication, the integration of subject areas of knowledge, the development of the ability to speak in a foreign language, and the acquisition of skills in the field of information and communication technologies. Absence of the corresponding linguistic environment gives an important role to the dictionaries of collocations, development of new information technologies, including electronic dictionaries and electronic text corpora. All these confirm the urgency of studying foreign languages when the language is the main carrier of information, and, accordingly, the demand for new teaching methods and techniques.

Keywords: Linguodidactics · Collocation and communicative competence · Combinatorial linguistics · Secondary linguistic personality

1 Methodology vs. Linguodidactics

In the course of language teaching we notice that the term "methodology" that we are accustomed to is gradually replaced by the term "linguodidactics". As the analysis shows, linguodidactics is a scientific field dealing with the problems of identifying the goals, selecting and organizing the teaching programs, developing of teaching aids. In addition, methodology develops the systems of learning activities or technology of teaching aimed at introducing foreign language learners to the content of the subject in specific learning conditions.

© Springer Nature Switzerland AG 2019
Z. Anikina (Ed.): GGSSH 2019, AISC 907, pp. 91–97, 2019.
https://doi.org/10.1007/978-3-030-11473-2_11

Khaleeva attributes linguodidactics to the field of methodological science which "proves the substantive components of education, learning, teaching in their inseparable connection with the nature of language and the nature of communication as a social phenomenon that determines the activity essence of speech products, which are based on the mechanisms of social interaction of individuals" [1].

Today, the anthropocentric approach prevails in science with the emphasis on the role of the human factor. In the modern methodological paradigm secondary linguistic personality acts as such a factor, which is understood as the totality of abilities and readiness of a person to intercultural communication in a foreign language, i.e. to the interaction with the representatives of other languages and cultures. At the present stage, linguodidactics takes the concept of "secondary linguistic personality" for the main category, describes it, as well as levels, mechanisms, conditions of functioning and formation of the linguistic personality.

One of the basic principles of teaching process with a secondary linguistic personality is, in our opinion, the principle of linguistic combinatorial theory which fundamentally affects the cognitive and psycholinguistic aspects of speech generation, because the combination of words in the linguistic perception of a person is closely related to the correlation of language and thinking.

2 Theoretical and Methodological Fundamentals of Combinatorial Linguodidactics

Theoretical fundamentals of combinatorial linguodidactics:

(1) The doctrine of the relationship between thought and speech from the standpoint of cognitive consideration of the language role in the formation of thought begins with the concept of Potebnya [2]. The basic theses of this theory are that the thought is derived from the word; the word is primary; the priority of the language to any other reflexible activity; the impossibility to think without words.

(2) These ideas are developed in the cultural and historical theory of mentality by Vygotsky [3]: the impossibility of thinking outside the sphere of verbal thought is based on the collocability of words. The words in the speech "throw the net of logical relations" to a certain reality revealing the its patterns and existing links between the phenomena of reality and between the phenomena of reality and words.

(3) Further on, these ideas are developed in the theory of language and thought of Katznelson [4]: The speech and thought process is realized by the mechanisms of speech thinking: language vocabulary and grammar + the notional aspect of speech. Sentences are formed with words and their meanings in the process of transforming deep syntactic structures into surface structures of generation and perception.

The syntagmatic idea – the laws of syntactic structures – formed the basis of the transformation-generative grammar of Chomsky [5]. In this theory language is viewed not as a set of units of language and their classes, but as a mechanism that generates correct phrases. Each sentence is a new syntagmatic combination of words, not previously encountered in people's speech practice. The system of collocability rules exists as the ability to generate and understand sentences where can be grammatically correct, but meaningless sentences as well as incorrect ones in both semantic and grammatical sense.

In the process of the syntagmatic transformation of thought into speech the operation of selecting the semantic and linguistic elements of a speech utterance is of greatest importance as it is regarded as a universal operation of speech generation. Certain semantic rules of words collocability exist for speech generation, and these rules become a sort of a "filter" that ensures the meaningfulness of the utterance. Thus, syntagmatics and combinatorics within a phrase or statement govern the process of speech generation.

3   Combinatorics vs. Syntagmatics

The word in human consciousness and memory does not exist in isolation: it is inevitably attracted toward others. The basis for the creation of any utterance in speech as a whole is the principle of linguistic combinatorics – or composition and study of word combinations that are subject to certain communicative tasks under given conditions of their implementation and which can be formed from a given number of words [6].

Syntagmatics is viewed as an aspect of language study which investigates the rules of linguistic units' collocability and their use in speech. Syntagmatics is manifested in two linguistic phenomena: (1) valency, which reveals itself at the level of the language and represents a potential combination of linguistic units, (2) collocability which emerges at the level of speech and represents the valency.

Considering syntagmatics and combinatorics as the linguistic bases of combinatorial linguodidactics, it is necessary to emphasize that syntagmatics involves the forming up of language units in a linear sequence according to the rules of their combinatorics. In turn, combinatorial limitations are conditioned by: (1) the solution of the communicative task (the given meaning), (2) the conditions for implementing this task, (3) a specific set of language units that express the given meaning.

Syntagmatics and combinatorics equally determine the collocability of linguistic units and are in relation of the equipollent opposition to each other.

4   Linguistic Bases of Combinatorial Linguodidactics. Collocation-Communicative Competence

The theoretical basis of combinatorial linguodidactics is combinatorial linguistics, which studies the syntagmatic relations of linguistic units and their combinatorial potential. Its emergence as a theoretical and applied branch in linguistics is dictated not only by the logic of the modern linguistic science development and the importance of studying the problems of linguistic combinatorics to describe the mechanisms of human speech activity, but also by purely practical goals – teaching foreign language and translation. The formation and development of this branch is conditioned by the need for a deep understanding of the language communicative function realization and features of the language functioning in various spheres of communication.

Combinatorial linguodidactics is aimed at developing the set of competences (communicative, linguistic, speech, cultural, collocational, etc.) of the secondary linguistic personality through the emphasis of the content, structural and instrumental components of teaching and learning in their inseparable connection with the nature of language and nature of communication, based on the mechanisms of social interaction of individuals. Thus, combinatorial linguodidactics contains a clearly expressed methodological aspect, since one of its main tasks is the development of methods and teaching techniques to form collocation and communicative ability to use foreign language vocabulary [7]. Thus, the main condition for learning in the framework of the combinatorial direction is the study of lexical units in their diversity and interconnection.

Speaking about the collocation competence [8], it should be emphasized that the most complex and vulnerable norms in the language belong to the sphere of word collocability. As a linguistic phenomenon, collocability is the main property of linguistic units, ensuring their functioning in speech, based on syntagmatic relations. The problem of words collocability is the most difficult area for studying the vocabulary of a foreign language, especially in terms of active use Knowledge of the native language does not always help when choosing the right lexical phrases in another language. For example, in Russian, you might say, разбить парк (to start a park) and гробовая тишина (deathly silence), but the English phrases *to break a park and *coffin silence are incorrect. The ability to correctly combine words and form collocations for the expression of thought is crucial for a secondary linguistic personality development. Taking into account the existing problems, it is necessary to look for new ways and approaches to teaching foreign languages.

The collocation and communicative competence means effective communication in a foreign language, and most importantly – the ability to combine words according to the language standards. According to the recent research, the formation and development of collocation and communicative competence is currently one of the most important tasks of a foreign language teaching. To achieve this goal, a special section of linguistics – corpus linguistics was created, which deals with "the development of general principles for the construction and use of linguistic corpus of texts with the aid of computer technologies" [9, p. 5].

5   Structure and Content of Combinatorial Linguodidactics

In terms of the goals of teaching and developing a secondary linguistic personality within the framework of combinatorial linguodidactics, it is very important to achieve the following goals.

Description of the general purpose combinatorial linguodidactics aimed at the formation of the mental lexicon of students by means of studying the combinatorial-syntagmatic potential of words of the general literary language.

The advantage of learning words in their relations with other words (i.e. in their environment) is confirmed by at least two statements. Firstly, the word, as a rule, is not used in isolation, therefore, it must be studied in typical and stable models of actual use. Secondly, in relation to communicative direction of learning, it is more expedient to memorize the whole word-combination or phrase, not the individual words constituting it. It should be noted that the semantics of words needs to be studied in comparison with the native language, since collocability in different languages may not coincide, deep grass – высокая трава; deep voice – грудной голос; deep forest – дремучий лес; deep drinking – беспробудное пьянство; deep enemy – заклятый враг; deep sin – страшный грех; deep arguments – веские доказательства; deep plans – далекоидущие планы.

English language, because of its idiomatic nature, is more expedient to be studied not through words, but through collocations or phrases. The word, in addition to the sound structure and conceptual-object content, has such an important communicative property as the ability to selectively combine with other words in a speech chain, i.e. collocability or syntagmatics, and form collocations.

Returning to the term "collocation", it must be emphasized that it takes one of the key places in combinatorial linguodidactics. In a broad sense, it is a combination of two or more words that tend to be "co-occurring". The phenomenon of semantic-grammatical interdependence of the phrase elements is the basis for most definitions of collocation [10].

Collocations in mastering a foreign language are of great importance, which is confirmed by several facts. Firstly, collocations exist in all natural languages; this is the main feature of any language. Secondly, collocation emphasizes and specifies the meaning of the words that form it; the exact meaning of the word in any context is determined by the word environment – its extension (collocates), or words that are united around it and form a collocation. Thirdly, the unification of words in collocations is fundamental to all language use. Words in the language are not arbitrary, since their choice is limited. For example, the noun drug – "an illegal substance that changes the body and mind" – has a predictable choice of correlated words: illegal, illicit, addictive, dangerous, etc.

In the situation of teaching a foreign language outside the corresponding linguistic environment or classroom bilingualism [11] the problem of teaching syntagmatic connections of words becomes of great importance. In these circumstances, an important role is played by dictionaries of collocations, development of new information technologies including electronic dictionaries. Now there is a rapid development of electronic text corpora, which contain millions of contexts, created specifically for the study of language.

The other means of learning the lexical connections of words is concordance which is a relatively simple type of applied program. Concordance is understood as the recording of all uses of a particular element of a language (mainly words or phrases) demonstrating its immediate context. Its function is to provide a constructive search for a large number of contextual examples of a specific word or phrase. When the keyword is located in the middle of the screen, its collocates are given to the right and left.

Computer dictionaries and concordances are important resources used in the learning process, as they are given one of the main roles in mastering a foreign language – assistance in the study of syntagmatic word associations. Within the context of combinatorial linguodidactics, teachers need to involve students in working with e-learning resources to fully develop their "lexical vision and understanding" of the language [12].

The branch of academic combinatorial linguodidactics develops ways and methods of learning the collocability of words for academic general scientific purposes. Presently, the English language has become the main means of international scientific communication. Every day the number of specialists and scientists grows as they need need to present the results of their research in English, prepae of reports for international conferences or apply for grants. In addition to lexicographic reference books, we need a carefully developed linguodidactic theory, including the theoretical foundations and practical developments to reach the scientific goals of the specialist.

Languages are arranged in different ways and literal translation of the Russian scientific text into English makes the text heavy and unreadable, moreover, often with semantic errors. The collocation and communicative competence in terms of scientific presentation is the high priority task for researchers.

However, the skills of using English for academic purposes cannot be fully realized without professional vocabulary. Consequently, there is the need in formation of the fundamentals of a specific combinatorial linguodidactics for the development of the collocation competence for the secondary linguistic personality in specific areas of knowledge (medicine, economics, radio engineering, electronics, nanotechnology, ecology, law, business, etc.). New branches belong to technical and natural sciences, so there are technical dictionaries, but there are no dictionaries on mechatronics, new fields of knowledge, sonocytology (the study of cell sounds), evolutionary developmental biology.

Nowadays, in the sphere of professional intercultural communication, great attention is paid to the translation on which is important for relations between partners in various fields of human activity. Professional-oriented oral and written translation links the professional activities of partners from different countries. Communication between them is carried out with the help of an intermediary language, which gives rise to certain difficulties and problems caused by linguistic interference. The psychological basis of linguistic interference is the transfer of certain elements from one language system to another as a result of language contact. Combinatorial linguodidactics can solve this problem, due to the study of "negative language material" and development of methods for preventing interference errors. From the perspective of vocabulary-oriented combinatorial linguodidactics, such concepts as "translation" and "interference" are very useful.

The ability of a secondary linguistic personality to understand lexical correspondences is extremely necessary. Firstly, mother tongue is the main base for studying any foreign language: all interpretation passes through it. Secondly, sometimes semantisation in a foreign language leads to misunderstanding of the meanings of new lexical units and misleads students (at the initial and intermediate levels of studying it is extremely important to take into account the insufficient level of language skills). Thirdly, the equivalents of idioms, set expressions and phraseological units need to be presented in the native language, otherwise they will either be misunderstood or 'missed' in general.

6 Conclusion

In conclusion, it should be said that combinatorial linguodidactics is a scientific field dealing with the problems of defining the goals, selecting and organizing the content of teaching, and also developing teaching tools relating primarily to the development of collocation and communicative competence of a secondary linguistic personality.

The theoretical fundamentals of combinatorial linguodidactics are, on the one hand, the cognitive and psycholinguistic components of a person's linguistic consciousness which facilitate the word combining and generation of speech, on the other hand, combinatorial-syntagmatic potential of the word is embedded in the very nature of language and linguistic units. The word is not an isolated element, it has combinatorial-syntagmatic properties – the ability to connect with other words logically and semantically, or to attract other words in speech.

In the process of studying foreign language and culture, combinatorial and semantic information accumulates in the language consciousness of a person and forms a cognitive basis which in turn favors the formation of a collocation ability in the secondary linguistic personality and ensures full communication in certain fields of knowledge and spheres of communication. In general, it follows that at the present stage the potential of combinatorial linguodidactics can be considered relevant and prospective.

References

1. Khaleeva, I.: Secondary language personality as a recipient of a foreign text. Language-system. Language-text. Language-ability. Institute of the Russian Language, Russian Academy of Sciences, Moscow (1995). (in Russian)
2. Potebnya, A.: Thought and Language. Labirint, Moscow (1999). (in Russian)
3. Vygotsky, L.: Thought and Word. Labirint, Moscow (1996)
4. Katsnelson, S.: The Typology of Language and Speech Thinking/SD Katznelson. Science, Leningrad (1972)
5. Chomsky, N.: Syntactic Structures. Mouton, The Hague (1957)
6. Makovsky, M.: Linguistic combinatorics: the experience of topological stratification of language structures. KomKniga, Moscow (2006). (in Russian)
7. Vlavatskaya, M.: Interdependence of word semantics and syntagmatics as problem of combinatory semasiology. In: Proceedings of the Conference on Cultural, Linguistic and Art Studies, pp. 88–93. Sibac, Novosibirsk (2018). (in Russian)
8. Lewis, M.: Implementing the Lexical Approach. Thompson Heinle, Oxford (2002)
9. Zakharov, V., Bogdanova, S.: Corpus Linguistics. Publishing house of St. Petersburg State University, St. Petersburg (2013)
10. Iordanskaya, L., Melchuk, I.: The meaning and Compatibility in the Dictionary: Monography. Languages of Slavic Cultures, Moscow (2007). (in Russian)
11. Frolova, G., Vishnevskaya, G.: Features of auditory bilingualism. http://elar.urfu.ru/bitstream/10995/40785/1/avfn_2016_10.pdf. Accessed 2 Sep 2018
12. Woolard, G.: Messaging: Beyond the Lexical Approach. Smashwords Edition, California (2013)

Self-instructed Distance e-Learning of Foreign Languages in the Far North of the Russian Federation

Ivan Artemiev$^{(\boxtimes)}$ ⓘ, Evgeniy Parfenov ⓘ, and Anatoliy Nikolaev ⓘ

North-Eastern Federal University, 677000 Yakutsk, Russian Federation
{ivanart,nickan07}@mail.ru, eap0l@inbox.ru

Abstract. In this paper, we describe the concept "self-development" in a context of philosophical, psychological and pedagogical science. The essence of the process of self-development appears in different philosophical and psychological scientific schools. Definition of self-development of the personality in distance learning of a foreign language is given. Efficiency of distance education technologies for self-development of the personality is shown by practical consideration. Vast territory, distant locality, undeveloped infrastructure and seasonal character of transport lines of communication are the most common features of foreign language teaching in the Far North regions of the Russian Federation, particularly in the Sakha Republic. In these circumstances, we consider Distance e-Learning to be one of the most optimal form of foreign language learning. The article considers Distance e-Learning as a combination of Distance Education and e-Learning which is characterized by the extensive use of Information and Communications Technologies (ICT) in the delivery of education and instruction as well as the use of synchronous and asynchronous online communication in an interactive learning environment or virtual communities, instead of a physical classroom, to bridge the gap in temporal or spatial constraints.

Keywords: Self-instructed Distance e-Learning ·
Learning of foreign languages · Self-development · Creativity ·
Personal development · Information and communication technologies

1 Introduction

Nowadays in conditions of scientific and technical progress, self-developing and self-educated people are necessary in our society. The teachers' purpose is to create necessary conditions for students to form their skills of self-instructed learning. Learning of foreign languages is one of the main means to develop a person in the modern world. A good command of foreign language provides the person's capacity, satisfies his/her social, cultural and professional needs.

Foreign language teaching in the Far North regions of the Russian Federation, particularly in the Sakha Republic, has its own features. For example, vast territory, distant locality, undeveloped infrastructure and seasonal character of transport lines of communication are the most common features. In these circumstances, we consider Distance e-Learning to be one of the most optimal form of foreign language learning.

© Springer Nature Switzerland AG 2019
Z. Anikina (Ed.): GGSSH 2019, AISC 907, pp. 98–104, 2019.
https://doi.org/10.1007/978-3-030-11473-2_12

The aim of our research is to define theoretically the efficiency of foreign language Distance e-Learning for self-instructed learning skills development. In our work, we follow the definition of self-instructed learning given by Dickinson [1, p. 47] which is: "situations in which a learner, with others, or alone, is working without the direct control of a teacher".

In our research, we use the following research techniques: analysis of theoretical research-related literature, Distance e-Learning process modelling, summarizing the research conclusions.

In our research, we based on our vast experience of development, use and implementation of Distance e-Learning courses for university students and schoolchildren. Thus we have developed teach-yourself packages - sets of a comprehensive e-Learning course materials. The findings of our research are published in scientific journals registered in the Russian Index for Science Citation.

In pedagogics, there is a concept of self-training as deeply conscious creative activity for mastering skills of informative, communicative and other kinds of activity, on this basis, developing necessary competence, and gaining the qualities required for personal self-development. However, specific character of self-training in psychological and pedagogical studies has not been fully revealed. Teaching and self-instructed learning are dialectically interconnected as means and components of personal development: with other people's assistance a person learns to show cogitative and verbal activity as well as to make efforts based on positive motivation. Accordingly, teaching acts as means of self-training skills formation and self-training represents a product, i.e. result of training [2].

2 Specific Features of Self-instructed Distance e-Learning of Foreign Languages in the Far North of the Russian Federation

Due to the mentioned above, it is possible to note that self-training and independent acquisition of knowledge by the student in distance learning of a foreign language is possible only with speech and cognitive activity on a target language and with created motivational and semantic kernel of the personality.

Thus, Trofimova and Eremina emphasize, the mechanism of self-training starts due to a contradiction between the formed, active informative interest and the level of development of the personality insufficient for satisfaction of his/her interest [2].

The creative pedagogics, which we consider as a project-based learning, is the mechanism of self-training in distance learning. The principles of creative pedagogics are a favorable condition for self-development of the personality. It is shown by the fact that one of the main tasks of an education system is training of creatively thinking experts with high intellectual potential.

Furthermore, as A.N. Kulik notes, that the education system always creatively thinking specialists in the education system, so prominent teachers and psychologists (Ushinsky, Leontyev, Vygotsky and others) developed and approved theories and methods of actuation of creative activity, which allow to increase students' potential [3].

In brief, we will define what we understand under the concept of "creativity". There are two various points of view on this problem. In a number of studies (Cox, May, Torrens etc.) creativity is considered as the highest form of thinking higher than usual cogitative acts. The turning point, which defines the nature of creative process, is existence of a product that is a solution of a task. In this regard the creative thinking is opposed to logical thinking and is considered only as an aspect of intelligence of the personality.

Other approach towards understanding the creativity as integrated phenomenon is widely presented in numerous studies of Russian scientists (Bogoyavlenskaya, Matyushkin, Yakovleva, Kulikova etc.) and in theories of foreign humanistic psychologists (Maslow, Rogers). The essence of creativity as manifestations of creativity comes to light on the basis of a combination of informative (general intellectual) abilities.

In Russian psychological science which we adhere in our research, creativity is considered as "creativeness" - some special property of a human to show socially significant creative activity [4]. Thus, an adequate unit when studying creativity, according to Bogoyavlenskaya, is "intellectual activity of the personality", reflecting "informative and motivational characteristics of the creative person in their unity" [5, p. 22]. Intellectual activity unites intellectual (mental) abilities and nonintellectual (personal, motivational) factors of intellectual activity, but it comes to neither of them. Mental capacities, in particular active speech and intellectual activity, are the bases of intellectual activity, and are shown not directly, but refracting through a motivational (motivational and semantic) kernel of the personality. Consequently, creativity is understood as "a derivative of the intelligence refracted through motivational structure which either brakes, or stimulates its manifestation" [5, p. 23].

Thus, we established that creative methods of training (for example, a method of project-based learning), promoting speech and cogitative activity as well as forming a motivational and semantic kernel of the personality, develop student's self-training skills, independent knowledge acquisition skills. In this regard, information and communication technologies represent a wide range of means and forms of education and creativity development of a student, including technologies of distance learning.

Use of modern information and communication technologies in educational process for personal growth is confirmed by the fact that Robert, investigating their role in education, explains the process of formation of the personality as follows: "Revelation of the opportunities and abilities of knowledge given by nature, creative initiative, their systematic development, improvement and timely realization is the way which each person should pass in the course of formation of the personality" [12, p. 480].

Modern social, psychological and pedagogical studies prove that the person who has realized his/her creative potential is a noncomplex personality. Thus, it is, at least, a necessary condition of comfortable existence for both the person, and the society as a whole for further movement on the way of improvement and self-improvement" [12, p. 205].

In our work, we rely on findings of the Russian researchers of distance learning, such as Andreyev, Panyukova, Robert etc. Under Distance e-Learning we understand the combination of Distance Education and e-Learning which is characterized by the extensive use of Information and Communications Technologies (ICT) in the delivery of education and instruction as well as the use of synchronous and asynchronous online communication in an interactive learning environment or virtual communities, instead of a physical classroom, to bridge the gap in temporal or spatial constraints.

In order to create our own model of Distance e-Learning process in the Far North of the Russian Federation we based on the following didactic principles: the principle of a value of an individual, the principle of definition being trained as an active subject of knowledge, the principle of orientation to self-development, self-training, self-education of the student, the principle of socialization of the student, the principle of subjective experience use of the student, the principle of the specific psycho-physiological features of the student, the principle of communicative abilities development of the personality.

In this regard, the didactic features of distance learning of foreign language provide new understanding and correction of the purposes of its introduction which we can designate as follows:

– development of abilities and skills of training and self-training that are reached by expansion and deepening of creative educational technologies and actions;
– stimulation of intellectual foreign-language speech and intellectual activity of students by means of definition of the purposes of the studying, dominating type of educational action;
– strengthening of educational motivation that is reached by accurate definition of the personally significant content of the teaching forming a motivational and semantic kernel of the identity of the student.

Considering the previously mentioned and having compared opportunities of the computer and computer telecommunication networks, we came to conclusion that the model of distance learning of a foreign language will include a system of training programs which:

– have to be substantial, have a quick search of necessary information, contain built-in reference books, built on completely authentic training material;
– have to combine the theory and practice that create prerequisites for development of language and speech skills on a conscious basis, to promote multichannel training (use of verbal, graphic, sound support);
– have to cause the interest stimulating to speaking, have a communicative focus, offer a fascinating interactive form of work which strengthens training effect and increases motivation as a whole;
– possess function of stage-by-stage management of foreign-language speech activity (detailed reasonableness of system of exercises and tasks in methodical, linguistic and psychological meanings), to achieve certain practical results;
– have adaptive character which allows to change the way of learning according to the feature of work of the student;
– are to provide instant feedback (to make comments on each answer of the student), to provide the necessary required help, to record statistics of the covered material and mistakes.

To achieve the aim of the research we have developed a model of Self-instructed Distance e-learning of foreign languages (Table 1).

Table 1. Model of Self-instructed Distance e-learning of foreign languages.

Educational activity		
Level of self-instructed learning skills	Dominant type of actions type of educational activity	Means of teaching
Formation of internal needs for self-organization of intellectual activity	Acquisition of linguistic knowledge Speech perception and speech generation exercises	Introduction to new information technologies Training within a basic level of foreign language skills with use of local multimedia computer training programs Presentations, video, audio etc
Shaping of independent activity skills in searching different ways of task solution	Shaping of speech skills Partially automated speech skills Reproductive and productive exercises	Work with interactive training. programs. Training in separate types of speech activity with use of multimedia tutorials in a local network (a chat session, e-mail, etc.) Interactive exercises
Formation of self-education competence	Development of speech skills Automated speech	Project work on studying of culture-specific aspects of a foreign language with use of multimedia tutorials on the global Internet Creative synthesis of communicative skills in the Internet

At the first stage, we form internal needs for self-organization of cognitive activity of the student. Here the communicative and informative need for self-training is actualized. For this purpose, foreign language is learnt within a basic level of foreign language skills with use of local multimedia computer training programs such as Hot Potatoes, Professor Higgins etc. At this stage of the Model students learn to perceive and remember a training material in perceptual and reproductive exercises.

At the second stage students' skills of independent actions on search of different ways to solve educational tasks are formed, personal sense of activity becomes relevant. Here profound learning of separate types of speech activity and aspects of a foreign language with use of interactive multimedia tutorials in a local network are offered (chat sessions, e-mail, etc.). Dominant type of educational action are reproductive and productive exercises.

The third stage is devoted to abilities of the rational organization of education in real actions of the self-education, which assumes formation of abilities on self-change. Students gain the automated speech actions and, having formed positive educational motivation, are able to perform projects on cultural aspects of a foreign language with use of multimedia tutorials on the Internet (Philamentality, MOODLE, TinCan etc.).

In the Model of distance learning we assess student's level of self-instructed learning skills by Bespalko's criteria [7]:

1. For the first stage of the Model: knowledge-introduction – recognition of objects, phenomena, processes, properties due to re-experience of earlier acquired information; knowledge-copies –re-productive activity and use of received information.
2. For the second stage of the Model: productive actions on the use of received information in the course of independent activity.
3. For the third stage of the Model: transformation-knowledge assumes possibility of creative use of received information by independent designing of student's own activity.

In the following work, we will submit the description and results of our experimental work on implementation of the Model of Self-instructed Distance eLearning of foreign languages in the Far North of the Russian Federation; we will develop recommendations about didactic support and actions for the organization of distance e-learning of foreign languages.

3   Conclusion

Following the results of research, the system of Distance foreign language e-Learning courses in High northern regions of the Russian Federation will be scientifically proved and developed. The system will promote self-instructed learning skills of students.

The system will allow to create proper conditions for:

– better schoolchildren training in foreign languages for the Unified State Exam to be enrolled in universities, particularly the North-Eastern Federal University;
– development of foreign-language communication skills of university students to prepare them for the international academic mobility programs and increase their professional competitiveness.
– language training of specialists of the organizations and the enterprises planning international cooperation.

References

1. Andreev, A.A.: K voprosu opredeleniya ponyatiya distancionnogo obrazovaniya [To a question on definition of Distance Education]. Distancionnoe obrazovanie **4**, 26–36 (1997). (in Russian)
2. Bespalko, V.P.: Osnovy teorii pedagogicheskih sistem: Problemy i metody psihologo-pedagogicheskogo obespecheniya tehnicheskih obuchayuschih system [Theoretical foundations of pedagogical systems: Problems and methods of psychological and pedagogical support of technical training systems]. Voronezh State University, Voronezh (1977). (in Russian)
3. Bogoyavlenskaya, D.B.: Puti k tvorchestvu [Ways to creativity]. Znanie, Moscow (1981). (in Russian)
4. Cox, B.F.: The Relationship Between Creativity and Self-Directed Learning Among Adult Community College Students. Ph.D. theses. University of Tennessee (2002)

5. Dickinson, L.: Self-Instruction in Language Learning. Cambridge University Press, Cambridge (1987)
6. Kulikova, L.N.: Problemy samorazvitiya lichnosti [Problems of personal self-development], 2nd edn. BGPU, Blagoveshchensk (2001). (in Russian)
7. Leontyev, A.N.: Automatizacia i chelovek [Automatization and a human being. Psychological researches]. Psychologicheskiye issledovaniya **2**, 8–9 (1970). (in Russian)
8. Maslow, A.: Samoaktualizaciya. Psychologiya lichnosti [Self-actualization. Psychology of personality]. Moscow State University, Moscow (1982). (in Russian)
9. Matyushkin, A.M.: Problemnye situacii v myshlenii i obuchenii [Problem situations in thinking and learning]. Pedagogika, Moscow (1972). (in Russian)
10. May, R.: Everyday Creativity: Our Hidden Potential. Everyday Creativity and New Views of Human Nature. American Psychological Association, Washington (2007)
11. Panyukova, S.V.: Koncepciya realizacii lichnostno orientirovannogo obucheniya pri ispol'zovanii informacionnyh i kommunikacionnyh tehnologii [Conception of realization of person-oriented teaching when using information and communication technologies]. IOSO RAE, Moscow (1998). (in Russian)
12. Robert, I.V.: Sovremennye informacionnye tehnologii v obrazovanii: didakticheskie problemy, perspektivy ispol'zovani [Modern information technologies in education: didactic problems, use prospects]. Shkola-Press, Moscow (1994). (in Russian)
13. Rogers, K.: Vzglyad na psihoterapiyu [View on behavioral therapy]. Progress, Moscow (1994). (in Russian)
14. Torrens, E.P.: The Nature of Creativity as Manify in the Testing. Cambridge University Press, Cambridge (1988)
15. Trofimova, N.M., Eremin, E.I.: Samoobrazovanie i tvorcheskoe razvitie lichnosti buduschego specialist [Self-education and creative development of the identity of future expert. Pedagogics]. Pedagogika 2, 42–47 (2003). (in Russian)
16. Ushinsky, K.D.: Rossiyskoye obrazovaniye [Russian education.]. Institut russkoy civilizacii, Moscow (2015). (in Russian)
17. Vygotsky, L.S.: Razvitie vysshih psihicheskih funkcii [Development of higher psychic functions]. Pedagogika, Moscow (1960). (in Russian)
18. Yakovleva, E.L.: Psihologiya razvitiya tvorcheskogo potenciala lichnosti [Development psychology of creative potential of a person]. Moscow Psychological Institute, Moscow (1997). (in Russian)

Research on Teacher Career Motivation in the Russian Pedagogical University Context

Maria V. Arkhipova$^{(\boxtimes)}$ ⓘ, Ekaterina E. Belova ⓘ,
Yulia A. Gavrikova ⓘ, and Olga A. Mineeva ⓘ

Minin Nizhny Novgorod State Pedagogical University, Novgorod, Russia
arhipovnn@yandex.ru, belova_katerina@inbox.ru,
y.a_gavrikova@mail.ru, mineevaolga@gmail.com

Abstract. In this article, we focus on the current issue of teaching career chosen by the young people in Russia. To explore the topic we present a survey aimed at analyzing students' motivation to choose teaching as their future profession. Our research was conducted on the basis of Minin Nizhny Novgorod State Pedagogical University (Russia), the survey involved 78 Bachelor and Master Degree students majoring in English as a foreign language. In the first stage we research students' motivation presented by its components, namely, motives, aims, emotions and interest. The findings demonstrate the cognitive motive as a leading one for 82% students, while 18% of students are socially motivated. 60% of students are persistent in reaching the goal, 25% is in need of extra motivation not to backtrack in case of failures, 15% - tries to find various reasons to refuse from a difficult task. 65% has a low level of overall emotional well-being. 33% is subject to lose interest in the activity facing difficulties. These results let us notice the change in students' motivation to choose teaching career throughout their years of studying. They show a constant decline in the number of highly-motivated students from the first year of studying being 75% with the absence of low level of motivation, with only 50% of senior and graduate students being highly-motivated and 10% considering teaching unattractive. There should be found reasons why highly-motivated students lose their enthusiasm to choose teacher career and means of fostering their motivation.

Keywords: Education · Career motivations · Prestige of the teaching profession · English language teaching · Motives · Level of motivation

1 Introduction

1.1 Research Problem Statement

The choice of a profession corresponding to the personality and abilities but also governed by the requirements and needs of the labor market for relevant specialists is a complex personal and social problem [1, 5, 7, 8, 12]. Scientific literature reveals a great body of extensive research on the prestige of different professions in the youth environment including the teacher education studies on the prestige of the teaching

© Springer Nature Switzerland AG 2019
Z. Anikina (Ed.): GGSSH 2019, AISC 907, pp. 105–115, 2019.
https://doi.org/10.1007/978-3-030-11473-2_13

profession and on the factors and reasons affecting the choices young people make to pursue teaching as a career [13, 14, 17, 22, 23].

Teaching plays a significant role in preparing younger generations of any society. The teacher is one of the basic figures of the educational environment who participates in the development and implementation of the educational process, who prepares a person for further education. Meanwhile, modern teaching is experiencing all the complexities and contradictions of the overall transformation of society and the processes that take place in the learning environment. Vorobyova indicated the social position of the professional group of teachers [25]. On the one hand, the society declares and recognizes the high social importance of pedagogical work, on the other hand, the social status and prestige of the teaching profession are decreasing which affect the lack of interest in the teaching profession among young people, teachers' shortage and outflow of specialists to other fields.

1.2   Literature Review

The Eurydice report (2015) analyzed the relation between the policies that regulate the teaching profession in Europe, and the attitudes, practices, and perceptions of teachers [3]. The analysis focuses on such aspects as primary teacher education, further professional development, transnational mobility, etc., but the most crucial areas for our research are teacher demographics and attractiveness of the profession. The statistics indicate that in Europe only one-third of teachers are aged under 40. While in Luxembourg, Malta, Romania and the UK over 50% of teachers are aged under 40, less than 25% of teachers in Bulgaria, Greece, Latvia, and Austria are aged less than 40 as well. Italy is the country with the oldest teaching population. Further, there is teachers' shortage, especially in some subject areas or particular geographical locations. It may be the result of a decline in prestige, deterioration in teachers' working conditions and their relatively low salaries compared with those of other intellectual professions.

Further research showed that Estonia is one of the countries with the oldest and still ageing teacher population [21]. The authors emphasized that young people do not find the teaching career sufficiently attractive to pursue. Young people's scarce interest in the teaching profession may be the result of the low prestige of the profession in the society. The results of the research intended to explore the students' perceptions and attitudes to the teaching profession in Estonia revealed that although the students highly value teachers' job (the reputation of the teaching profession is above average having a mean of 3.14 on a 5-point scale), they consider it a hard, underpaid and low-challenging job.

A study by Gomes and Palazzo in Brazil concluded that the numbers of students entering teaching degree courses, especially in the fields of Chemistry, Physics, Mathematics, and Biology are insufficient to meet the needs of the education system, thereby showing that too few young people in Brazil pursue a career in teaching [7]. One noticeable finding in the study indicated that a great number of pre-service teachers choose a teaching career as a result of the impossibility of building some other career for which they had a real vocation. As a result, it can lead to a teacher shortage in the not too distant future.

Kane and et al. investigated the perceptions of teaching in New Zealand from different aspects and inferred that the majority of senior secondary school students who participated in the study did not consider teaching to be an attractive career [10]. In their opinion, teaching was underpaid, stressful and too ordinary. For most senior students, the familiar, predictable and difficult job of teaching, paled into insignificance in the face of other, more appealing careers which offered higher salaries, more esteem and enhanced opportunities for advancement in salary and status.

In Turkey, the teaching profession is cited as one of the less reputable professions. The social status of the profession has fallen due to low teachers' income. Yüce et al. maintained that "an abundant increase in the number of teachers, and appointment of unqualified teachers have all reduced the quality of teaching and the prestige of teachers in the eyes of the public" [26].

By contrast, in their study of Singapore pre-service teachers, Low et al. introduced "the Singapore context" and reported that Singapore differed from some other countries as teachers in Singapore were well-paid and enjoyed a relatively high social status [16]. Hence, Singapore succeeded in recruiting and maintaining a high-quality teaching force. Moreover, they found that Singapore's pre-service teachers were mainly motivated by altruistic and intrinsic factors to enter teaching, and least by extrinsic factors.

One earlier study on motivation to teach among Hong Kong and mainland Chinese pre-service teachers found that teaching seemed to be more popular in Hong Kong than in the mainland China [15]. Secondary students regarded the teaching profession as the occupation which they "most wanted" and "most respected" among 20 listed occupations. In addition, extrinsic motives, such as the relatively favorable initial pay level, job security, service conditions and long holidays, were among the main reasons for joining teaching in Hong Kong. Moreover, it was argued that in mainland China and Hong Kong both, teaching attracted more students of low academic standard and from families of lower socioeconomic status. A high proportion of mainland Chinese pre-service teachers enrolled in teacher education programs in Hong Kong, however, they came from economically better-off families.

In the Russian context studies on teacher career motivation are growing in momentum. Findings by Zadonskaya revealed that one of the main social mechanisms regulating the process of choosing a profession as well as the level of vocational education and an institute is attractiveness of professions [27]. The role of the social status of a profession in the course of young people's professional self-determination lies in the fact that prestige determines attractiveness of professions and their popularity among the youth thereby orienting young people towards preferences for some professions over others. It is emphasized that the prestige of the profession is changeable in time: the hierarchy of professions changes in the public consciousness and depends on changes in the sphere of professional stratification and mobility. Didkovskaya stated that the professional self-determination is influenced not only by the esteem of the profession but also by the prestige of the university [6]. Kalimullin and Vinogradov conducted a research on the state of schoolchildren's vocational guidance in Russia and discovered that the most popular majors among the youth are economic theory, law, and management of the organization [9]. About 42.5% of school-leavers want a career in those branches. According to Vershinin and Delaryu, the trend towards a decline in the traditionally prestigious professions (teachers, engineers, and scientists) begun 10-

15 years before remained [24]. The data obtained from the survey of 2069 high school students (junior high school students and senior high school students) conducted in Volgograd in December in 2013 showed that the most prestigious professions for young people in Russia were lawyers (45.4%), doctors (39.7%), while teaching attracted too few young people (21.3%). Similarly, Denisenko relying on the data of the Russian Public Opinion Research Center (VTsIOM) claimed that the most prestigious professions in Russia are lawyers (20%), doctors and economists (12% each). Bankers rank next (7%) with only 4% of the respondents believe that it is prestigious to be a teacher [5]. However, despite teaching profession being not one of the most popular professions, today it is in great demand on the labor market in Russia.

Karpukhova in a small scale study including 60 young people (aged from 14 to 30) from Smolensk Region, Russia, found that, on the one hand, 100% of the participants noticed the importance of the teaching profession. On the other hand, 75% of them was persuaded that the teaching profession was not desirable in the modern society because of low salary and lack of respect for teachers in society, government's low regard for education, poor students' knowledge of the subjects, etc. The question "Would you like to choose teaching as your future career /second (third) major?" 64% answered in the negative and gave their grounds, some of the reasons being low salary, their wish to make a career in the other field, admitting that their personality does not suit the teaching profession, that the teaching profession is morally and physically exhausting, and they do not like the profession. 26% answered affirmatively and the most cited reasons were love for teaching and love for working with children. 10% had difficulties to answer the question [11].

Other lines of research have reported that the teaching profession in Russia has failed to attract young people. Actually, pre-service teachers often join the teacher education program, not entering the profession per se or because they have come to the education faculty accidentally or they drop out of teaching after a short period of time [2, 23]. A survey of pre-service teachers by Poyarova et al. showed that 35.1% of the prospective student teachers entered the teacher's training college only to have any higher education degree [19]. The experimental results of Regus and Ermilova revealed that 37% with a bachelor degree in pedagogics did not intend working as teachers, 12% have not decided yet [20]. According to Popova the most attractive professions in Russia include the field of programming and IT, followed by economy, finance, loans and banking. Engineers, economists and doctors are also mentioned in the ranking, while teacher's profession was not in the list [18].

1.3   Basic Assumptions

Professional orientation of students primarily depends on their academic motivation which includes the following components: aims, motives, interest and emotions, which, in their turn, greatly influence the process of studying and further profession-centered orientation.

We present a survey to better understand students' motivation to choose pedagogical university in Russia, and the following questions constituting the foundation of our study: 1) What are the perceptions and motives for choosing teaching profession by the modern generation of pedagogical university students? 2) How does the level of motivation to choose teaching career by students change throughout the years of studying?

To reach the aim we used the following methods: theoretical literature analysis, diagnostic method of observation and questionnaire poll. The authors developed the questionnaire basing on the following underlying principles: a) voluntary participation of respondents in the research, their privacy and anonymity; b) no offensive, discriminatory language in the questionnaire and interview; c) the highest level of objectivity analyzing the research data [4]. The questionnaire clarified the aim of the survey, contained an instruction and a list of questions of open and closed types.

2   Materials and Methods

Our research was conducted in Minin Nizhny Novgorod State Pedagogical University (Russia) involving 78 students – 1–4-year students of the bachelor programme "Foreign language and primary school education" and 1–2-year students of the master course "Foreign language (English)".

The first stage of our research is focused on motivation components. The tested students were given a questionnaire with 8 statements regarding their professional orientation to assess students' motivation for studying the English language, the testees were to agree or disagree with the offered statements. The questions refer to 4 different components: components of motives, aim, emotion and interest:

1. The motive component may be illustrated by the statements:
   "If possible, I try to use my groupmates' materials and notes on the subject and ask for somebody's assistance",
   "I consider all the knowledge on the subject valuable".
2. The aim component is evident from the statements:
   "I am usually persistent in reaching the goal, do not backtrack, even in case of obstacles and failure",
   "While studying this subject the knowledge I get at the tutorials is enough for me".
3. The emotion component is represented by the statements:
   "This very subject does not come easy to me and I have to make myself do the training tasks",
   "If I miss tutorials in this subject due to my illness and for some other reasons I get very much upset".
4. According to the interest component,
   "I consider that complex theoretical issues in this subject can be neglected"
   "My interests and hobbies in my leisure time are related to the subject".

In the second stage of our research the survey results were studied and decoded in the way that a certain score demonstrates a certain level of motivation – high, medium and low.

3   Research Data

Motivation is based on motives. There exist two major groups of motives: cognitive and social. The findings of the mentioned questionnaire demonstrate the presence of cognitive motives of 82% students. 41% students have self-educational motives as part of cognitive motives: these students study additional materials independently, make an effort to solve a very complicated situation, do the tasks independently and do not like to be advised, work successfully without external control. The rest 18% of students is socially motivated, they fully recognize the duty and the social significance of studying, but are dependent on relationship with the teacher (13%) who demands certain output. 5% students strive to be recognized by their environment, to assert their role and position with the help of good marks, giving much importance to marks.

One of the learner's motivation components is ability of goal setting. The students' ability to define goals at a certain stage speaks about the maturity of their activity. The questionnaire contained issues referring to goal-setting in the professional sphere. As the survey shows, 60% of students gave affirmative answers to the statement "*I am usually persistent in reaching the goal, do not backtrack, even in case of obstacles and failure.*" 25% of students is not very persistent, some of them requiring strict monitoring for extra motivation. 15% tries to find various reasons to refuse from a difficult task, that certainly means their low level of motivation and lack of aims in the process of studying at a pedagogical higher educational institution.

Emotions are also of great importance for academic motivation. By means of generating positive emotions in the process of studying it is possible to influence students' attitude not only to the process of studying but also to the prospective professional orientation. Emotions are closely connected with motives for activity, and each step of the educational process is followed by emotions. Being limited in time during a certain type of activity a person can be overwhelmed by emotions and, consequently, may work much worse and considerably more slowly, but sometimes under the same circumstances the outcome may be quite opposite. The productive students whose learning activity is supported by emotions while working with new material, feel excited, curious and absorbed in the subject. 35% students work with optimism, hope for success, they are very active and prone to take the initiative. The students perceive their abilities to solve complicated problems while studying. However, negative emotions while studying are inevitable, and 65% students experience frustration, deep emotional pain, or complete indifference to studying, namely, foreign languages, being ready to reconsider their failures.

The positive attitude to activity results in interest as an important reason for an action. The cognitive interest is reflected through students' emotional attitude to the object of cognition. 67% of students are interested in studying the foreign language, read extra literature extensively, have a lot of discussion with their groupmates, have their hobbies connected with the languages. While 33% is subject to lose interest in the activity very soon, especially having faced a failure.

The component analysis of motivation of 1–4-year students of the bachelor programme "Foreign language and primary school education" (hereinafter 1–4 year) and 1–2-year students of the master course "Foreign language (English)" (hereinafter 5–6 year) as well as the additional survey of the results enabled us to estimate accurately students' motivation for studying foreign languages, effort or lack of effort to master the profession and succeed in it.

Motivation is the most essential incentive for students' interest in learning English and intellectual activity.

First-year students are highly motivated to study English, 75% and 25% representing a high and a medium level of motivation respectively. The absence of students with a low level of motivation confirms the idea of the mature choice of the University to study foreign languages as their profession and vocation, being a good indicator for a pedagogical institution.

The situation is different with second-year students. There is a 15% decline in high level of motivation and, correspondingly, 15% increase of a medium level of motivation, thus leading to 60% and 40% respectively.

Possessing certain skills and competences in the chosen profession, the third-year students tested have already acquired their own view on the profession and they did not find it difficult to respond to the questionnaire issues. As the survey shows, the majority of the participants (77%) demonstrate a medium level of motivation, while 23% of students is still highly motivated to acquire the chosen profession. The medium level of motivation can be observed even from the first issues, where they state that the tasks they do while studying do not appeal to them. They do not take time to go deep into the subject but enjoy the knowledge obtained during the lessons as they do not find the material interesting and the tasks effective and significant. The third-year students also showed fear of failure that considerably changes students' motivation degree and their productivity in studying and work. Only 16% possesses motivation for success that implies that obstacles challenge them and increase the appeal of a task and attractiveness of the chosen route. They are insistent in reaching the goal and plan their future prudently and well in advance.

The fourth-year students are 50% highly-, 43% medium-, 7% low-motivated.

The same questionnaire was given to the 1- and 2-year master's degree students of the "Foreign language (English)" major. The first-year master students (hereinafter 5th year) divided evenly showing high and medium level of inner motivation.

The motivation indicators are demonstrated immediately from the initial reactions, where the students were to agree with the statements "*Studying the subject will enable me to learn a lot of important things and reveal my abilities*", "*I find the subject interesting, I would like to learn as much as possible*". The 100% answers are positive. Inner motivation can be easily observed in the number of positive reactions – 80% – to the statement "*If a difficulty arises, if something goes wrong, I try to do my best to get to the bottom of the case*". The majority of students tries to work independently, does not want any assistance. Their intense reading in a foreign language outside the classroom in their free time gives them a lot of advantages in their future profession and tells a lot about their aspiration to study, develop and master the language on the way to their future profession, competence and self-development. 58% of students is planning to choose teaching languages as a career and are ready to overcome obstacles and find ways of development in this profession. The high score, absence of students with fear of failure and low motivation characterize the students' choice as mature and thoroughly thought over. Studying for the master's degree, students have often already acquired a certain experience of working as an English teacher or a private tutor. They already have an idea regarding the skills and abilities required for the profession.

The master degree graduates (indicated as 6th year) clearly understand what the teaching profession implies due to their own experience and perceive its difficulties and disadvantages. The students are still motivated to study the foreign language, recognizing its importance and feasibility in the modern world. 50% are highly motivated to continue their education, are willing to improve their knowledge, study something new, reach their goal and succeed. The medium level of motivation was demonstrated by 40% of students. Thus, 10% of students is not interested in the subject and they are often very much indifferent to studying in general.

Basing on the data obtained from the students of a pedagogical institution, the experiment reveals the following indicators of students' motivation dynamics during the entire period of studying (6 years) (see Fig. 1).

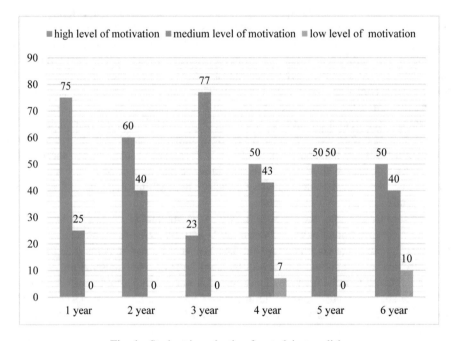

Fig. 1. Students' motivation for studying english.

The above-mentioned diagram clearly shows the actual dynamics of students' motivation for studying English. The blue column in the diagram illustrates a decline in the number of highly-motivated students, the most considerable one (by 37%) being among 2nd and 3rd-year students. The decline and the surge may be connected with the fact that studying foreign languages is no easy matter and it is followed by thorough regular exercises. The effort, intelligence and patience involved and the time consumed turn students away from the process of studying no matter what the chosen profession is, how mature their choice was and how much well-paid their job is expected to be.

High motivation remains unchanged (50%) in the other courses of studying, which is rather a favourable indicator for an educational institution.

Medium motivation shown with the red column increases up to the third year and then decreases by 34% being still represented by 40–50% of students. The growth of medium motivation up to the third year can be accounted for by the difficulties awaiting mainly the second- and third-year-students on their prospective way to success and appropriate salary in the future. As one of the students appeared to presume, the surge can be connected with the changes in the curriculum leading to studying all the difficult subjects in one year.

The low level of motivation is presented by bachelor and master graduates only. When the students are at the final stage of their educational program before they get a job according to the education they have acquired or they change their mind for a different job.

4   Conclusion

Today the list of prestigious professions in Russia has changed significantly compared with that of ten years earlier. Over the past decades the top of the most prestigious professions in the world are headed by programmers and IT-specialists. Computerization and rapid development of technology made this profession highly-paid and in demand for many years. The profession of the teacher in Russia used to be prestigious, and teachers served as role models. Nowadays the teaching profession in Russia is no longer considered to be attractive.

As a rule, when choosing a profession or direction in a university, the person already has a certain idea and knowledge about it. However, after graduation only a small number of people stay in the profession. Often the percentage of graduates, working by profession is very small. There is an urgent need to find ways of stimulating and motivating students to take a teaching career after graduating. It is motivation that forms educational trajectories, on the basis of which professional preferences of young people are formed and a real choice is made.

The limitation of our study is that we have examined the career motivations of students majoring only in English as a foreign language and studying in one pedagogical University. Thus, the findings do not reflect the full picture on the topic examined and similar studies, including other majors and universities, are necessary to be conducted for a more accurate, detailed and comprehensive view of career choice.

We see the prospects for our further research in finding the reasons why students throughout their years of studying at a higher school lose their enthusiasm, interest and motivation to choose teacher career, and the means promoting well-being and greater job satisfaction.

References

1. Berdova, M.V.: Gotovnost' starsheklasnikov k vyboru budushchey professii kak social'no-pedagogicheskaya problema [The readiness of senior pupils to choose their future profession

as a socio-pedagogical problem]. Probl. Mod. Pedagog. Educ. **48**(3), 43–49 (2015). (in Russian)

2. Bicheva, I.B., Filatova, O.M.: Perspektivy professional'noj podgotovki pedagogov: aksiologicheskij podhod [Prospects of professional training of teachers: axiological approach]. Vestn. Minin Univ. Prof. Educ. **6**(2), 3 (2018). (in Russian)

3. Birch, P., Balcon, M.-P., Borodankova, O., Ducout, O., Sekhri, S.: The Teaching Profession in Europe: Practices, Perceptions, and Policies. Eurydice Report. Publication Office of the European Union, Luxembourg (2015)

4. Bryman, A., Bell, E.: Business Research Methods, 2nd edn. Oxford University Press, Oxford (2007)

5. Denisenko, S.E.: Prestizh i vostrebovannost' professiy - vazhnejshie usloviya professional'nogo samoopredeleniya molodyozhi v usloviyah transformacii v mirovye globaliza-ciobbye processy [Prestige and demand of professions the most important conditions of youth's professional self-identification in the conditions of transformation into the world globalization process]. Mod. Probl. Ways Their Solut. Sci. Prod. Educ. **1**(1), 111–115 (2016). (in Russian)

6. Didkovskaya, Y.V.: Professional'noe samoopredelenie molodezhi: sociologicheskij analiz [Professional self-identification of youth: sociological analysis]. GOU UGTU-UPI, Ekaterinburg (2004). (in Russian)

7. Gomes, C. A., Palazzo, J.: Teaching career's attraction and rejection factors: analysis of students and graduates perceptions in teacher education programs. Ensaio Aval. Pol. Públ. Educ. **25**(94), 90–113 (2017)

8. Gordienko, I.V.: Professional'naya orientachiya i pedagogicheskoye soprovozhdenie profil'noy i predprofil'noy podgotovki obuchayushihsya k vyboru pedagogicheskoy professii: regional'nyi aspect [Professional orientation and pedagogical support of the profile and pre-profile training of the pedagogical professions learning to the election of the pedagogical professions: regional aspect]. Azimuth Sci. Res. Pedagog. Psychol. **6**(4/21), 50–53 (2017). (in Russian)

9. Kalimullin, A.M., Vinogradov, V.L.: Professional'naya orientaciya shkol'nikov: sostoyanie problemy i puti resheniya [Schoolchildren's vocational guidance: state of the problem and solutions]. Educ. Self-Dev. **6**(34), 148–155 (2012). (in Russian)

10. Kane, R.G., Mallon, M.: Perceptions of Teachers and Teaching. Ministry of Education, New Zealand (2006)

11. Karpukhova, M.I.: Prestizh professii uchitelya v sovremennoj molodyozhnoy srede [Prestige of the teacher's profession in the modern youth environment]. Youth Sci. Acute Probl. Pedagog. Psychol. **2**, 65–70 (2017). (in Russian)

12. Kaznacheeva, S.N., Bondarenko, V.A.: Uchyot osobennostej kontingenta studentov-zaochnikov s cel'yu razvitiya poznavatel'noj aktivnosti pri obuchenii inostrannym yazykam [The peculiarities of extra-mural students contingent for the purpose of development of cognitive activity in learning foreign languages]. Vestn. Minin Univ. Gen. Educ. Issues **4**, 5 (2017). (in Russian)

13. Kobzareva, I.I., Volobueva, E.V.: Problema vybora budushchey professii v kontekste sovremennogo profil'nogo obucheniya [The problem of choosing a future profession in the context of contemporary profile training]. World Sci. Cult. Educ. **2**(57), 268–270 (2016). (in Russian)

14. Krivtsova, N.S.: O formirovanii polozhitel'nogo obraza professii u starsheklassnikov: osnovnye podhody i rezul'taty [On the formation of a positive image of the profession among high school students: basic approaches and results]. Balt. Hum.Itarian J. **6**(3/20), 204–209 (2017). (in Russian)

15. Lai, K., Chan, K., Ko, K., So, K.: Teaching as a career: a perspective from Hong Kong senior secondary students. J. Educ. Teach. Int. Res. Pedagog. **31**(3), 153–168 (2005)
16. Low, E., Ng, P., Hui, C., Cai, L.: Teaching as a career choice: triggers and drivers. Aust. J. Teach. Educ. **42**(2), 28–46 (2017)
17. Nemova, O.A., Svadbina, T.V., Zimina, E.K., Kostyleva, E.A., Tsyplakova, S.A., Shevchenko, N.A.: Professional orientation of youth: problems and prospects. J. Entrep. Educ. **20**(3), 6, 1 (2017)
18. Popova, E.S.: Motivacija i vybor v obrazovatel'nyh strategijah molodezhi [Motivation and future choice in educational strategies of youth]. High. Educ. Russ. **1**, 32–37 (2014). (in Russian)
19. Poyarova, T.A., Nosova, N.V., Fokina, I.V.: Pedagogicheskoe obrasovanie: motivy vybora professii sovremennoj molodezh'yu [Teacher education: the motives of choice of profession modern youth]. Probl. Mod. Pedagog. Educ. **53**(4), 390–395 (2016). (in Russian)
20. Regus, L.A., Ermilova, E.E.: Professional'nyj vybor vyposknikov bakalavriata kak reshenie prognosticheskoj zadachi [The professional choice of graduates with a bachelor degree as a prognostic task solution]. Educ. Sci. J. **19**(8), 75–89 (2017). (in Russian)
21. Saks, K., Soosaar, R., Ilves, H.: The students' perceptions and attitudes to teaching profession, the case of Estonia. In: Proceedings of 7th International Conference on Education and Educational Psychology, pp. 470–481. Future Academy, Rhodes, Greece (2016)
22. Serebryakova, T., Morozova, L., Kochneva, E., Zharova, D., Kolarkova, O., Kostina, O.: Social and psychological adaptation of higher school students: experimental study. Man India **97**(9), 151–162 (2017)
23. Shabanova, T.L., Tarabakina, L.V.: Issledovanie ehmocional'noj zrelosti u studentov pedagogicheskogo vuza [A study of emotional maturity of students of a pedagogical university]. Vestn. Minin Univ. Pedagog. Psychol. **6**(1), 13 (2018). (in Russian)
24. Vershinin, E.G., Delaryu, V.V.: Prestizhnost' professij kak refleksija cennostnyh orientacij podrostkov g. Volgograda [Occupational prestige as reflection of the youth value system in Volgograd]. Sociol. City **4**(24), 27–36 (2014). (in Russian)
25. Vorobyova, I.V.: Prestizh professii uchitelya [Prestige of the teaching profession]. RSUH/RGGU Bulletin. "Philosophy. Social Studies. Art Studies" Series **4**(126), 247–254 (2014). (in Russian)
26. Yüce, K., Sahin, E.Y., Kocer, Ö., Kana, F.: Motivations for choosing teaching as a career: a perspective of pre-service teachers from a Turkish context. Asia Pac. Educ. Rev. **14**(3), 295–306 (2013)
27. Zadonskaya, I.A.: Rol' institutov professii i obrasovaniya v regulyacii processa professional'nogo samoopredeleniya studencheskoj molodezhi [Role of institutes of professions and education in regulation of process of professional self-identification of student's youth]. Tambov Univ. Rev. Ser. Humanit. **1**(105), 173–179 (2012). (in Russian)

Multilingual Regional Contest of Student-Created Videos: Pilot Project

Marina Bovtenko$^{(\boxtimes)}$ and Maiya Morozova

Novosibirsk State Technical University, Prospekt Marksa 20,
630073 Novosibirsk, Russian Federation
bovtenko@is.nstu.ru, morozova@corp.nstu.ru

Abstract. The article presents the Novosibirsk State Technical University's (NSTU) pilot project on regional contest of student videos in foreign languages «My University» in the context of current approaches to new literacies—digital, media, multimodal. The study is based on analysis of 60 video projects in four languages created by students of 28 universities from 15 cities of Russia and Kazakhstan and on the feedback from students and teachers.

According to new literacies concepts, video is considered as a multimodal message, which requires a wide range of competences to create it - from functional digital and media to linguistic, audio, video, gestural, spatial. The contest procedures and contest specific rubric developed for summative assessment are described; the strong and weak points of the projects and current level of students' media competences revealed are presented. Recommendations are offered regarding further contest development to be a valuable resource for improvement of students' and teachers' media and multiliteracy competences.

Keywords: Student video · Media literacy · Multiliteracy ·
Video project contest · Foreign languages

1 Introduction

Films and videos are valuable resources for language learning and teaching. The efficient ways of their use for development of language skills and cross-cultural competence are presented in great number of studies, practice projects, language classes' observations, resource books for teachers [11, 13], and professional development programs.

Video creation is a task which becomes more and more common and popular in language classes for students of different ages and levels of language proficiency. Variety of modern devices, software and applications easily allow users to film and edit videos, upload videos to social media and cloud platforms, comment and evaluate other users' videos online, and share created videos through a variety of e-communication channels. Video is an integral part of not only everyday life but also of business communication in such genres as video CV, job interview video, purpose statement video, project proposal video, video report, etc. Contests on customers' created visuals, mostly spreading through social media, is common practice in company's customers loyalty programs.

© Springer Nature Switzerland AG 2019
Z. Anikina (Ed.): GGSSH 2019, AISC 907, pp. 116–123, 2019.
https://doi.org/10.1007/978-3-030-11473-2_14

Participation in video creation contests of any level and types—class, school, university, city, regional, national, international; academic, business, customers'—motivates students to develop creativity, allows them to demonstrate their achievements, acquire new knowledge, skills and experiences, provide opportunities for independent assessment, self-promotion and portfolio development.

Current approaches to exploring phenomenon of student-created videos in language learning and teaching imply concepts of new literacies—digital, media, multiliteracy [12, 16]. Media literacy's key components include understanding the role of media in a society, media critical evaluation, engagement with media, and media production in different formats [5, 24, 26]. One of the latest studies on media literacy proves the necessity "to re-think media literacy in the age of platforms" [4, p. 11] as traditional notions of media literacy are challenged by expanded engagement with media through audience-generated content in social media. "This new engagement includes more active participation by individuals, but also more influence from platforms and media creators, raising questions about responsibility and control" [4, p. 14]. Multiliteracies as it is defined by New London Group [19] deals with multimodal information design in digital age which includes linguistic, visual, audio, gestural, and spatial modes. The efficiency of students' multiliteracy development in foreign language teaching through a variety of tasks and genres—presentations, multimedia compositions, web-texts creation, digital/video storytelling, video projects,—are widely discussed and explored [27, p. 29]. However, it is also noticed that though students are active users of the great number of digital technologies it doesn't guarantee high level of their media and multimodal competences [4, 7].

Among the research questions arising in multiliteracy exploration are the issues of genres modifications [2, 3, 10, 14], pedagogical implications of multimodality [19, 20], and assessment of multimodal artifacts [21, 23, 25]. The researchers pointed out that "teachers may be consciously or unconsciously working with "a paradigm of assessment rooted in a print-based theoretic culture" [cf. 21, p. 4].

The purpose of this work is to reveal current level of university students' new literacies demonstrated in pilot regional contest of student video projects in foreign languages and discuss the ways of making the contest a valuable resource for further development of students' and teachers' media and multiliteracy.

2   The Study

2.1   Contest Rules and Procedures

The pilot regional contest of student video projects in foreign languages "My University" was held in Novosibirsk State Technical University (NSTU) in March–April, 2018 within the framework of the university's program on development of students, teachers and staff foreign language proficiency. The contest topic was selected for pilot project as it is commonly included in Russian university's foreign language curricula. Moreover, it provided the basis for the integration into classroom activities and projects.

Students, undergraduates, and postgraduates of non-linguistic majors of Russian universities were invited to take part in the contest with the projects on three subtopics:

"My University/faculty/major";
"Research in my university/my research";
"Students' life (projects, events, etc.)."

Video could be applied as individual or group project in one of six languages—English, German, French, Chinese, Japanese, and Russian as a foreign language.

The requirements to the projects included: genres (non-fiction and non-commercial), duration (no more than 4–7 min); obligatory audio text, recorded by students; title and credits.

The number of projects from one university was not limited. Submitting the application, participants guaranteed compliance with copyrights and agreed to personal data processing and possible use of projects in non-commercial purposes. The projects were available on the university's YouTube channel.

Independent jury consisted of universities and foreign languages centers teachers from Novosibirsk and other cities; the regional representatives of DAAD and Macmillan Publishing, digital video specialists, advertising and public relations officers, representatives of the university educational department and international services. As multilingual contests are not common practice for language contests and olympiads [28] it was a new experience to manage the work of the jury, students and teachers in a distance mode.

The list of projects and the winners were published on the NSTU website, all students and teachers received official confirmation of participation in the contest, and the winners were awarded with diplomas.

2.2   Data Collection and Analysis

Contest Participants. 60 projects—59 projects from Russia and a project from Kazakhstan competed in the contest. They represented 28 universities from 15 cities. Total number of students participated in the contest was over 140. Although the competition was announced for students of non-linguistic majors—natural science, engineering, business, medicine, architecture, agriculture, and humanities—five projects were created by students of teachers' training, regional studies, international relations departments. Most of the projects were made by teams and only 3 projects applied were individual. Some teams consisted of students of the same academic groups, and there were also teams of students of different years of study and majors.

Foreign Languages. The video projects were created in 4 languages: English (40 projects), German (17 projects), French (2 projects), and Japanese (1 project) (Fig. 1).

The level of participants' foreign language proficiency varied from elementary to upper intermediate.

Projects Topics. The topic "My University" was selected for 29 projects, "My faculty"—7, "My major"—5, one project was devoted to research, and the topic

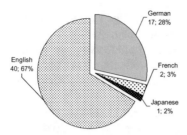

Fig. 1. Distribution of the projects by languages.

"Students' life" was presented by 18 projects. Among the projects in "My University" section there was a video about a Chinese university created by NSTU postgraduate student who was in China at that time as an exchange student (Fig. 2).

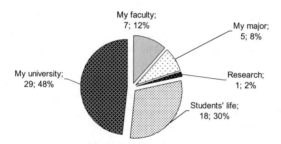

Fig. 2. Distribution of the projects by topics.

Video Assessment. The winners were defined in each language and topic. The jury graded the videos using the rubric developed for the contest. It included the following groups of criteria: content, foreign language, video, design, copyright. The members of the jury could also add their own criteria.

Contest Evaluation. Feedback forms were collected from students, teachers, members of jury and organizing committee through e-mail and discussions during post-event seminars for teachers.

2.3 Results and Discussions

Video Project Rubric's Focuses. The rubric developed for the projects' assessment was based on customized Rubistar Multimedia and Oral Projects rubric templates [22] and on the analysis of more than 30 rubrics developed for video projects and available through open educational resources web-sites and teaching practices as well as research on educational video-projects. [1, 8, 9, 12, 16, 27]. The rubric was focused on the following students' media and multimodal competences: video genres, content, video production and editing, foreign language writing, pronunciation, speaking skills, sound mixing, graphic design, branding and copyright issues.

As student presented their final products for the contest, such typical video project rubrics criteria as team- and pre-production work were excluded. The design, branding and copyright issues were limited to a university's name and logo, copyright statements and references.

The weight of each criterion in total score was the following: content (informative, interesting, corresponding to the genre)—20%, foreign language (adequate (written/spoken) style; use of academic terms; grammar, intonation, pronunciation, spelling, punctuation)—33%, video (locations; video, sound, graphics combinations, remixing—if used)—20%, design (titles, credits, a university brand components)—20%, copyright statements and references—7%. The members of the jury could add their own criteria—maximum +20%. Among the criteria offered by members of jury were: the sense of humor, creative ways of video shooting and editing, translation of interviews in a foreign language.

Students' Competences Revealed. Though the contest participants had different level of foreign language proficiency and experience in creating and editing video, the projects had a lot of common both in their strong and weak points. The information genres requirements could not meet the only project. Such linguistic literacies as proper use of academic terms, grammar, intonation, pronunciation, and spelling in most cases were often awarded with high marks. Lower marks were given to the audio texts of commentaries, interviews, voice-over narrations as students often used written style. Combination of informative and original content, dynamic video, and students' engagement were the features of the winners' projects. However, approximately half of projects didn't have interesting, creative, and catchy titles, and they had to be numbered in all lists of projects. Though the branding, anti-plagiarism and copyright issues are supposed to be well-known to all the students it is turned out that such university brand components as logo, full and abbreviated university title, as well lists of references for remixed video or audio were not used in about a third of projects. Some videos were marked by the YouTube Content ID system as "includes copyrighted content" [6] and could not be uploaded to NSTU YouTube Channel as the authors used copyright-protected music.

Video Creation. Integration Projects into Curricula. The projects were a part of students' out-of-class independent work and were integrated into foreign language courses. In both cases there were some extreme approaches. In the case of out-of-class work, the teacher of a foreign language could not take part in the preparation of the project. Another option was that the video was considered as a real project for the university promotion and the project was supervised by a teacher of a foreign language and university's international department. In case of projects integrated into language learning curriculum, the work on the content and production issues were considered as independent—every week one class was devoted to video pre-production work, while all technical issues of video creation were assigned as independent students' work.

Video Creation. Teachers' Comments. As the most challenging moments in the process of working on the project the teachers mentioned that many students had no experience in using video equipment, shooting and video editing, writing a video script, communicating to find information, interviewing, feeling confident on a camera, working in a team. The teachers also assumed that video creation process is rather time-consuming.

Contest Evaluation. Students' Feedback. Positive students' feedback was associated with the opportunity to participate in such a contest, improve foreign language skills, show creativity, learn new skills related to creating and editing videos, and teamwork:

"This format of studying a foreign language is cool! Thank you for the opportunity to participate in the competition of video projects!"

"Thank you for the opportunity to show the abilities which I did not even suspect to have. The motivation for learning the language has definitely improved:)."

"It was not so easy to work in the team, but we did it. And I'm glad that I managed to improve my phonetics!"

"I did not expect I would manage the editing video... Thanks for the help and support!"

At the same time, students would like to see a different format for summarizing the results of the competition—publishing information about the winners on the web-site was not enough for students from Novosibirsk, and though the request for ceremony of awarding the winners was taken into account, the participants of the ceremony offered to turn it to a real festive event.

Contest Evaluation. Teachers' and Members of Organizing Committee Feedback. Teachers as well as students gave very positive feedback to the idea of the contest, and though practice of student video projects varied in different universities, both students' teams and teachers asked for holding the contest on regular basis. As the main positive output of the contest the teachers mentioned the students' involvement, changing perception of video production process, and improvement of foreign language skills—regardless of whether they had won the contest. The teachers appreciated the official letters of gratitude from organizing committee for their help in promotion and recognition of the work done in their universities.

The organizing committee highly evaluated the efficient work of participants and members of jury in distance mode. As for assessment of the projects as multimodal product it was noticed that not all members of the jury were ready to evaluate students' projects according to the set of criteria offered—foreign language teachers paid special attention to students' level of foreign language proficiency, while representatives of public relations and educational departments tended to consider the students' videos created in the framework of foreign languages learning curricula as ready products for promoting the universities and assessed students' video projects as professional promos.

3   Conclusion

The pilot regional contest of student video projects in foreign languages "My University" attracted students of various majors with different levels of foreign language proficiency not only from Siberian Federal District, but also from other regions of Russia—Central, Northwestern, Ural, Volga Federal Districts, and Kazakhstan. The most valuable contest outcomes are development of new competencies of students, expanding the scope of language learning projects, independent assessments of students' achievements, focus on multimodal productive and receptive skills, cooperation between universities, language centers and education, public relations and video production professional communities.

The analysis of projects and feedback from students, teachers, members of jury and organizing committee allowed to reveal current level of students' media and multiliteracy and to offer recommendations for students, teachers, and the contest's team. For student video projects to be efficient, special attention should be paid to such competences as story writing and creation of storyboard, video interviewer's and interviewee's skills, voice-over narrations, basics of video shooting, editing, remixing, design of graphical elements, sound mixing, brand presentation, copyright for social media. It is also important to focus on students' awareness in varieties of video genres [15, 17, 18], negotiation skills in the context of video creation both in and outside the classroom, ethics and aesthetics issues of video production [29]; post-production video promotion, including event management skills. Taking into account the video project assessment rubric created with special focus on multiliteracy [27: 36–37] it is possible to customize it for the contest format. Development of educational materials, conducting workshops on video production, a web-site for sharing experience, revision of the contest terms and student evaluation criteria will make the contest a valuable resource for improvement of students' and teachers' media and multiliteracy competences.

References

1. Abramova, I., Sherekhova, O., Shishmolina, E.: Student created videos in English language teaching. Vysshee obrazovanie v Rossii **6**(213), 36–43 (2017)
2. Baldwin, K.M.: Multimodal assessment in action: what we really value in new media texts, Doctoral Dissertation (2016), https://scholarworks.umass.edu/dissertations_2/851. Last accessed 15 Aug 2018
3. Bowen, T., Whithaus, C. (eds.): Multimodal Literacies and Emerging Genres. University of Pittsburgh Press, Pittsburgh (2013)
4. Bulger, M., Davison, P.: The promises, challenges and futures of media literacy. J. Media Lit. Educ. **10**(1), 1–21 (2018)
5. Chelysheva, I.: Strategii razvitiya rossiickogo mediaobrazovaniya: tradicii i innovacii. Mediaobrazovanie **1**, 71–77 (2016)
6. Copyright on YouTube. Rights enforcement with Content ID.: In: YouTube Creators Academy. https://creatoracademy.youtube.com/page/course/copyright?hl=en. Last accessed 15 Aug 2018
7. Fedorov, A.: Levels of university students' media competence: analysis of test results. Magister Dixit **1**, 17–56 (2011)
8. Forester, L.A., Meyer, E.: Implementing student-produced video projects in language course. Unterrichtspraxis **2**(48), 192–210 (2015)
9. Gaparyan, L.A.: Metod videoproektov kak sredstvo formirovaniya umenii inoyazuchnogo dialogicheskogo vzaimodeistviya. Pedagogicheskoe obrasovanie v Rossii **3**, 80–86 (2012)

10. Godwin-Jones, R.: Technologies digital video revisited: storytelling, conferencing, remixing. Lang. Learn. Technol. **16**, 1–9 (2012)
11. Goldstein, B., Driver, P.: Language Learning with Digital Video. Cambridge University Press, Cambridge (2015)
12. Gremler, C., Wielander, E.: The benefits of student-led video production in the language for business classroom. In: Xiang, C.H. (ed.) Cases on Audio-Visual Media in Language Education, pp. 155–193 (2018)
13. Keddie, J.: Bringing Online Video into the Classroom. Oxford University Press, Oxford (2014)
14. Kibrik, A.A.: Multilodalnaya lingvistika. Cognitivnue issledovaniya **4**, 134–152 (2010)
15. Kolesnikova, N., Ridnaja, Ju: Zhanrovaja kompetencija kak komponent mezhkul'turnoj inojazychnoj kompetencii magistrantov. In: Kuznetsova, D.I., Almazova, N.I., Valieva, F.I., Haljapina, L.P. (eds.) Innovacionnye idei i podhody k integrirovannomu obucheniju inostrannym jazykam i professional'nym disciplinam v sisteme vysshego obrazovanija, International school-conference, pp. 62–64. Peter the Great St. Petersburg Polytechnic University, St. Petersburg (2017)
16. Leontyeva, T.: Videotechnologii v processe podgotovki stydentov izuchaucshih inostrannue yazyki k krosskukturnoy kommunikazii. Jazyk i kul'tura **4**, 81–88 (2008)
17. Millar, D.: Promoting genre awareness in the EFL classroom. Engl. Forum **2**, 2–10 (2011)
18. Mogoş, A., Trofin, C.: YouTube video genres. Amateur how-to videos versus professional tutorials. Acta Universitatis Danubius. Communicatio **9**(2), 38–48 (2015)
19. New London Group: A pedagogy of multiliteracies: designing social futures. Harv. Educ. Rev. **66**(1), 60–93 (1996)
20. Paesani, K., Allen, H.W., Dupuy, B.A.: Multiliteracies Framework for Collegiate Foreign Language Teaching. Pearson, London (2015)
21. Ross, J., Bell A., Curwood J.S.: Assessment in a digital age: rethinking multimodal artefacts in higher education. Learning & Teaching Conference 2018 Outputs. University of Edinburgh, Edinburgh (2018)
22. RUBISTAR. Customizonable Rubrics. http://rubistar.4teachers.org/index.php?screen= NewRubric&section_id=1#01. Last accessed 15 Aug 2018
23. Sills, E.: Multimodal assessment as disciplinary sensemaking: beyond rubrics to frameworks. J. Writ. Assess. **9**(2), 1–27 (2016)
24. Thoman, E., Jolls, T.: Literacy for the 21 Century: An Overview & Orientation Guide to Media Literacy Education. A Framework for Learning and Teaching in a Media Age. Center for Media Literacy, Ontario (2005)
25. Wierszewski, E.: Something old, something new: evaluative criteria in teacher responses to student multimodal texts. In: McKee, H.A., DeVoss, D.N. (eds.) Digital Writing Assessment & Evaluation, pp. 389–411 (2013)
26. Wilson, C., Grizzle, A., Tuazon, R., Akyempong, K., Cheung, Chi K.: Media and Information Literacy Curriculum for Teachers. United Nations Educational, Scientific and Cultural Organization, Paris (2011)
27. Yeh, H.-C.: Exploring the perceived benefits of the process of multimodal video making in developing multiliteracies. Lang. Learn. Technol. **22**(2), 28–37 (2018)
28. Zelenina, T., Butorina, N.: Mnogoyazychnaya olimpiada mladshih shkolnikov "Yunuy poliglot-2017" (v proektnom setevom vzaimodeystvii). Pedagogicheskoe obrazovanie v Rossii **12**, 153–157 (2017)
29. Zettl, H.: Sight, Sound, Motion: Applied Media Aesthetics. Wadsworth Cengage Learning, Boston (2011)

Multilingual Pedagogy in the Russian North: Theoretical and Applied Issues

Maria Druzhinina$^{(\boxtimes)}$ ⓘ and Inga Zashikhina ⓘ

Northern (Arctic) Federal University named after M.V. Lomonosov,
163000 Archangelsk, Russian Federation
{m.druzhinina, i.zashikhina}@narfu.ru

Abstract. The article highlights the issue of multilingualism in multidisciplinary perspective. The authors aim at solving the problem of defining the multilingualism notion, as compared to such notions as plurilingualism and "mnogoyazychie", which is a Russian term for the given context area. A five-stage research has been conducted. As a result, an innovative concept of multilingualism for educational pursuits, based on ten methodological principles, is suggested. Multilingualism is viewed as an actively developing integrative and synthesising research area, which shows positive dynamics results of its innovative conceptual ideas application. The research has resulted in publishing a book "Culture in the North. Pedagogy of Multilingualism" which was written in six languages and incorporated the ideas of multilingualism pedagogy. A thorough analysis of the authors' experience of working with the book, gained in educational practices at the Northern (Arctic) Federal University, Russia, is presented. The study focuses on the period of 2009–2018 and includes more than 6000 participants of the academic process and 3000 students in extracurricular activities.

Keywords: Multilingualism · Polycultural process · Education · Language pedagogy

1 Introduction

To introduce the topic of multilingualism, we find it appropriate to recall the story of the Tower of Babel. We believe that the Tower of Babel parable describes real events. The Tower remains were found and explored by modern scientists. There were times when all people spoke the same language—Sanskrit. After the Great Flood the Tower collapsed, and people started to degenerate. As a result, they stopped understanding each other, and lots of languages and dialects appeared. However, the Sanskrit roots are found in all of the world languages. Allegorically, the Tower of Babel depicts scientific and technological progress. People believe in intellectual methods, in technologies, in logical constructions and institutions. For a long time the Tower grows above the Earth, rising higher and higher. However, the Golden Age ends, and Kali-Yuga (the Iron Age) comes and the Babylonian creation collapses. People cease to understand each other. People value only their own life and comfort. They begin to quarrel and make wars. This is the allegory of mixing languages. The Book of Genesis reads: "And the whole

© Springer Nature Switzerland AG 2019
Z. Anikina (Ed.): GGSSH 2019, AISC 907, pp. 124–135, 2019.
https://doi.org/10.1007/978-3-030-11473-2_15

earth was of one language and one speech. And so it happened, when they traveled from the east that they found a plain in the land of Shinar; and they settled there. And they said to each other, "Let us build ourselves a city and a tower, the top of which can reach heaven, and we will create a name for ourselves." And the Lord went down to see the city and the tower that the children of men had built for themselves. And the Lord said: ... Let's go down there and mix their language so that they will not understand ... each other .... The Lord there mixed the language of the whole earth; and from there the Lord scattered them across the face of the whole earth" [1].

The article aims at solving the problem of defining the "multilingualism" notion in science, in language education, and in multilingual pedagogy. This issue has enjoyed a dynamic development on the turn of the 20th century and a significant experience has been accumulated in theoretical and practical education [2, 3]. However, we have to state that the notion of multilingualism is not fully represented in a multidisciplinary context. Also, the trends and perspectives of multilingual and polycultural educational environment are poorly investigated. The existing contradiction between the available polycultural context of the global community and spread of Anglo-Saxon culture in education creates a major problem of inadequate understanding and ineffective communication in all areas of human life: in everyday practices, business, professional and academic endeavours.

The purpose of the study defines its objectives. First of all, we would like to produce a thorough analysis of the multilingualism notion from a multidisciplinary perspective. Second, we'd like to investigate how multilingualism is defined in science and society today; what tendencies in the multilingualism development are observed in the world; what processes occur when one language influences another, and if the multilingual pedagogy has a future. The objective that features practical application of the investigation on multilingualism phenomenon and its manifestation in pedagogy is connected to the concept of the multilingual study aid "Culture in the Russian North. Multilingual Pedagogy". This is the objective to describe the results of its use in university education practices.

The concept of "Multilingualism" is very complex, multidimensional, multilevel, versatile, interdisciplinary. The notion of multilingualism synthesises a large spectre of scientific knowledge and data of multiple scientific fields: philosophy, language pedagogy, applied linguistics and educational politics, philology and others [4–13]. The phenomenon is multifaceted, which makes its ideas and implications a subject to generalisation and systematisation.

In the article we analyse the literary sources on the given topic and describe the results obtained in the process of implementation of the multilingual pedagogy concept in the educational process. It should be noted that in the course of the study our research position underwent a significant modification. At the beginning of our research we understood multilingualism as the use of different languages and dialects, switching from one language to another depending on the situation of communication. Today we realise that multilingualism integrates a wide conceptual area which makes it a source for reflection and systematisation to define its future in terms of its convergent application in language education.

So, the logic of the article implies the following:

- Research of the integrative notion of multilingualism;
- Study of the problems of language educational environment to define tendencies and theoretical perspectives of multilingualism;
- Analysis of the authors' experience gained in the educational practices at the Northern (Arctic) Federal University.

2 Literature Review

Many definitions of the concept under the study emphasize that multilingualism is a command of several languages and regular switching from one language to another, depending on the situation [14]. Other scientists believe that understanding multilingualism only as an individual ability does not cover the whole essence of this phenomenon, since this term is also used to describe interregional and interstate relations, which were historically conditioned and became especially pronounced in the 20th century [15].

There are works which use the term plurilingualism, understanding it as "the concept of plurilingualism as distinct from multilingualism, explaining the advantages of the former over the latter in such contexts, and analyzes possible synergies between plurilingualism and creativity through the lens of complexity theories and the theory of affordances, with the related concepts of 'affordance spaces' and landscape of affordances" [16]. Researchers from Donetsk Republican Institute of Additional Pedagogical Education argue that "plurilingual approach in ESL teaching often overlap the notion of multilingualism" and define these two terms as identical [17].

However, plurilingualism is different from multilingualism, as it is synonymous of a definite intercultural competence, comprising three basic parts: cognitive, emotional and behavioural. The society consisting of plurilingual members mainly and existing within the framework of cross-cultural dialogue is usually called plurilingual. The term multilingualism is generalizing, and it does not have any specific framework. In terms of language pedagogy, the plurilingual aprroach is more effective and recommended as the key principle of second (third, etc.) language acquisition. According to Ponomareva, the term "mnogoyazychie" (literally "many languages" in Russian, which is different from both "multilingualism" and "plurilingualism") comprises the notions of "multilingualism" and "plurilingualism" [18].

Multilingualism is associated with the language environment in a geographical area, whereas plurilingualism reflects a person's ability to speak several languages [18]. In his turn, Piccardo introduces the term plurilanguaging as "a dynamic, never-ending process to make meaning using different linguistic and semiotic resources" to streamline discussion of a potential synergy between plurilingualism and creativity [16]. But once you start anylising literary sources, you discover that all three notions are interconnected. The problem of definition appeared because all three terms started to be actively used in the documents on language policy of the European Council. In these documents there are no commonly known definitions of plurilingualism, multilingualism or "mnogoyazychie" [19]. So, in this article we are going to choose the term multilingualism as it is most often used in the English language literature. We should also mention that in our research we pay a special attention to development of a polycultural personality able to speak several languages. This process is a mandatory part of educational policy of universities and other educational institutions, e.g. colleges, schools, etc. We do not deny the notion of plurilingualism, but include it into the conceptual field of "mnogoyazychie".

Now we would like to move to scientific contexts which feature the multilingualism phenomenon. First of all, multilingualism is the subject of attention of political contexts, and that is proved by multiple researches [20, 5, 21]. Political discourse of multilingualism issue has its own hierarchy. To begin with, the multilingualism state, process and perspectives of its development are studied within the framework of language or educational language policy [20, 10, 22, 16]. The policy level is also taken into consideration, and then the question of multilingualism is relevant at the global, European, state, regional or institutional level. Normative legal documents of various political levels come as a separate block, for example, the European Council documents that set multilingualism development direction are profoundly used by researchers [23].

Multilingualism is construed as a society development strategy, as a mechanism of advancement, as a goal of polycultural society welfare, as a navigator directing from monolingualism to polylingualism [24, 18]. To achieve this, attempts of system analysis and research methodology design for multilingualism are made [10, 16, 25].

Multilingualism, as we can see, is very multifaceted. Not only in everyday life, but also in scientific literature, multilingualism is referred to as a factor, a principle, a phenomenon, a norm and even a symbol [4, 14, 26]. However, if we analyze a bulk of literature, we will find that scientists share common ground on the following: the variety of languages is connected with the originality of different cultures. This, in turn, actualizes the problem of not only linguistic, but also cultural knowledge, necessary for building both the national identity of an individual and a multicultural personality [2, 3, 22, 11, 27, 28]. Multilingualism is regarded as the highest level of linguistic consciousness [19], because in the process of communication in multicultural situations, thinking, creativity, feelings, knowledge and abilities are mobilized [29, 22, 30, 16].

More often, however, scientists investigate multilingualism as an issue of language education. There exists a far-reaching interest to various aspects of multilingualism, to scientific and educational creativity and unexpected discoveries in pedagogy and language teaching in particular [31, 2, 22, 32, 11, 33, 34, 35]. Language pedagogy treats learning foreign languages for professional purposes as a means for multilingualism development [36].

3 Basic Assumptions/Research Question

Based on the studied literary sources, theories of Russian scholars and foreign authors, research works of this article's authors and their rich experience of teaching languages at the university, we have made an assumption that the concept of multilingualism which we are investigating both in theory and in practice from 2009, is as follows:

- Stable
- Motivating for students of various levels and directions (bachelor, master, post-graduate, in additional professional education, studying Russian as a foreign language)
- Fostering creativity
- Encouraging respect towards native and foreign cultures
- Forming the communicative sociocultural competence.

Our work has brought to plentiful results which, however, have not yet undergone the statistical analysis. In compliance with our assumption, we have reviewed the concept of multilingualism; edited it, aiming at more accuracy; systematised the concept, and highlighted its criteria. We have also set the scope for our research, focusing on the time period of 2009–2018 and bringing attention of the research to a definite target group of students at various educational levels.

4 Methodology

Following the subject, aim and objectives of the research, at the initial stage of our work 70 literary sources were analysed. Majority of authors belong to international research groups. Most of papers were published during last three years. At the same time we find Russian scientific works to be equally important. At the next stage of our work the theoretical data on the multilingualism concept, its directions and trends, as well as existing research results have been analysed and presented in the article. The third stage included a study aid design which incorporated our educational perspective on the main principles of language pedagogy based on the multilingualism concept. The book "Culture in the North" was created by a team of teachers who worked as authors and translators with the help of native speakers and reviewed by experts. At the fourth stage the book was introduced into N(Ar)FU's educational practice. The course book has been tested during 9 years of teaching in 6 languages. It has been used by more than 300 student groups what means more than 6000 participants of the academic process. The book was helpful for 3000 students in extracurricular activities.

Here we present the principles of the multilingual concept which the book is based on, as they reflect the methodology of our research:

1. Resources of the multilingual study aid should feature the culture of the Russian North (Arkhangelsk and the Arkhangelsk region, Russia). Based on our rich experience of international collaborations, we state that a person who studies foreign cultures needs to be able to discuss his/her own culture.
2. The topic and information on cultural phenomena in the Russian North should be of personal importance for the authors of the book. We mean that authors need to express their individual perceptions of cultural events and phenomena, for example, on literature, personalities, excursions and sightseeing.
3. All of the topics offered in the aid are autonomous which allow learners to choose from the book's chapters those that present the biggest interest for them and motivate them to study foreign languages and cultures.

4. Each chapter is a subject for editing, contains recommendations for future work on the topic, and encourages further investigation into it.
5. Texts should be fairly short, informative, and refer to various text genres, e.g. an interview, a fairy tale, a dialogue, a description, a report, an excursion, etc., what is very motivating for students and other readers.
6. Interpretation of the sources implies felt-through imagery, authors' artistic and spiritual interests, and their cultural content input. At the same time the simplicity of presentation is important for motivating learners. Images of the book go through a thorough selection and refer to the authors' personal experience.
7. Readers are to go into the depth and uniqueness of culture in the Russian North, to understand the reason for the saying which goes as "one's soul rests" and "one's soul recovers in the Russian North". Such a motif is not common and contains a creative element, encourages authors to begin a dialogue with the reader which adds to the setting of multilingual and polycultural communicative environment.
8. Assignments of the study aid should be various, miscellaneous and not tiring. A reader will get interested in the content which is amusing and easily absorbed.
9. All the tasks should facilitate further learning, so it is vital to formulate them according to the pragmatics of a specific language. The assignments encourage reflection on linguistic and cultural phenomena.
10. The study aid presents its content in 6 languages: Russian, English, German, French and Norwegian, and Chinese in the third edition. This allows having a look at the foreign language without even the knowledge of its alphabet. Thus, a reader is given an opportunity to compare texts in different languages, making his/her first steps towards a multilingual world.

The fifth stage of our research was devoted to analysis of key criteria and indicators of the educational process, such as 1) language competence dynamics (A–C); 2) the ability to present the learner's own culture (in speaking and writing practices); 3) the growth of creativity (the quality of creative works); 4) the ability to communicate with native speakers (the time period of effective communication without an interpreter); 5) self-assessment of language skills, and 6) assessment of a lesson or event quality (surveys).

5   Data Analysis

As it has been mentioned above, the present research and its results were conducted in five stages. The first three stages were theoretical and can be divided into three time periods:

1. 2001–2007—introductory theoretical investigation of the multilingualism problem, study of foreign experience, design of the basic issue of the course book in five languages and its publishing under the title "Culture in the North: Towards Multilingualism" in 2004, obtaining of testing results from using the new textbook in the educational process.

In the course of the textbook testing the students' and instructors' perception of the book were analysed in various programmes (Fig. 1). In general, the survey results from 300 participants appeared positive. Critical feedback was given in terms of some inaccuracy in translation, insufficient quality of images, and drawbacks of layout. The content, pre-text and after-text assignments, motivation towards discussion of the presented information, tasks variety, perspectives of further work on the resources were positively assessed as well.

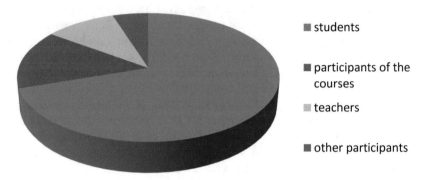

Fig. 1. Structure of participants of the survey "Your Opinion on the Course Book" (2001–2007).

2. 2007–2009—clarification of the concept, preparation and publishing of the reviewed course book edition, continuation of educational process with the book, further study of students' and teachers' opinions which showed satisfactory dynamics.
3. 2009–2018—renewal of the course book concept, publishing the third edition under the renewed title "Culture in the North. Pedagogy of Multilingualism" in 2017, which was added with three new chapters on the Arctic topic and translation into

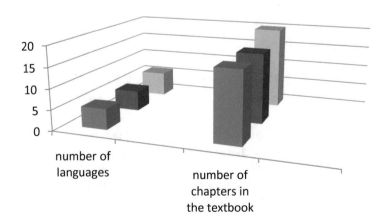

Fig. 2. Languages and chapters in the Course Book editions in 2004, 2009, 2017.

Chinese, continuation of the educational process with a more numerous student audience from N(Ar)FU and Hainan College of Information Technology (China) (Fig. 2).

We have analysed a number of groups and students of various study formats, as well as a number of participants of extracurricular events, organized in 2007–2018. When we speak of study formats, we mean full-time and part-time programmes, additional programmes, additional professional programmes, and specialised educational programmes for various purposes. Speaking of extracurricular activities, we include the course book presentations, round tables, conferences, exhibitions and other kinds of scientific, popular-scientific and cultural educational events with the course book where discussions and disputes on the lingua-cultural content and layout of the course book were discussed and tested. Our work resulted in the following data:

- Number of individual participants grew from 2000 to 6000 students
- Number of group participants grew from 100 groups to 300 groups
- Number of participants of extracurricular activities grew from 50 to 3000 (Fig. 3).

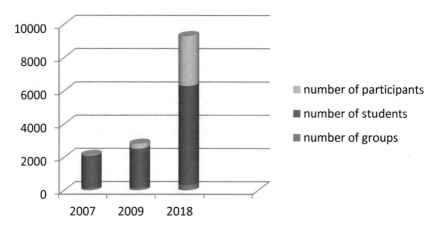

Fig. 3. Growing numbers of individual and group participants and participants of extracurricular activities in 2007–2018.

Now we are moving from quantitative to qualitative analysis of the following parameters: the ability to present native culture in a foreign language, improvement of lingua-cultural creativity, ability to communicate with native speakers on the topics of culture without communicative breakdowns, advancement of language competence, self-assessment, and evaluation of the activities quality. According to the criteria, we have selected benchmarks that show the results dynamics, where the first three benchmarks are basic and the last three benchmarks come as additional. In our calculations we have used average statistical indicators for the results of 400 students, presenting target audience for the article authors. Dynamics of qualitative indicators is assessed as positive (Fig. 4).

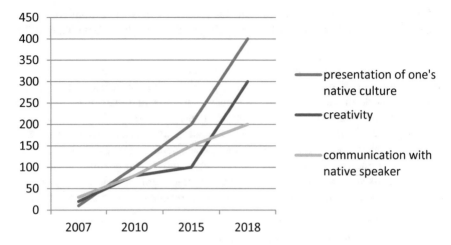

Fig. 4. Growth of qualitative indicators of the Course Book use in 2007–2018.

We have made an assumption that use of the course book's conceptual ideas, as well as regular learning sources in the educational process, will lead to formation of language competence by students; development of their metacompetence, ability of critical assessment in the process of work with learning sources both in their native language and in a foreign language. Moreover, evaluation of knowledge and activities quality will allow students and teachers advance their learning and professional activities. In the process of experimental practices we received the following results on the additional parameters of the quality of the book's concept assessment:

- Positive dynamics of language competence which showed advancement from A1 to A2-B1 levels (70%) with international online testing in language skills. There were also cases of advancement to B2 (20%) and even C1 levels (10%);
- Growth of critical assessment skills up to 36%;

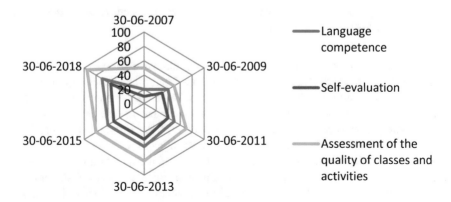

Fig. 5. Percentage of dynamics of language competence, critical assessment skills, knowledge and activities quality while teaching with the course book.

- Usefulness, rationality and future perspectives of the course book application in language teaching have grown as assessed by 87% of the survey participants (Fig. 5).

Accuracy and validity of the experimental results is proved by the experiment duration, large number of experiment participants and a careful selection of research methodology for the investigation.

6 Conclusion

To conclude, we would like to summarise the obtained theoretical and practical results of our research. The notion of multilingualism is multidisciplinary and connected with the fields of business, education, sociolinguistics, applied linguistics, cultural studies, and many other. "Mnogoyazychie", multilingualism and plurilingualism are interconnected. "Mnogoyazychie" comes as the most general term, depicting multilingual and polycultural societal, educational and scientific context. Multilingualism phenomenon is vividly expressed in the areas of politics, science and pedagogy. Educational trends in multilingualism are language and cultural politics, interrelation of different languages within one geographical area, multilingualism pedagogy, multilingualism concept in the language education and other professional areas. Multilingualism pedagogy as an integrative and synthesising research area is actively developing on the turn of the centuries and shows positive dynamics results of its innovative conceptual ideas application. It has been experimentally proved in the process of approbation of the multilingual course book "Culture in the North: on the Way Towards Multilingualism" ("Culture in the North. Pedagogy of Multilingualism"). Qualitative and quantitative results presented in this article prove positive dynamics of the book's concept realisation in Northern (Arctic) Federal University and have a rich potential for future development.

References

1. Genesis 1–11: The Primeval Story. http://barrybandstra.com/rtot4/rtot4-04-ch1.html. Last accessed 02 Aug 2017
2. Druzhinina, M.: Formirovanie yazykovoj obrazovatel'noj politiki universiteta kak faktora obespecheniya kachestva professional'noj podgotovki sovremennyh specialistov [Building language educational policy of the university as a means of ensuring quality of the modern professional education]. Pomorskij universitet, Arhangel'sk (2007). (In Russian)
3. Gal'skova, N.: Teoriya obucheniya inostrannym yazykam. Lingvodidaktika i metodika: ucheb.pos. dlya stud. lingv. un-tov i fak. in.yaz. vyssh. ped. ucheb. zav-j [Theory of foreign language teaching. Language pedagogy and methods of teaching. Study aid for the students of linguistic universities and faculties of foreign languages of pedagogical educational institutions]. Izdat. centr «Akademiya», Moscow (2008). (In Russian)
4. Baryshnikov, N.: Didaktika mnogoyazychiya: teoriya i fakty. Inostrannye yazyki v shkole [Didactics of multilingualism: theory and facts. Foreign languages at school] 2, 22–24 (2008). (In Russian)

5. Daniels, Sh, Richards, R.: Equitable multilingualism? The case of Stellenbosch University Writing Laboratory. Stellenbosch Pap. Linguist. Plus **53**, 59–77 (2017)
6. Davila, L.: Ecologies of heritage language learning in a multilingual Swedish school. J. Lang. Identity Educ. **16**(6), 395–407 (2017)
7. Dedic, N., Stanier, C.: MLED_BI: a new BI design approach to support multilingualism in business intelligence. TEM J. Technol. Educ. Manag. Inform. **6**(4), 771–782 (2017)
8. Fernandez Juncal, C.: Innovation for multilingualism: e-lengua. Caracteres. Estudios culturales y criticos de la esfera digital **6**(2), 282–297 (2017)
9. Flores, F., Jose, A.: Past, present and future of the original languages of Mexico. Zeitschrift fur romanische philology **133**(4), 973–997 (2017)
10. Hiss, F.: Workplace multilingualism in shifting contexts: a historical case. Lang. Soc. **46**(5), 697–718 (2017)
11. Kruchinina, A.: Prepodavatel' inostrannyh yazykov v usloviyah mnogoyazychiya: problemy i perspektivy [The teacher of foreign languages in multilingualism conditions: problems and prospects]. Inostrannye yazyki v ehkonomicheskih vuzah Rossii: Vserossijskij nauchno-informacionnyj al'manah **8**, 6–15 (2009). (In Russian)
12. The council conclusions of 12 June 1995 on linguistic diversity and multilingualism in the European Union. http://www.vestnik.mgimo.ru/sites/default/files/pdf/026_filologiya_02_rekoshkh.pdf. Last accessed 07 Jan 2011
13. UNESCO Universal declaration on cultural diversity. November 2, 2001. http://ec.europa.eu/avpolicy/docs/ext/multilateral/gats/decl_fr.pdf. Last accessed 10 Sept 2008
14. Kovtun, A.: O perevode traktata Karsavina L.P. «O sovershenstve» [Multilingualism as a Constant of L.P. Karsavin's Linguistic Consciousness. Translation of L.P. Karsavin's Tractate "On Perfection"]. Stud. Lang. **13**, 38–43 (2008). (In Russian)
15. Smokotin, V.: Mnogoyazychie i obshchestvo [Multilingualism and the society]. http://www.lib.tsu.ru. Last accessed 07 Jan 2011 (In Russian)
16. Piccardo, E.: Plurilingualism as a catalyst for creativity in superdiverse societies: a systemic analysis (2017). https://www.ncbi.nlm.nih.gov/pmc/articles/PMC5743751. Last accessed 02 Aug 2017
17. Mul'tilingvizm VS Plyurilingvalizm. http://www.donippo.org/2017/04/21/%D0. Last accessed 02 Aug 2017 (In Russian)
18. Ponomareva, O.: Strategiya mul'tilingvizma i izucheniya inostrannyh yazykov v Evrosoyuze [Strategy of multilingualism and foreign languages acquisition in European Union zone]. Vestnik TGU **2**(2), 23–28 (2015). (In Russian)
19. European charter of plurilinguisme. http://www.aplvlanguesmodernes.org/IMG/pdf/charte_plurilinguisme_fr.pdf. Last accessed 02 Aug 2017
20. Baptista, M.: On the role of agency, marginalization, multilingualism, and language policy in maintaining language vitality: commentary on Mufwene. Language **93**(4), E298–E305 (2017)
21. Mehmedbegovic, D.: Engaging with linguistic diversity in global cities: arguing for 'Language Hierarchy Free' policy and practice in education. Open Linguist. **3**(1), 540–553 (2017)
22. Hurst, E., Mona, M.: "Translanguaging" as a socially just pedagogy. Educ. Chang. **21**(2), 126–148 (2017)
23. Communication from the Commission of 22 November 2005—A new framework strategy for multilingualism. https://eur-lex.europa.eu/LexUriServ/LexUriServ.do?uri=COM:2005:0596:FIN:EN:PDF. Last accessed 02 Aug 2017
24. Mori, J., Sanuth, K.: Navigating between a monolingual utopia and translingual realities: experiences of American learners of YorA (1) ba as an additional language. Appl. Linguist. **39**(1), 78–98 (2018)

25. Singh, M.: Post-monolingual research methodology: multilingual researchers democratizing theorizing and doctoral education (2017). http://www.mdpi.com/2227-7102/7/1/28. Last accessed 02 Aug 2017

26. Obshcheevropejskie kompetencii vladeniya inostrannym yazykom: Izuchenie, prepodavanie, ocenka [General European competences of foreign language proficiency: study, teaching, assessment]. MGLU, Moscow (2003). (In Russian)

27. Druzhinina, M.V.: Kul'tura na Severe. Pedagogika mnogoyazychiya [Culture in the North. Multylingualism pedagogy]. SAFU, Arhangel'sk (2017) (In Russian)

28. Van Kerckvoorde, C.: Bildungsziel: Mehrsprachigkeit. Towards the Aim of Education: Multilingualism. Unterrichtspraxis-teaching German 50(1), 106–107 (2017)

29. Evdokimova, N.: Formirovanie sposobnostej k izucheniyu inostrannyh yazykov [Building the ability of foreign language acquisition]. Vysshee obrazovanie segodnya 9, 89–93 (2009). (In Russian)

30. Kyratzis, A.: Peer ecologies for learning how to read: exhibiting reading, orchestrating participation, and learning over time in bilingual Mexican-American preschoolers' play enactments of reading to a peer. Linguist. Educ. 41, 7–19 (2017)

31. Court, J.: I feel integrated when I help myself': ESOL learners' views and experiences of language learning and integration. Lang. Intercult. Commun. 17(4), 396–421 (2017)

32. Kordt, B.: Affordance theory and multiple language learning and teaching. Int. J. Multiling. 15(2), 135–148 (2018)

33. Lee, Ch., Curtis, J., Curran, M.: Shaping the vision for service-learning in language education. Foreign Lang. Ann. 51(1), 169–184 (2018)

34. Moeller, A.J., Abbott, M.G.: Creating a new normal: language education for all. Foreign Lang. Ann. 51(1), 12–23 (2018)

35. Thi, H.N.N.: Divergence of languages as resources for theorizing (2017). http://www.mdpi.com/2227-7102/7/1/23/htm. Last accessed 07 Jan 2017

36. Merino, J.A., Lasagabaster, D.: CLIL as a way to multilingualism. Int. J. Biling. Educ. Biling. 21(1), 79–92 (2018)

Using LMS Moodle for Mastering English Skills as an Interactive Competition Tool

Ksenia A. Girfanova[1,2(✉)] 📧, Inna A. Cheremisina Harrer[2] 📧,
Liudmila V. Anufryenka[3] 📧, and Alena V. Kavaliova[4] 📧

[1] Department of Russian Language and Special Disciplines for International
Students, Tomsk State University of Architecture and Building, Solyanaya
Square 2, 634003 Tomsk, Russian Federation
ksenia_astra700@mail.ru
[2] National Research Tomsk Polytechnic University, Lenin Avenue 30,
634050 Tomsk, Russian Federation
{ksenia_astra700,gwhcher}@mail.ru
[3] Polotsk State University, Blokhin Street 29, 211440 Novopotsk,
Republic of Belarus
liudmila.anufryenka@gmail.com
[4] Mozyr State Pedagogical University named after I.P. Shamyakin,
Studencheskaya Street 28, 247760 Mozyr, Republic of Belarus
alena.kavaliova@gmx.de

Abstract. The paper outlines the use of Learning Management System Moodle
as a tool for running an international competition in English. Since 2012 partner
universities from Russia, Belarus and Ukraine have initiated an international
student competition in English based on the facilities and resources of the
Moodle platform. The participating students from partner universities were
exposed to interactive activities and networking resources with the aim to
demonstrate and master their English and communication skills. The use of
virtual learning environment was intended to assist students in developing their
digital competence as well. The competition included several rounds based on
different Moodle features and services and enabled the participants to demonstrate their skills in listening comprehension, reading comprehension, use of
English, creative writing, and public speech. The participation of students from
different universities and managing the English competition by an international
team of teachers are described as challenging experiences that resulted for both
sides in promoting their personal and professional development.

Keywords: Learning management system moodle · English and
communication skills · Interactive competition

1 Introduction

Globalization of economy and constantly developing information technologies demand
from modern specialists a variety of skills that will allow them to adapt to the changing
world market and respond to its challenges. The employers need professionals with the
so-called core skills and competences. Modern pedagogy names them also as 21st

© Springer Nature Switzerland AG 2019
Z. Anikina (Ed.): GGSSH 2019, AISC 907, pp. 136–144, 2019.
https://doi.org/10.1007/978-3-030-11473-2_16

century skills or deep learning skills. Briefly, they include the following subgroups of skills: ways of doing jobs by means of active communication and cooperation; critical thinking and problem-solving, innovation and metacognitive skills; ability to use digital tools, in other words, demonstrating information and communication literacy; global citizenship and civic responsibility including intercultural competence [1–4]. Learning English will assist students promote the core skills and competences in a variety of ways, including realistic tasks in the virtual English-speaking environment.

2 Research

2.1 Communication Skills and Digital Literacy as Core Skills

In the modern world where nations do not have boundaries to interact and learn about different peoples, the ability to communicate and express oneself in one's native and second language plays the most important role. That is why the main task of any institution is to teach students how best to engage with other cultures and how to lead this involvement. English is an important mediator in the context of global communication, and the ability to use English for professional and personal purposes complements to the status of the modern specialist.

In a general sense, communication is a conveyance by ways of exchanging the facts of a culture, a nation, a civil society, etc. which is being described and disclosed as to how its features and characteristics are. The communicator is the facilitator as well as the transmitter of the ideas. The communication skill is the ability to convey or share ideas, news and facts effectively. In addition to the communication skill(s), one more skill is regarded as most vital and up-to-date. It is digital literacy.

As it is difficult these days to imagine any sphere of public life without the use of digital systems, like one cannot imagine the educational process without computer appliance anymore, the application of digital hardware needs proper preparation and knowledge as well. Digital literacy is commonly referred to as Internet literacy and multimedia literacy, cyber literacy or online literacy, and thus can be defined as the capacity supporting the user in engaging in social and cultural activities through the utilization of various media [5].

It is considered that digital literacy helps people to be active in all social affairs and provides them with the possibility to easily express their opinions [6], as well as the ability to search or understand the information needed and therefore better comprehend with others [7].

The main goal of any English educator is to develop the communication skills in every student, and in so doing, one will naturally find digital literacy to help decisively and it actually exercises the communication skills, being wise to use computer technologies in the process of English language learning.

2.2 Methodology

We based our research on the data collected during the English competition performed with the support of a virtual learning environment. The methods we used included observation method, discussion method, method of interview, and the questionnaire survey. We focused our observation on the student cohort participating in the English competition. The students were observed in the natural setting and it was also important for us to analyze how the natural setting (in our case, free access to the competition materials and the personal decision of when and where to do the competition rounds) could play a role and influence the results the students had achieved. The teaching team from the Polotsk State University in Belarus interviewed 20 students who were taking part in the second round of the competition. At that stage of the research the teachers from the organizing team of the Polotsk State University were also interviewed by their peers. The teacher cohort included 18 senior and junior staff members. The data from both cohorts were analyzed by means of narrative inquiry, focusing on how the interviewees constructed and narrated their experience while doing or monitoring the assignments of the second round.

The method of group discussion proved its effectiveness at the stage of designing English competition materials and evaluation forms. The participating universities from Russia and Belarus agreed on the structure and content of the complex English test and assignments for creative writing with a focus on key topical areas and an account of the level of English competence of participating students. The members of the international teaching team were in charge of designing sets of evaluation criteria and reviewing forms for evaluating written essays and spoken presentations. Both the materials and evaluation forms had been regularly changed and improved in the course of the English competition as a result of joint group discussions.

3 LMS Moodle as an Attractive Tool for Learning English

3.1 Origin of the Moodle-Based English Competition

In the era of globalization, online educational programmes and competitions gain their popularity. They have influence on international students' cooperation in the sphere of education. Such activities help to master the English language and get a new learning experience.

Learning Management System Moodle (Modular Object-Oriented Dynamic Learning Environment) has become an efficient software platform in the world with 103063 currently active sites registered in 233 countries [8]. Due to its dynamics and flexibility, Moodle is used at the National Research Tomsk Polytechnic University, Russia, for managing the learning process in a variety of courses including English [9]. The use of LMS Moodle at TPU is seen as an educational strategy for digital re-engineering of traditional courses, introducing more on-line courses and involving teaching staff in the development of open educational resources and modules for undergraduate educational programs [10].

Apart from analyzing the role and place of computer technologies and learning management systems in educational establishments, researchers also focus on the importance of using modern information technologies for the purpose of fostering self-learning, self-control and correction of English skills as well as learner autonomy [11–16].

In 2012, the Department of Foreign Languages of the Energy Institute at the National Research Tomsk Polytechnic University hosted the International Internet

English Competition (IIEC) using the LMS Moodle software platform [17]. The idea to bring together students from partner institutions to share and demonstrate their knowledge and skills in English by implementing an interactive competition mode was supported by the Polotsk State University, the Republic of Belarus, National Technical University of the Ukraine, Kiev Polytechnic University, and later by Mozyr State Pedagogical University named after I.P Shamyakin, the Republic of Belarus. 1877 students participated in the competition over the period from 2012 to 2018.

Participation of students in the competition was seen as a challenging activity, organization and managing of the competition was a challenge for teachers as well. We set the aim of the competition as expanding students' language skills, promoting the English language and developing learners' self-confidence in mastering English.

3.2 Organization and Technical Issues

To take part in the competition, students were invited to fill out an online registration form in order to get via e-mail an authorized access to the materials that involved a personal login and a password. During the whole period of the English competition, which normally took about 4 weeks, the registered participant stayed in contact with the administrator of the competition web-site in case of urgent questions, procedure difficulties or technical connection failures. The communication between the administrator of the English competition web-site and the participating students was performed via Moodle Forum and e-mail.

Often the students faced difficulties because of typing incorrect login or password when entering the system. Our experience revealed that to manage such an event without serious problems we need at least two administrators in charge—one being responsible for the technical functioning of the English competition web-site, including communication with the participants, and the other—administering the content layout of the competition and providing feedback to students when necessary. Prior to the English competition, the teachers from the participating institutions were given the webinar 'Basic features and resources in the Virtual Learning Environment Moodle' in order to be familiar with the procedure and management of the learning platform.

Supported by the Moodle features and taking into account available interactive activities and networking resources, we designed a personalized learning environment consisting of three competition rounds. Thematically, we agreed to focus on the following burning topical areas - 'Environment protection and green issues', 'Digital media in the modern world', and 'Planning a career and managing professional self-development'. The above areas were continually retrieved in the course of the three competition rounds: firstly, in the complex language and communication test, secondly, in writing an essay, and thirdly, in making a presentation.

3.3 Sequence of Competition Rounds

The first round of the English competition was organized with the help of Moodle activities in the form of tests. The participants were to demonstrate their knowledge and skills in listening comprehension involving audio and video resources, reading comprehension, and use of English. The tasks were considered from the viewpoint of their authenticity, students' level of language acquisition, availability and applicability. According to Avanesov, tests are considered to be one of the main ways of checking the language level. Therefore, it is important to choose proper material and arrange it in the right way: test tasks should be short, forms as well as the content should be correct, phrasing should be logical; rules for test evaluation and instructions for the participants should be identical [18].

The following types of tests were used—multiple choice, matching activities, cloze tests, word-building tasks, etc. The most commonly used kind of test in our materials was a multiple choice test, which offers a variety of answers for a participant with only one answer being correct. The test typically contained questions for tenses in active voice, passive voice constructions, gerund, article, prepositions, etc. For example, in the sentence *"The American bald eagle ... off the endangered register in the next two years"* the optional answers may be as follows: *(a) will probably take, (b) will probably be taken, (c) is probably taking, (d) takes probably*. As you see, different grammatical forms are offered as distractors, they are correct but only one is suitable. Topical vocabulary may be also checked, as well as word-building skills, for example, in the sentence *"Thousands of people participated in direct action to stop the destruction of ... wildlife sites"* the possible distractors are *(a) irreparable, (b) irreplaceable, (c) irrevocable*, which are correct in form but different in meaning.

Listening Comprehension test provided the authentic listening material with 5% unknown words that could be easily guessed while listening. The number of attempts was not limited and if the students had enough time, they had a possibility to listen to the audio materials for several times. Reading test was also a part of the competition and the texts for reading were aimed at checking reading comprehension. Like in the listening test, there were a number of unknown words, but not more than 5% from the whole text. The test level could be defined as low-intermediate—intermediate, which corresponded to the university curricula for language learning intended for non-linguistic students.

In the first round, all registered participants were exposed to the same set of tasks to be done within a fixed time limit but in a self-paced mode. Before starting the test, the participants were to read the instructions that guided them in the process of doing the tasks and submitting the completed work. The students' works were assessed automatically by the Moodle software. The evident advantage was that, in doing so, it eliminated personal involvement, occasional mistakes, which resulted in high objectivity. 60 best participants of the first round were able to take part in the second round of the competition.

The second round of the competition involved writing an essay on one of the suggested topics. At that stage, the students were able to choose a topic according to their personal interests. They were provided with the instructions of doing the task and were set a time limit for essay writing. The topics were perceived as both interesting and educative for students and focused on the following issues—the power of education for changing the world, education in the era of digital technologies, globalization and brain drain trends, integration into the labor market in a host country, the role of robots in human life, computers in education and everyday life, etc.

In our opinion, essay writing is an excellent way to check the vocabulary on a given topic, check grammar, and master the ability to express personal opinion using higher-order reasoning skills in 200 words.

The essays were evaluated with the help of the reviewing form being attached to the individual work of each second round participant within the Moodle system. The evaluative criteria included the category of the main idea (correspondence to the subject area of the topic chosen by the student, sufficient number of words, essay structure, presenting several points of view, logical reasoning, etc.), the category of language idea (use of English, adequate vocabulary and grammar), and the category of additional points (logically built paragraphs and correctly highlighted issues). The evaluation of the essays was realized as peer review by the teachers from the participating institutions. It was agreed that the students' works from one participating university were evaluated by the teachers from another participating university. The idea itself was progressive, in our opinion. On the one hand, the evaluation was free from any biased conclusions, and, on the other hand, the teachers were able to analyze and compare English communication skills of the students from partner institutions. As a result, 30 best participants of the second round were invited to the final third round.

The third round was organized with the help of the Moodle networking resource webinar. The participants were asked to make a presentation on the topic of their choice so that it should reveal their specialization or demonstrate their research experience. All the presentations were uploaded to the Moodle system by the technical administrator of the English competition in advance. In order to run the webinar smoothly and avoid serious problems during the students' presentations, we also installed the additional software Adobe Acrobat Connect Pro Meeting and performed several test connections between the participating institutions. During the webinar mode, we sometimes faced certain difficulties with a connection quality or sound failures, but they did not interfere with the students' performance.

The third round of the English competition could be considered as the most difficult one because students not only prepared their online presentations, but they had to interact with the jury. Therefore, it was necessary to pay attention to that point since, having a brilliant presentation, some students were not able to give examples on the topic they presented and answer the questions of the jury.

The evaluation of students' presentations was also performed as a peer review by the jury from another participating university. The reviewing criteria were given to the jury in advance and included the following categories: layout of the presentation (structure of content, number of slides and references), quality of slide design (information sufficiency, language economy, visuals and their design, text layout in slides), presentation (relevance to the presentation topic, text cohesion, correct use of language means including the variety of lexical and grammatical means, pronunciation and phonetic aspects, intonation), level of content awareness (i.e. being able to speak on the topic without supportive notes), and interaction with the jury and ability to answer questions logically providing reasonable points. Taking into account the results of all three rounds, each participating university was able to designate the winner and medalists of the English competition individually.

4 Opinions and Evaluation

It should be taken into consideration that the students participating in the English competition were of civic expertise and the usage of English was based on the language for specific purposes. On the one hand, that helped the learners realize how much they knew, and, on the other hand, feel how many diverse language patterns they could use in speech.

In general, the feedback of the students who were engaged in the English competition was that the acquired experience helped them realize how important it was to know the language and to share the knowledge with people from other countries. Actually, the knowledge of the language enabled them to express their own thoughts, attitudes and ideas, as well as to present their own achievements.

Additionally, the students were able to understand their strengths and weaknesses in English. At the same time, communication and interaction with other countries enhanced their confidence in their own abilities and demonstrated that the fear to speak and use the language was the main obstacle that prevented smooth and easy information exchange. Since the main power in developing communication skills is the communication itself, such a competition provides a real opportunity to use a language and motivates students for further development.

With regard to the English competition, it should be noted that it plays a critical role both for students and educators. Facilitating a competition reveals the innermost potential of every supervisor and organizer. For example, being a scientific supervisor, an educator has to show all his/her knowledge and experience in order to help students with the competition process. Each educator has to monitor the procedure of rounds, inform about the results, provide reasoning if students are in doubts or need justification of the results. The most important qualities of an educator who acts as a competition supervisor are as follows:

- good knowledge of the topic of students' interest;
- being supportive at all stages of a competition both technically and personally;
- allowing students to make their own decisions about a competition, for example, when choosing topics for presentation.

As an organizer, a teacher should have a number of leadership qualities in order to be able to monitor the whole competition process. These qualities may include quick-thinking, strategic thinking, display of strong business acumen, critical thinking, strong interpersonal skills and clear message sending, interactional and decision-making skills.

All in all, these qualities will help to arrange such competitions, assure students that they will be guided properly, and will help participants from partner universities to run a competition in the best possible way.

5   Conclusion

LMS Moodle is on the ways of becoming wide-spread in Belarus. Therefore, it is important to employ this tool to create new possibilities and develop the educational model that will increase the proportion of online learning within a traditional educational course, reaching the required 30% of its syllabus. A definite advantage is that this tool is accessible to both students and teachers whenever they need it. The competition forum is also of great help in case of difficulties or questions that arise throughout a competition.

In 2018, eight Turkmenistan students took part in our competition as members of the Mozyr State Pedagogical University team. The event presented a real challenge for them, since, initially, they underestimated their skills, but after the competition they felt more confident in their English classes. A lot of students have taken part in the competition twice, and they realize their progress and share their experience with the peers. Interactive assessments are important from the viewpoint of self-evaluation, students may compare their presentations with others, they are exposed to different ideas that will be necessary in their future profession.

Such international competitions are highly beneficial for both the participants and organizers since they provide students with an opportunity to evaluate their level of English, analyze mistakes, and work better on their essays and presentations in the future, and teachers—with an opportunity to effectively cooperate, discuss and share their competence at international level.

References

1. Cisco Homepage. http://www.cisco.com/c/dam/en_us/about/citizenship/socio-economic/docs/ATC21S_Exec_Summary.pdf. Last accessed 02 July 2018
2. British Council Homepage. https://www.britishcouncil.vn/en/programmes/education/connecting-classrooms/core-skills-for-learning-work-and-society. Last accessed 22 July 2018
3. Fullan, M., Langworthy, M.: A Rich Seam. How New Pedagogies Find Deep Learning. Pearson, Toronto (2014)
4. New pedagogies for deep learning Homepage. http://npdl.global. Last accessed 22 July 2018
5. Younghee, N.: A study on the effect of digital literacy on information use behavior. J. Libr. Inf. Sci. **49**(1), 26–56 (2017)
6. Hobbs, R.: Digital and Media Literacy: A Plan of Action. The Aspen Institute, Washington (2010)
7. Kwon, S., Hyun, S.: A study of the factors influencing the digital literacy capabilities of middle-aged people in online learning. Korean J. Learn. Sci. **8**(1), 120–140 (2014)
8. Moodle Homepage. http://moodle.net/sites. Last accessed 16 July 2018
9. Tomsk Polytechnic University Electronic Learning Homepage. http://portal.tpu.ru/eL. Last accessed 22 July 2018
10. Rymanova, I., Baryshnikov, N., Grishaeva, A.: E-course based on the LMS Moodle for English language teaching. Development and implementation of results. Soc. Behav. Sci. **206**, 236–240 (2015)

11. Kazatchyonok, V.V., Mandrik, P.A.: Primenenie IKT v vyschem obrazovanii Respubliki Belarus [Use of computer technologies in higher education in the Republic of Belarus]. In: Viktorov, A., Ovodenko, A. (eds.) Primenenie IKT v vyschem obrazovanii stran SNG i Baltii: tekushchee sostojanie, problemy i perspektivy rasvitija [Use of information computer technologies in higher education in the CIS and Baltic States: state-of-the-art, challenges and prospects for development], pp. 41–54. GUAL publishing centre, St. Peterburg (2009) (in Russian)

12. Robert, I.V.: Sovremennye informatsionnye technologii v obrazovanii [Modern information technologies in education]. Shkola-Press Publishers, Moscow (2004). (in Russian)

13. Khamitseva, S.F.: Komp'uternye tekhologii v obuchenii inostrannym jazykam [Computer technologies in teaching foreign languages]. Molodoj uchenuj **8**, 1057–1059 (2015). (in Russian)

14. Zubov, A.V.: Metodika primenenija informatsionnyh technologij v obuchenii inostrannomy jazyku [Methods of using information technologies in teaching a foreign language]. Akademia Publishers, Moscow (2009). (in Russian)

15. Dichkovskaya, E.A.: Ispolsovanie komp'uternyh tekhnologij v obuchenii inostrannym jazykam. [Use of computer technologies in teaching foreign languages]. Teorija i praktika professionalnogo obuchenija inostrannym jazykam. In: Shimanskaya, O.Y. (ed.) Teorija i praktika professionalnogo obuchenija inostrannym jazykam [The theory and practice of professional foreign languages teaching], Conference 2013, pp. 28–31. MITSO, Minsk (2013) (in Russian)

16. Dofs, K., Hobbs, M.: Autonomous language learning in self-access spaces: moodle in action. Stud. Self Access Learn. J. **7**(1), 72–83 (2016)

17. International Internet English Competition Homepage. http://web.tpu.ru/webcenter/portal/iyaei/mio?_adf.ctrl-state=f0rsl33vi_21. Last accessed 22 July 2018

18. Avanesov, V.S.: Kompozicija testovyh zadanij. [Composition of tests]. Centr testirovanija Publishers, Moscow (2002) (in Russian)

Development of Students' Polycultural and Ethnocultural Competences in the System of Language Education as a Demand of Globalizing World

Nadezhda Almazova[iD], Tatjana Baranova[iD],
and Liudmila Khalyapina[✉][iD]

Peter the Great Saint-Petersburg Polytechnic University,
Saint-Petersburg 195251, Russian Federation
almazovanadial@yandex.ru, baranova.ta@flspbgpu.ru,
lhalapina@bk.ru

Abstract. The study examines the increasing role of the process of the growing demand of globalizing world for the development of polycultural and ethnocultural competences of students as a means of preparing them for global (not international) successful cooperation in different economic spheres in the conditions of a growing global market and as a means of avoiding intercultural conflicts. The requirements for the education system lie in the field of searching and suggesting new approaches and technologies to solve this task. The present paper is devoted to the theoretical and experimental study of different models that integrate knowledge and skills lying in such aspects of cognitive linguistics as the analysis of universal cultural concepts. Three new ideas are proposed by the authors: reference to the ideas of applying cognitive linguistics in a broader sphere of language education – for technical university students; reference to the interpretation of universal cultural concepts both from the viewpoint of their similarities in different cultures (for the development of polycultural competence) and differences (for the development of ethnocultural competence); reference to internet resources (forums and chats) for the investigation of the peripheral layer of cultural concepts. The findings reveal a growth in various types of competences necessary for global communication and cooperation.

Keywords: Globalization · Global communication · Multiculturalism ·
Multicultural education · Cultural concepts ·
Interpretation of universal cultural concepts

1 Introduction

The development of the processes of globalization and internationalization of economy and business has set a new goal in higher education - training of professional staff capable of working effectively in the changed conditions of the global market. For this purpose, new ideas and approaches in the field of language and culture education should be proposed in foreign language teaching methodology to replace those which

© Springer Nature Switzerland AG 2019
Z. Anikina (Ed.): GGSSH 2019, AISC 907, pp. 145–156, 2019.
https://doi.org/10.1007/978-3-030-11473-2_17

were actively developed and applied in previous decades: sociocultural, cross-cultural, intercultural.

In the globalizing world, it is not enough to prepare students for intercultural communication, it becomes more important to prepare them for effective global communication where many countries and cultures are making contacts simultaneously.

The most important characteristics of a modern global society is multiculturalism defined as the preservation and integration of the cultural identity of the individual in a multinational society. Multiculturalism determines the development of tolerant relations between different nationalities, the formation of a culture of interethnic communication [1], as well as interaction of all types of local cultures, which, in turn, allows creating conditions for the development of cultural tolerance.

In this sense, of special interest is a multicultural education including the issues of mastering knowledge, experience and behavioral norms, the problems of grasping the phenomenon of the development of trainees' individual identity, their humane qualities, deep knowledge of innate properties, development tendencies, abilities, needs, opportunities for self-development [2, 3]. It is the multicultural education that becomes the basis that enables civil and ethnic identity development by educating a person capable of effectively living in a multinational setting, acquiring a special "planetary thinking", feeling not only as a representative of the national culture but also as an active citizen of the world, the subject of the dialogue of cultures.

The development of this interpretation of global communication is directly associated with the problem of searching for new approaches and methods to prepare students for this type of communication. Our suggestion is to integrate some aspects of cognitive linguistics and foreign language teaching methodology in order to develop students' ability to penetrate deeply into cultural similarities and differences of people belonging to different cultures.

The idea of our approach is based on the new goal in the system of language education: development of new types of competences – polycultural and ethnocultural – within the multicultural education system. These competences have been defined while investigating the demands of a new global society. Thus, we have developed a new methodological system of foreign language teaching, which includes three models recommended to be used in sequence.

The first idea of the research is to evaluate the theoretical position of two interconnected scientific spheres (global communication and possible models of interpretation of universal cultural concepts for a better understanding of people from the global world), and the second idea is to analyze the outcomes of the experimental teaching of students.

The paper presents the preliminary results of the experimental research aimed at developing the level of students' polycultural and ethnocultural competences.

The paper has the following structure: a literature review providing different viewpoints on global communication and ideas of multicultural education; description of the procedure and results of the experimental education and their interpretation; summary of the findings.

2 Polycultural and Ethnocultural Competences as the Components of Multicultural Education

Multicultural education has been considered by many Russian and foreign authors. In particular, the general issues of multicultural education are studied in the works of Aydin and Tonbuloğlu [5], Baker [6], Grant and Chapman [7], Kearney [8], Kramsch [9], Krummel [10], Bekova [11], Griva [12], Dubinina and Konzhiev [13], Zvyagint-seva and Valiakhmetova [14], Krechetova [15], Loseva [16], Mutsalov [17], Novolodskaya [18], etc. Features of business communication training in the context of multicultural education are presented in the studies of Özturgut [19], Datsyuk [4], Demina [20], etc.

The research into intercultural and interpersonal communication, which included the works of such well-known specialists as Bennett [21], Lasting and Koester [22], Sitaram and Prosser [23] among many others, helped us to arrive at the idea that an interdisciplinary investigation should be performed for the purpose of creating a model or models for the development of polycultural and ethnocultural competences required in a global society.

The fact is that many scholars agree on the idea that the main reason of misunderstanding between people that belong to different cultures is not in the sphere of language (it is not difficult to overcome this problem), but in that of mentality, a conceptual picture of the world. The investigation associated with this issue covers such areas as intercultural communication, linguistic education, sociology, and psychology.

Sociology studies of globalization as a social theory and global culture and of multiculturalism as its result have proved the importance of the development of a new group of competences. University graduates are expected to be able to build bridges with representatives of various countries and cultures.

Intercultural communication and linguistic education (especially cognitive linguistics) studies have proved the idea of incorporating the theory and practice of conceptual analysis into the system of language education so that students could be taught the procedure of analyzing different language material. Two types of competences – polycultural and ethnocultural – can be developed in this case. Knowledge and skills that comprise these competences are connected with the conceptual analysis technique at three levels of investigation: word etymology+ dictionary analysis (nuclear components), broader context analysis (interpretation components) and analysis of forums and chats (peripheral meaning of the concept, received from real native speakers via the internet). One of the main goals in this case is to teach students to realize that the universal concepts have only partially common features in different cultures, and that in many aspects their interpretation in different cultures is different, i.e. universal cultural concepts possess both polycultural issues typical of the majority of cultures and ethnocultural issues typical of a particular culture. Thus, we propose a new idea in the system of the universal cultural concepts analysis, since all the previous studies [24, 25] focus their attention on teaching students to use universal cultural concepts for only one purpose – to interpret similar features in different cultures. According to Wierzbisca, the combination of a set of semantic primes represent a different basic concept, combination of those concepts into meaningful messages constitutes a natural semantic prime language, or natural semantic metalanguage [24]. Methodological system created by Khalyapina is also based on the idea of 4 level concept spheres where universal concepts refer only to common features [25]. Based on our extensive experience, we have concluded that from the viewpoint of language education, reference to the theoretical ideas of universal cultural concepts interpretation possesses double-aimed possibility for increasing the level of students' readiness for global communication: it can become an interesting didactic resource for teaching common features typical of different cultures, but at the same time it can help students realize that, besides similarities (polycultural issues), all these universal concepts are characterized by their inner (ethnocultural) differences. To be aware of these differences, rather than similarities, is more important for effective global communication. Thus, on the basis of the three level interpretation of universal cultural concepts we have proved the idea of the development of two competences necessary for global communication – polycultural and ethnocultural.

3   Method and Procedure

3.1   Participants

The study was carried out at the Polytechnic University in Saint-Petersburg. The participants of the experiment were 1st- and 2nd-year technical students. There were two groups of each year of studying: (1) the control group without intervention and (2) the experiment group in which the participants were taught the English language according to our special program.

3.2   Procedure

3.2.1   Empirical Stage

At the empirical stage of our study, we conducted an associative experiment of the Russian and foreign respondents studying at Peter the Great Saint-Petersburg Polytechnic University.

The objective of the associative experiment was to find out whether the empirical stage reflected the scientific knowledge concerned with similarities and differences in the interpretation of well-known universal cultural concepts within representatives of different cultures.

Thus, we conducted the associative experiment with 100 students as participants (see Fig. 1): 31 Russian students, 69 foreign students (from such countries as China, Kazakhstan, the Republic of South Africa, Turkey, Uzbekistan, Jordan, Iran, Albania, Mongolia, Moldova).

For quantitative and qualitative analyses, we suggested the questionnaire entitled "Cultural and comparative analysis of interpretation of universal cultural concepts among Russian-speaking and foreign students".

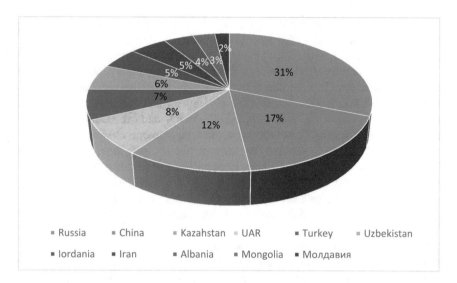

Fig. 1. Percentage of participants from each country.

The questionnaire included a question task:

"What associations appear in your mind when you hear the following concepts: "friendship", "love", "family", "happiness", "mother", "help"?"

The results allowed us to demonstrate to the students participating in our experimental study a number of factors that prove the general scientific idea about similarities and differences in the interpretation of one and the same universal cultural concepts known in all cultures.

For example, free associations of the "family" concept revealed the existence of the following semantic nets:

Family "substantive and metaphorical" - love (8); mutual understandings (4); happiness (4); help (3); unity (2) trust (3); force (2) union (2) support (2) cell of society (2); marriage; budget; relationship; life together; friend, friendship; unity; associations; responsibility; couple; understanding; joy; native corner of the earth; heart, connection; I.

Family "attributive" - large (3); respect (2) only (4); friendly (4); unity; darling; small; inseparable; unity; happy; sincere; others.

Family "comfort; protection" - heat (7); coziness (6) tranquility (4); protection (2) wait for each other; center; family center; hearth; care; comfort; durability; small country of love; beacon; world; for me, it is good to be with them; we are good together; you will forgive all; back; I love them and Murka.

Family "microsociety" - children (16); mother (10); father (10); family (8); relatives (6) parents (5); person (5); sister (4); we (3); wife (3); brother (5); relatives (3); people relatives; father-mother-child; darling; all; I am a person; family; niece; two; it is a lot of; grandmother; mine; the.

Family "tool" - raising children in love; for her it is worth living; friendly company; will always come to the rescue; cohabitation of man and woman; you always want to come back to them.

Family "injuring" - duties (3); I was; I had her; clan; alcoholic husband; unhappy; divorce; cross; quarrel; heavy burden; difficult; chain.

Family "real" - the house (20); housing; apartment; cat; kitchen.

Thus, the universal cultural concept "family" in the interpretation of both Russian and foreign students is considered today mainly in terms of "substantive and metaphorical", "attributive", "comfortable-protective" and "microsocial", and much more rarely – in terms of "tool", "injuring", "real", and "recreative". Among the Russian students "injuring" was used more often than among the foreign ones.

The semantic similarity between the concepts "family" and "house", "families" and "genetic" was also observed. Yet, another conclusion that we made at that level of our investigation was that the empirical stage was not sufficient for students' deep understanding of differences between cultures. For that purpose, it was necessary to develop special models for polycultural and ethnocultural competence development.

3.2.2 Models for Competence Development

As a result of the performed theoretical and practical research, we developed a set of three models aimed at teaching students to understand similarities and differences of various cultures (students dealt with three languages and cultures: Russian, English, French).

Types of the models are:

1. Model No. 1. Analysis of definitions and microcontexts.

The aim of the first model was to train students to interpret the nuclear value of a concept on the basis of lexicographic sources. The descriptive analysis of dictionary definitions is an important and necessary stage of studying the structure of a concept, as noted by many scientists. A lexical system is presented in various types of dictionaries and reflects specific features of the language picture of the world.

Exercises on comparison of semantics of the words representing a concept were the exercises of this stage.

For example, having analyzed the interpretation of concepts stated in English, Russian and French dictionaries, students were to choose the most suitable one. Examination of the nuclear part in the structure of a concept and the following creation of the nominative field of a concept were carried out on the basis of lexicographic sources (thesaurus, synonymic, phraseological dictionaries – analysis of idioms).

2. Model No. 2. Contextual interpretation of concepts.

For example, while analyzing paremia or contemporary e-periodicals in Russian, English, and French, students could find out some additional shades of the core meaning of one and the same concept, and compare them. The aim of the second model was to teach students to interpret the peripheral value of a concept on the basis of context research (e-periodicals, proverbs, etc.).

Verification of the obtained cognitive description (additional meanings of a concept) was carried out by verification of the data obtained while studying peripheral signs of a concept. These additional conceptual signs were often very ambiguous, or even opposite to core meaning, but they helped students understand the differences lying on ethnical level.

At this stage, the students' attention was on the differences (ethnocultural ideas) in the interpretation of universal cultural concepts.

3. Model No. 3. Analysis of forums and chats.

Exercises on a broader contextual interpretation of concepts with the help of such Internet resources as forums and chats.

At this level, the interpretative field of a concept represented its periphery and contained estimation and interpretations of the core meaning of a concept.

Students were free to search for additional information on the present day understanding and explanation of a cultural concept which could be given by different native speakers, people of different gender, age, education level, etc.

The objective of the third model was to teach students to generalize and conclude on similarities and differences of the investigated universal cultural concepts.

4   Results

In order to analyze the effectiveness of multicultural approach, determine its specifics, and reveal the level of the development of polycultural and ethnocultural competences of students, some test and essay statistics were calculated.

Before the experiment, we estimated different types of knowledge referring to polycultural and ethnocultural competences. The results were evaluated on the basis of a questionnaire. The results (see Fig. 2) revealed that when estimating the cognitive aspect of polycultural competence, most students demonstrated an average level of the development of the system of polycultural knowledge in the field of communication. The knowledge was of general nature, it did not rely on the fundamental ideas about modern multicultural society and the processes of its formation; the stereotypical vision prevailed. Students demonstrated their uncritical attitude to the information they received; most of them did not reveal any independent activity to obtain information about multicultural world since they did not recognize the need for doing this.

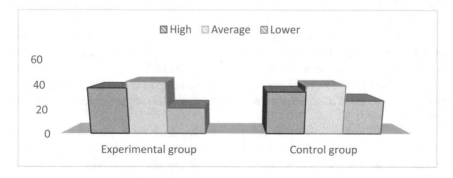

Fig. 2. Results of the pre-experimental studies.

Within all the three models, students could observe by themselves both similarities and differences in the perception and interpretation of universal cultural concepts (friendship, money, family, etc.).

The second stage of the statistics was performed after the experimental study by means of our three models. The results of the study are presented in Fig. 3; the summary data are given in Table 2.

The results were evaluated on the basis of final investigation essays.

To evaluate the polycultural competence, the following skills were analyzed: to determine cultural characteristics of a communication partner and to convey one's own culture; to underline cultural features of a communication partner; to define similar meanings in the interpretation of universal cultural concepts.

In order to evaluate the ethnocultural competence, another group of skills was estimated (Table 1).

Table 1. Assessment of polycultural and ethnocultural competences.

Types of competences	Skills
Polycultural competence	- to determine cultural characteristics of a communication partner and to convey one's own culture - to underline cultural features of a communication partner - to define similar meanings in the interpretation of cultural concepts
Ethnocultural competence	- to define specific ethno-cultural content of universal concepts - to interact with representatives of different cultures taking into account the knowledge of both similarities and differences

These skills were identified via students' essays and screenshots of their contributions to forums and chats. For example, in his final essay one of the students wrote: "Regarding the microcontext, e.g. idioms and proverbs, it is notable that in English and French we can frequently meet idiomatic expressions with similar meanings but different ways of expressing them in words. In English we most frequently see the noun "money", while in French the most frequent is the verb "payer" (to pay) and its derivative "prix" (price): e.g. hush money – prix du silence (*price of keeping silent); the money's good – c'est bien paye (* it's paid well).

From this we can conclude that for the English money is valuable as a substance, while for the French it isn't, it is a means of payment: e.g. put your money where your mouth is – passez a la caisse (*go to the cashier's stand)".

We defined three levels of the competence development – high, average, lower, which were estimated in accordance with the level of skills demonstrated in essays and screenshots (Table 2).

With regard to the levels and indicators for the analysis of ethnocultural competence, it is necessary to note that they were developed and evaluated using the same scheme, but in that case the students were to demonstrate the ability to determine specific features of cultural concepts. This type of ability was evaluated at three levels which were defined while analyzing the results of students' work for three different models with three types of representation: complete, partial and superficial.

The experimental group students demonstrated only high and average levels of polycultural and ethnocultural competence development.

Statistical processing and objective interpretation of the obtained results yield the following conclusion: multicultural approach to foreign language teaching allows the formation of a multicultural environment as a basis for more effective and more productive human interaction with representatives of other cultures of the global world (Table 3).

Table 2. Levels and indicators for the analysis of polycultural competence.

Levels	Indicators		
	Ability to determine common features of cultural concepts at the level of the 1-st model	Ability to determine common features of cultural concepts at the level of the 2-nd model	Ability to determine common features of cultural concepts at the level of the 3-d model
High	Complete representation of the results of comparative analysis of common features, typical of contacting cultures (model 1) – 3 points	Complete representation of the results of comparative analysis of common features, typical of contacting cultures (model 2) 3 points	Complete representation of the results of comparative analysis of common features, typical of contacting cultures (model 3) 3 points
Average	Partial representation of results (model 1) 2 points	Partial representation of results (model 2) 2 points	Partial representation of results (model 3) 2 points
Lower	Superficial representation of results (model 1) 1 point	Superficial representation of results (model 2) 1 point	Superficial representation of results (model 3) 1 point

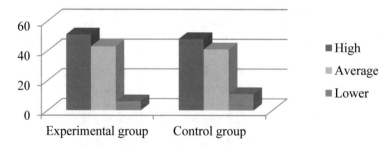

Fig. 3. Results of the experiment of the post - experimental studies.

Table 3. Final data on the level of polycultural and ethnocultural competence development.

Level of polycultural and ethnocultural competence development	Traditional approach		Multicultural approach	
	Experimental group, %	Control group, %	Experimental group, %	Control group, %
High	37	34	51	48
Average	42	39	49	41
Lower	21	27	0	11

5   Conclusion

The experimental training has proved our research hypothesis. The research results correlate with the goals stated in the introduction of the article. Thus, these results are as follows:

1. Our globalizing world requires the development of new types of multicultural competences among students of different non-linguistic universities as far as they should be well prepared for international and even global cooperation in various economic spheres. We define these types of competences as polycultural and ethnocultural.
2. Development of polycultural and ethnocultural competences would take place while teaching foreign languages on the basis of the special methodological system that incorporates three models connected with the development of students' skills and abilities to define common or specific features while analyzing universal cultural concepts.
3. As a result of the implementation of our approach in the educational process, students have demonstrated the following types of knowledge and abilities:

   - knowledge of universal cultural concepts;
   - ability to compare cultural concepts of a universal character, defining similarities between representatives of different countries and cultures;
   - ability to reveal and interpret specific features or differences between different cultures;
   - ability to realize that universal cultural concepts possess various interpretation and semantic values in various cultural groups, and, in accordance with this realization, to be more attentive while establishing relations for positive cooperation with representatives of a global polycultural world.

In addition, multicultural approach to education promotes the development of well-known and crucial qualities for global communication and cooperation including tolerance of views and judgments of people, recognition and development of cultural pluralism in societies, recognition of equal rights, duties, and opportunities for cooperation between all the countries in the world.

These results have proved the effectiveness of our system in foreign language education.

References

1. Matis, V.I.: Polikul'turnyj podhod v vospitanii kak metodologiya formirovaniya polikul'turnoj lichnosti [Polycultural approach in upbringing of polycultural personality]. Sibirskij pedagogicheskij zhurnal V, 162–172 (2007). (in Russian)
2. Ziatdinova, F.N.: Polikul'turnoe obrazovanie uchashchihsya mladshih klassov v nacional'-noj shkole [Polycultural education of young learners at national schools]. IPK BGPU, Ufa (2006). (in Russian)
3. Shum, S.K., Yazdanifard, R.: The significance of intercultural communication for businesses and the obstacles that managers should overcome in achieving effective intercultural communication. Glob. J. Manag. Bus. Res. XV(IV), 6–12 (2015)
4. Dacyuk, V.V.: Obuchenie studentov vostochnogo fakul'teta argumentativnym strategiyam angloyazychnogo delovogo diskursa [Teaching of the Eastern department students to argumentative English language business discourse]. Saint-Petersburg State University. Publishing company "Admiral", Saint-Petersburg (2017). (in Russian)
5. Aydin, H., Tonbuloğlu, B.: Graduate students' perceptions on multicultural education: a qualitative case study. Eurasian J. Educ. Res. 57, 29–50 (2014)
6. Baker, C.: Foundation of Bilingual Education and Bilingualism, 4th edn. Multilingual Matters, Bristol, UK (2008)
7. Grant, C.A., Chapman, T.K.: Challenging the myths about multicultural education. In: Grant, C.A., Chapman, T.K. (eds.) History of Multicultural Education, vol. 2, pp. 316–325. Routledge, New York (2008)
8. Kearney, E.: Intercultural Learning in Modern Language Education: Expanding Meaning-Making Potential. Multilingual Matters, Bristol (2016)
9. Kramsch, C.: Teaching foreign languages in an era of globalization: introduction. Mod. Lang. J. 98(1), 296–311 (2014)
10. Krummel, A.: Multicultural teaching models to educate pre-service teachers: reflections, service-learning, and mentoring. Curr. Issues Educ. 16(1), 1–8 (2013)
11. Bekova, M.: Osobennosti realizacii polikul'turnogo obrazovaniya v sovremennyh usloviyah [Peculiarities of realization of polycultural education in modern conditions]. Azimut nauchnyh issledovanij: pedagogika i psihologiya 4(13), 22–24 (2015). (in Russian)
12. Griva, O.A.: Mezhkul'turnye kommunikacii kak resurs kul'turnoj identichnosti lichnosti v usloviyah polikul'turnosti [Intercultural communication as a recourse of cultural identity of a person in conditions of polyculturalism]. In: Usovskaya, E.A. (ed.) Nacional'nye kul'tury v mezhkul'turnoj kommunikacii, pp. 99–108. Kolorgrad, Minsk (2016). (in Russian)
13. Dubinina, N.N., Konzhiev, N.M.: (2015). Vospitanie i obrazovanie v polikul'turnom prostranstve Russkogo Severa [Upbringing and teaching in the polycultural environment of the Russian North]. In: Shirokov, O.N. (ed.) Mezhdunarodnoj nauchno-prakticheskoj konferencii vospitanie dobroty i gumannosti u detej i molodezhi v polikul'turnom prostranstve, pp. 55–56. Interactiv plus, Cheboksary (2015). (in Russian)
14. Zvyaginceva, E.P., Valiahmetova, L.V.: Fenomen polikul'turnosti, ego idei i principy v obrazovatel'nom prostranstve sovremennoj Rossii [Polyculturalism fenomenon, its ideas and principles in educational environment of modern Russia]. Zhurnal nauchnyj publikacij aspirantov i doktorantov 1(91), 130–134 (2014). (in Russian)
15. Krechetova, G.A.: Specifika polikul'turnosti kak faktor ehffektivnogo sovremennogo obrazovaniya [Peculiarities of polyculturalism as factors of effective modern education]. Sovremennye innovacii 1(15), 77–79 (2017). (in Russian)

16. Loseva, A.A.: Polikul'turnoe obrazovanie: dialog kul'tur i bilingval'noe obuchenie [Polycultural education: dialogue of cultures and bilingual education]. Molodoj uchenyj **7** (5), 52–53 (2016). (in Russian)

17. Mucalov, S.H.: Polikul'turnost' i polikonfessional'nost' kak harakteristiki rossijskogo obshchestva [Polyculturalism and polyconfessionalism as characteristics of Russian society]. Obrazovanie i vospitanie **2**, 68–71 (2016). (in Russian)

18. Novolodskaya, S.: Istoriya i sovremennost' polikul'turnogo obrazovaniya: monografiya [History and modernity of polycultural education]. ZIP SibUPK, Chita (2015). (in Russian)

19. Özturgut, O.: Understanding multicultural education. Curr. Issues Educ. **14**, 2–11 (2011)

20. Demina, I.N.: Mesto i rol' kommunikacii v biznes-processah [Place and role of communication in business process]. Izvestiya Bajkal'skogo gosudarstvennogo universiteta **2**(82), 202–206 (2012). (in Russian)

21. Bennet, M.J.: Towards ethnorelativism: a developmental model of intercultural sensitivity. In: Paige, R.M. (ed.) Education for the Intercultural Experience, pp. 21–71. Intercultural Press, Yarmouth (1993)

22. Lusting, M.W., Koester, J.: Intercultural Competence. Interpersonal Communication Across Cultures. Addison Wesley Longman, California (1999)

23. Sitaram, K.S., Prosser, M.H.: Civic Discourse: Multiculturalism, Cultural Diversity and Global Communication. Ablex Publishing Corporation, London (1998)

24. Wierzbisca, A.: Cross-Cultural Pragmatics: The Semantics of Human Interaction. Mouton de Gruyter, Berlin/New York (1991)

25. Khalyapina, L.P.: Metodicheskaya Sistema formirovaniya polikulturnoi yazykovoi lichnosti [Metodolochical system of polycultural language personality formation]. Publishing company "Kuzbassvuzizdat", Kemerovo (2006). (in Russian)

Structure of Professional and Business Communication of Graduates Majoring in Innovation

Dmitry L. Matukhin[1]([✉]) ⓘ, Alexander V. Obskov[1] ⓘ,
Maria A. Vikulina[2] ⓘ, Nikolay A. Kachalov[1] ⓘ,
and Olga I. Kachalova[3] ⓘ

[1] National Research Tomsk Polytechnic University,
Tomsk 634050, Russian Federation
mdlbuddy@mail.ru, alexanderobskov@hotmail.com,
kachalov@tpu.ru
[2] Linguistics University of Nizhny Novgorod,
Nizhny Novgorod 603005, Russian Federation
centr_nid@lunn.ru
[3] The Russian Presidential Academy of National Economy and Public
Administration, Tomsk 634050, Russian Federation
kachoi@mail.ru

Abstract. Verbal activity is always directed, i.e. the subject of communication determines its scope. This pattern conditions both choice and arrangement of the teaching material, i.e. its content and structure, as well as development of verbal skills, its motivation (e.g., cognitive interest), and acts primarily in the field of the content of training material and organization of the pedagogical process and, accordingly, methods, goals and means of instruction.

The article attempts to establish, firstly, the subject of professional and business activities of graduates majoring in *Innovation* and to describe it; secondly, based on the characteristics found, to outline the scope of professional activity; thirdly, to establish the subject of professional and business communication of graduates in the specialty given and, consequently, to determine the scope of professional and business communication. In addition, the authors consider the situational structure of professional and business communication. Hence, the approaches suggested can be used as a basis to develop methods for training such skills as speaking and writing, and allow revealing the content of communicative competence of graduates majoring in Innovation.

Keywords: Situations of professional and business communication ·
Speaking and writing · Foreign language communicative competence ·
Scope of professional activities · Basics of banking and computer science

1 Introduction

Nowadays, not only those universities where foreign language (FL) training has always been part of the curriculum intensify their work on its deepening, but also those in which until recently FL teaching was of general educational origin. Among the latter,

© Springer Nature Switzerland AG 2019
Z. Anikina (Ed.): GGSSH 2019, AISC 907, pp. 157–164, 2019.
https://doi.org/10.1007/978-3-030-11473-2_18

undoubtedly, are non-linguistic universities, which emphasize the general mastery of a foreign language and try to make up their language training programs with regard to professional major in a particular field of knowledge.

Modern English is a complex and multi-sided phenomenon, thus, teaching it, considering the professional training of graduates, requires specification of the training content. We find it expedient to suggest as a basis for teaching English an independent course "Computers and Banking", the thematic scope of which corresponds to the content of professional training of graduates majoring in *Innovation*.

This course was chosen due to a number of unbiased reasons. Today, regardless of occupation, every person faces the need to understand the principles of economy. An ordinary person needs this knowledge when opening a bank account or transferring money. An employee of any enterprise, responsible for the timely execution of cash payments with business partners, encounters the same necessity. Finally, knowledge of the economy laws, in our case of its banking, is necessary for those specialists whose professional training plan comprises it as an obligatory component. It is also obvious that the study of banking fundamentals as a specific field of general economic knowledge presupposes the mastery of economic laws as a whole, which is of great demand nowadays. Since an economic component is included in the training program of graduates in this specialty, they can certainly be regarded as those for whom the social significance of economic knowledge is inextricably linked with FL professional communication training.

The social importance of computer skills today cannot be overemphasized. Actually, we observe that a person who does not know how to use a personal computer is illiterate, and, at least, is certainly isolated from a huge information field, in which, the information is likely to be new, up-to-date, and directly related to his/her professional activities. However, sound training of graduates in the field of information technology is unthinkable today without simultaneous training in English. This need is due to the fact that the acquisition of computer technology by students is closely linked to another problem, namely, the English language origin of most programs entering the Russian market. In addition, the statistics show that today 85% of the world's information stored in computer memory is written in English, 90% of web pages are in English, and finally, 75% of Internet users have English as their language of communication.

It is no secret that English is the means of international communication of specialists in various fields of activity. Nowadays, graduates of Russian universities are provided with great opportunities for professional growth in a variety of ways, e.g., exchanging professional experience with international specialists. In particular, students can participate in joint international projects, take internships abroad, and conduct scientific research in their specialty with the involvement of foreign materials or/and in the territory of other countries. Yet, the implementation of all these and other opportunities for professional development is limited by the insufficiently developed skills of FL professional and business communication, primarily speaking and writing skills, of graduates of non-linguistic universities including graduates majoring in *Innovation*.

2   Theoretical Background and Influences

Domestic science considers communicative competence as a choice and implementation of verbal behavior depending on the communicant's ability to interact in a situation adequately to the topic, task, and communicative attitude that the communicants have before and during their conversation. The provisions of communicative approach [4, 9–12] allow considering the FL communicative competence as a set of knowledge, language and verbal skills, communicative skills, that affect the ability and readiness of a person to communicate in the scientific, technical, professional, and general cultural fields.

The analysis of professional and business communication is based on the works of Russian and international scholars in the field of linguodidactics, theory and methodology of foreign language teaching [1–3, 6, 7, 14].

Awareness of such problems as the subject of professional activities of specialists majoring in engineering and economics, scope and subject of their professional and business communication, is based on the research conducted by Berezina [3], Erastov [8], Samayeva [13], and others.

The analysis of existing English coursebooks for non-linguistic universities confirms the fact that today there is no course capable of combining these two topical areas: the basics of banking and computer science course intended for *Innovation* graduates with a similar content of professional training. In addition, the lack of a training manual in the English language aimed at graduates in this specialty with the relevant methodological specialization, namely the development of productive verbal skills, makes designing the "Computers and Banking" course completely justified.

3   Methods

A special place in the work is given to the use of the situational approach as a linguodidactic strategy for implementing the communicative principle in teaching professional and business communication to *Innovation* graduates when determining the level of their communicative competence in speaking and writing.

The greatest importance for professional and business communication is the creation of lines of verbal communication. The following components are considered: formulation of the task, modelling of the motive and purpose of speech action; description of the situation (conditions and participants of communication, modelling of such components of a speech situation as situational afferentation and action subject). Our observations reveal that these general provisions are concretized in relation to the nature of the speech act, which is subject to modelling at a certain stage of the study.

Training for professional business communication involves both training in speech actions under typical, repetitive conditions, which develops readiness for action, and the accumulation of experience in self-orientation in various oral and written speech situations that require the elements of creativity, self-expression, and the vision of the situation. Only the combination of these two aspects of teaching speech interaction in the foreign language classroom, as can be seen from the survey conducted by graduates, can provide practical knowledge of the target language.

The communicative task of the upcoming communication should be a stimulus for self-expression. The wording of the task should include a communicative-psychological and practically (professionally) oriented message. The task of using the situational approach in FL teaching as a linguodidactic strategy for teaching oral and written language is seen in anticipating natural speech situations in the future professional activity of undergraduates in the Innovation specialty and, thus, preparing them for real speech communication, ensuring the transfer of speech actions from the conditioned exercises into speech practice.

4   Scope of Professional Activities

Characteristics of verbal interaction of communicants are determined, firstly, by the goals and subject of professional activity. Therefore, it is first required to establish the subject of professional activity of the graduates majoring in *Innovation*.

The graduates majoring in *Innovation* shall be trained in the field that requires comprehensive usage of knowledge of two sciences: engineering, namely information technology, and economics. Consequently, the subject components of their professional activity are economics as a set of production relations [2] and information technology of various applications. However, if we focus only on this notion of the subject of professional activity, it will be, undoubtedly, deprived of another key component.

Following the execution of the social order to integrate Russian higher education into the international educational environment, professionally-focused language training becomes an integral part of the curriculum. In this context, we try to determine the FL special part in the structure of students' professional activity.

Although FL itself is a dependent component of the professional activity subject (as it is explained above, the basis is economics and information technology), it often becomes an integral part of these two elements of the graduates' professional activity. This provision is confirmed by the fact that FL often acts in the educational activities as a tool for mastering a particular major. It becomes an intermediary between students and information, which is actually the subject of professional and business communication.

This is exactly the role that English plays in delivering the "Computers and Banking" course. English becomes not a separate subject of professional activity but a means of mastering the main components of the subject – economics (fundamentals of banking) and information technology (computer science). With regard to teaching students within the considered course, we may essentially talk of the mutual integration of the engineering, economic, and linguistic aspects of training graduates majoring in *Innovation*, which corresponds to our definition of the professionally-focused FL training as an integral aggregate of linguistic and cognitive knowledge [10].

The next step for us is to determine the scope of professional and business communication of the graduates, to outline the scope of their professional activities. Firstly, we conclude that the subject of professional activity determines its scope, which, in turn, will condition the scope of professional and business communication between the graduates.

Since we have found out that the subject of professional activity in our particular case is economics as a set of production relations, on the one hand, and information technology, on the other hand, the scope of professional activity shall assume application of the professional skills acquired as a result of mastering the professional activity subject.

Consequently, the authors state that the practical tasks, which graduates face, are resolved in various fields of activity including production, management, finance, and banking. This justifies the diversity of modes and content of training graduates in various specialties: management, marketing, economics, computer science, etc. The proposed mode of FL teaching (Computers and Banking) with its thematic content affects two fields of professional activity of the graduates majoring in *Innovation*: finances, namely banking fundamentals, and information technology.

The adequacy of this statement is also confirmed by the fact that finances (Fundamentals of Banking) and information technology (Fundamentals of Computer Science) correspond to the subject of professional and business relations of the *Innovation* graduates– economics and information technologies.

Hence, having established the scope of the graduates' future professional activities, we have approached the notion of the scope of professional and business communication.

5   Scope of Professional and Business Communication

Firstly, we should clarify what we understand by the "scope of communication" in general. Based on the definition given by Berezina [3], by the scope of communication we mean an array of topics that make up the content of communication in the relevant fields of social interaction. Proceeding from this definition, we state that the scope of business communication for the graduates majoring in *Innovation* will be the aggregate of topics that make up the content of students' communication in the fields of economics and information technology.

It should be noted that each field of communication has its own set of typical communicative situations, and each communicative situation includes the social roles typical of the communicants. This is mainly due to the relationship between language, thinking and reality, social conditioning of communication, and consideration of the learning process as a form of management of cognitive and FL verbal activity [4].

Since linguistic and extra linguistic factors interact in communication, it affects teaching of the FL communication, i.e. teaching of the main types of FL verbal activity is thematically and situationally conditioned. Since in our "Computers and Banking" course we arrange training on the basis of the situational approach, it seems appropriate to give its brief justification.

Researchers explain the necessity for organizing training based on situations differently. Some believe that teaching can be efficient only in the "situational presentation" system, since it triggers a verbal action. The stimulating role of the situation is also emphasized by Apelt [1] who recommends bringing the learning situations closer to real life. West [14] considers that teacher's mastership is to adapt real life situations to linguistic needs. Actually, the same idea is emphasized by Greyser [6] who states that any verbal unit shall be mastered only in the situations typical of it.

According to several researchers, the use of situations allows better implementation of the didactic principles: firstly, high activity is achieved since the learner feels the need for the skills; secondly, it corresponds to the principle of consciousness since it facilitates understanding of the speech content; thirdly, the principle of visibility [8] is implemented.

Apparently, there are many reasons for the high significance attributed to the situation. Firstly, speech situationality, both oral and written, is recognized as one of its main characteristics. Secondly, the situational approach allows us to reconsider most methods of teaching and arranging of the topical material. In particular, the so-called thematic organization of the material has ceased to satisfy nowadays. Instead, researchers suggest that situations be the initial component. Thirdly, this is one of the reasons that contributes to the success of the functional approach. The concept of the functional approach is well presented by Gurrey [7], who believes that in the classroom it is necessary to use situations to guide the students to the relevant thought and, having elicited the thought, to seek means for its expression.

Hence, from a psychological viewpoint, the situation provides the condition for the development of verbal skills and serves as a stimulus to speech.

Considering the above arguments in favor of the situational approach, it becomes evident that the approach should be used in the design of our course.

Recent analysis of the methodological references and manuals on teaching FL business communication allows us to distinguish two groups of communicative situations of business communication: situations of written and oral communication [3]. These situations yield communication the product of which is either an oral dialogue or monologue, or a written text.

Genre characteristics of these texts are determined by the communicative intentions of the speaker or listener, their attitude to other communicants and the subject of communication, the conditions for a certain situation of professional and business communication. The communicative tasks in the situations of business communication between graduates can be performed both in native and foreign language. The present study only considers the situations of professional and business communication in English.

It should also be noted that the suggested structure of situations present both communicative situations of professional and business communication in general, and the situations typical of applying knowledge gained in the "Computers and Banking" course.

Based on the references analysis [5, 6, 9, 11–13] and the results of our own observations, we determine the scope in which the graduates majoring in *Innovation* first feel the need to apply the acquired skills in the field of professionally-focused FL training. As mentioned above, each field of communication has its own set of typical communicative situations; therefore, in the structure of these fields we consider certain typical situations of professional and business communication.

1. Employment, e.g. at Russian enterprises that have professional business contacts with international partners in order to perform professional and business communication (both written and oral) with international company representatives.
2. Professional growth, e.g. participation in exchange programs for specialists with internships at an international company.
3. Research, e.g. participation in international programs for obtaining a scholarship (grant) to conduct scientific research in Russia.

These fields of professional and business communication between graduates refer to the subject of their professional activity in general. However, among the mentioned fields, it is also possible to single out a number of specific communicative situations, when the *Innovation* graduates can apply the knowledge obtained within the course of Computers and Banking for the first time. These situations are as follows:

1. Situations related to the employment at Russian banks and other financial institutions constantly cooperating with international partners.
2. Situations associated with the professional development: taking live online exam for a diploma in one of the specialties that are part of the professional competence of graduates.

When developing the method for the delivery of the "Computers and Banking" course, we found it expedient to organize the training material in such a way that, while mastering it, the students can also master their professional and business communication, both orally and in writing, in each of the listed fields of professional and business communication, as well as in their typical situations. This specifies the thematic content of the course. It should presuppose teaching fluent FL communication to students, first, in the fields related to employment; then, in those fields that require professional growth, and, finally, in the fields associated with research in their specialty. However, with regard to the latter point, it should be noted that we intend to create the basis for conducting any research and developing a steady interest in the course subject.

Each type of the communication situation has its own characteristics: motives, content, structure and conditions of execution. In addition, each type of the situation involves a set of typical social roles taken by the participants in professional and business communication.

6 Conclusion

To sum up, the conducted analysis of the structure of professional and business communication of the graduates majoring in *Innovation* allows us to determine the level of communicative competence in speaking and writing that students should achieve when mastering the "Computers and Banking" course. The structure of professional and business communication conditions the special features of the generated oral or written utterances, affects the development of verbal skills, determines the FL communicative competence. Actually, we assume this level of language proficiency to be the main goal of the entire learning process.

Based on the analysis of the typology of areas, situations and roles of professional and business communication of graduates majoring in *Innovation*, it will be possible to determine the criteria for the selection of language and factual material, its methodological organization, organization of the exercise system and method for controlling the level of the development of productive speech skills based on the "Computers and Banking" course.

References

1. Apelt, V.: Rol' situativnoi grammatiki v prepodavanii inostrannogo yasyka [The role of situational grammar in the foreign language teaching]. Inostrannye yasyki v shkole **2**, 3–27 (1967). (In Russian)
2. Baghdasaryan, M.E.: Obuchenie professionalno-orientirovannomu obschetniyu na osnove nauchno-populyarnykh tekstov [Teaching professionally-focused communication based on science-fiction texts]. State Linguistic University, Moscow (1990). (In Russian)
3. Berezina, N.E.: Obuchenie pismennym formam delovogo obscheniya v situatsiyakh vkhodzdeniya v spheru professionalnoi deyatelnosti [Teaching written business communication in situations of the professional activity introduction]. State Linguistic University, Nizhny Novgorod (1998). (In Russian)
4. Bim, I.L., Makarova, T.V.: Ob odnom iz vozmodznykh podkhodov k sosyavleniyu programmy po inostrannym yasykam [About one of the possible approaches to designing an educational program in foreign languages]. Inostrannye yasyki v shkole **5**, 3–5 (1992). (In Russian)
5. Gorobets, L.N.: Phormirovanie professionalnykh kommunikativno rechevykh yumenij stunentov ne philologov [Development of professional verbal communication skills of non-linguistic students]. Herzen State Pedagogical University, St. Petersburg (1998). (In Russian)
6. Greyser, A.V.: Ob usvoenii rechevykh edinits v tipichnykh situatsiyakh [On mastering of verbal units in typical situations]. Inostrannye yasyki v shkole **5**, 12–17 (1987). (In Russian)
7. Gurrey, P.: Teaching English Grammar. Longman, London (1961)
8. Erastov, N.P.: Protsessy myshleniya v rechevoi deyatelnosti [Thinking processes in verbal activity]. State University of Psychology & Education, Moscow (1970). (In Russian)
9. Kakepoto, I.: Perspectives on oral communication skills for engineers in engineering profession in Pakistan. Int. J. Appl. Linguist. Engl. Lit. **1**(5), 176–183 (2012)
10. Matukihn, D.L., Kachalov, N.A., Fedorenko, R.M.: Peculiarities of teaching translation of scientific and technical papers to engineering students. In: MATEC Web of Conferences, vol. 92. https://doi.org/10.1051/matecconf/20179201041 (2017). Last accessed 11 Nov 2018
11. Nakatani, Y.: Developing an oral communication strategy inventory. Mod. Lang. J. **90**, 151–168 (2006)
12. Rahman, M.M.: Teaching oral communication skills: a task-based approach. ESP World 1 (9). http://www.esp-world.info/Articles_27/Paper.pdf (2010). Last accessed 10 Oct 2018
13. Samayeva, T.P.: Formirovanie professionalno-metodicheskikh umenij rechevogo vzaimodejstviya na inostrannom yasyke [Development of professional and methodical skills of foreign language verbal interaction]. State Linguistic University, Moscow (1990). (In Russian)
14. West, W.: Catenizing. Engl. Lang. Teach. J. **5**(6), 147–151 (1951)

Teaching A2-B2 Learners to Understand and Use Reported Speech

Lyudmila A. Milovanova[1]([⊠]) [ID], Valentina A. Tsybaneva[2] [ID],
Marina S. Kalinina[1] [ID], and Elena F. Grebenyuk[1] [ID]

[1] Volgograd State Socio-Pedagogical University, Lenina 27, 400066 Volgograd,
Russian Federation
{ludmilamilovanova, dimar23}@yandex.ru,
elena_boldareva@mail.ru
[2] Volgograd State Academy of Postgraduate Education, Novodvinskay 19a,
400012 Volgograd, Russian Federation
valentinatsybaneva@yandex.ru

Abstract. The purpose of this paper is to underscore some common difficulties such as comprehension, word order, reporting verbs and pronunciation, that A2-B2 learners face in learning English reported speech. Looking through some specific analysis of reported speech, particularly its meaning, form and usage, the paper will consider some issues for teaching that will lead to selection of activities in which students will need both to use and understand reported speech, and will also suggest teaching activities that can help to improve students' communication skills. Selected activities focus not only on practicing form and use of reported speech, but on development of learners' independence and their ability to work in pairs or groups. Moreover, with the help of learning on how to use reported speech in English, A2-B2 learners will expand their knowledge and understanding of how the language works. They will benefit from learning something new and from gaining a new perspective on the understanding of reported speech.

Keywords: Reported speech · Learning problems of reported speech · Suggested teaching issues

1 Introduction

Teaching reported speech in English is an essential lesson for all students. They need to be aware of the fact that, when speaking English, they should also use the past tense in the secondary clause. However, they interpret the meaning of the sentence through the filter of their native language. Students should be given detailed explanations of the contextual problems they are facing as native speakers of their mother tongue. This will provide them with a deeper understanding of the phenomenon which will help them remember how to use correctly the tenses when they turn direct speech into reported speech.

As Lewis noticed, "remoteness is a full and systematic explanation of the choice of verb forms appropriate after "reporting verbs" [7, p. 72]. Batstone considers it to be

© Springer Nature Switzerland AG 2019
Z. Anikina (Ed.): GGSSH 2019, AISC 907, pp. 165–172, 2019.
https://doi.org/10.1007/978-3-030-11473-2_19

"distance" [1]. When and how to follow the rules makes reported speech complicated for our learners and they try to avoid it. To encourage our students to use reported speech in everyday spoken and written communication, we are going to reconsider the learners' awareness of backshifting and non-backshifting "rules" of reported speech, learners' problems with use, form and pronunciation of reported speech and reporting verbs and choose some classroom activities that can help the A2-B2 students.

2 Literature Review

(1) Meaning and use
Reported speech is used when we are interested in the words that someone has said, in the essential information they convey. It is used in reports, mostly in the written form. The following Table 1 shows the distinction between direct speech and reported speech

Table 1. The distinction between direct speech and reported speech.

Direct speech	Reported speech
She said, "The shop's shut"	She said (that) the shop was shut

is clearly demonstrated in these two sentences [3, p. 89]:

The choice of past or present forms can be influenced by "subjective perspectives on events" [1, p. 20]. If we use present forms and do not backshift the tense, we feel psychologically that they are still close and relevant to us. If we use past forms and backshift the tense, the experiences we refer to are complete and no longer relevant. It is called "psychological distance" (ibid). Thornbury shows this principle on the diagram (see Fig. 1) [10, p. 7]:

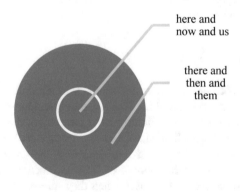

here and now and us

there and then and them

Fig. 1. The principle "psychological distance"

(2) Form and use (for some points)

(a) *Backshifting*

A "rule" for the "sequence of tenses" has "the distance effect" [4, p. 152]. If the verb in the main or reporting clause is in the Past Tense, it is usual for a verb in the reported clause to be backshifted. There are two possible types of backshift: Present → Past and Past → Past Perfect [5, p.106]. It is shown in the Table 2.

Table 2. The rule for the "sequence of tenses". Statements.

Direct	Indirect
Will – would	
The exam will be difficult	They said that the exam would be difficult
Present Simple – Past Simple	
I need help	He thought he needed help
Present Progressive – Past Progressive	
My English is getting better	I knew my English was getting better
Present Perfect – Past Perfect	
This has been a wonderful day	She told me that it had been a wonderful day
Past Simple – Past Perfect	
Ann grew up in France	I found out that Ann had grown up in France
can - could	
I can fly!	He thought he could fly
may - might	
We may come back late	They said they might came back late

In reported questions the subject normally comes before the verb and auxiliary (if needed) [6, p. 135]. The following rule is shown in the Table 3.

Table 3. The rule for the "sequence of tenses". Questions.

Direct speech	Indirect speech
"Do you live here?"	She asked him if (or whether) he lived there
"Where are the President and his wife staying?"	I asked where the President and his wife were staying

Yes/no questions are reported with if or whether [9]:

- in a formal style, whether is usually preferred in a two-part question with or: The Directors have not decided whether they will recommend a dividend or reinvest the profits.

Pronouns

A change of speakers may mean a change of pronoun:

Bill says, "**I** need a car". → Bill said **he** needed a new car.

Deictic expressions

A change of place and time may mean changing words like here → there, this → that/ the, now → then, today → that day.

Punctuation

In indirect speech, we do not normally separate the reporting clause from the reported clause by punctuation when the reporting clause is first [2, p. 222]: The lorry driver simply <u>said that</u> is was meat from another delivery. When the reporting clause comes second, a comma is used to separate the clauses (ibid): It had been painted with love, he said. There is no question mark in indirect questions: He asked when she would come.

(b) *No backshifting*

Past modals are usually left unchanged in indirect speech:

It would be nice if we could meet. → He said it would be nice if we could meet.

After present, future and present perfect reporting verbs, tenses are usually the same as in the original:

I don't want to play any more. → He says he doesn't want to play any more.

After past reporting verbs, conditionals are usually unchanged:

It would be best if we started early. → He said it would be best if they started early. However, if-sentences that refer to 'unreal' situations can change as follows: If I had any money I'd buy you a present. → She said that if she had had any money she would have bought me a present.

Speeches expressing "general truth" (proverbial or scientific nature) are likely to keep to the Present Tense even in indirect speech [5, p. 106]:

"Virtue is knowledge." → Socrates said that virtue is knowledge.

(c) *Reporting verbs*

Speech relating to actions (promises, agreements, orders, offers, requests and advice) is often reported with infinitives: He promised to phone. He agreed to wait for her.

Object + infinitive is common with ask, advise, tell and order (but not with promise or offer): I told Ann to be careful. I advise you to think again before you decide.

Reporting verb + ing is common with deny, apologise for, etc: I didn't take your bag. → He denied taking/ having taken my bag.

Some reporting verbs can be followed by that-clause: You're always late. → She complained that I was always late.

(d) *The passive structure*

The passive structure with verbs like think, feel, believe, know, etc, the object + infinitive is common, and often occurs in news reports [9].

He is believed to be dangerous. It is considered to be the finest cathedral in Scotland.

(3) Pronunciation

That is not stressed and has a weak form: /ðət/. As there is no comma in before that, there is no pause before *that*. Indirect questions are pronounced with falling intonation.

3   Learning Problems and Teaching Issues

3.1   Problems with Comprehension

A2-B1 learners are sometimes confused by the tense and pronoun changes that can occur in reported speech [8], particularly if their first language does not involve making similar or parallel changes. For example, they may understand Mary said I was ill to mean that Mary was ill. Moreover, a passive structure is problematic for B1-B2 learners: which subject to choose – she is believed to be or it is believed.

Deictic expressions are particularly challenging for A2-B1 learners when they are to re-situate text – "move text in either space or time" [11, p.124]: She said she was going to leave <u>tomorrow</u>.

3.2   Problems Caused by Word Order

In many languages there is no difference between the order of words in statements and questions. Learners can over-use correct word order in English questions when they report them: *She asked them did they like the film. He wanted to know were there any more people to come.*

B2 learners consider word order of questions beginning with *who/ what/ which + be* asked for a subject or a complement difficult as they have different word order (WO) [9].

Ask for a subject	
Who is the best player?	She asked me who **was** *the best player* (indirect WO)
	She asked me who *the best player* **was** (direct WO)
Asks for a complement	
What's the time?	She asked me what *the time* **was** (only direct WO)

3.3   Problems with Reporting Verbs

B1-B2 learners need not only to understand the meaning of the verbs they use, but they also need to know the construction of the clauses which follows each verb. Learners of a higher level often consider this to be the biggest problem with reported speech. *She said me she had to go. She explained me how to do it. She told she was ill.*

Say and *tell* are commonly used by A2-B2 students and sometimes they are used to report questions: *He said what was the time.* A2 and B1 students are confused by *say* (to somebody) and *tell somebody*. They can use *"say"* with a personal object but without to: *She said me that she would be late*; or they can use *to* after *tell*: *I told to her that I would come.* The meaning of *tell*, which is 'instruct' or 'inform', is not obvious for learners *He told them, 'Good morning'*. There are also some expressions with tell which learners have to learn: *tell a lie/ the truth*, etc, and they usually have difficulties in recollecting them.

Learners tend to over-use *say*, especially when they change conversation into reported speech: *She asked Robert if he wanted something to eat and he said that he didn't. He said that he would prefer a cup of coffee and she said that would be fine. He said thank you.*

Suggest is the most complicated reporting verb for B2 learners. *Suggest* is followed by *that-clause* and *–ing* structures, but students usually use *suggest* as they use *offer*: *Her uncle suggested that she to get a job in a bank* or *Her uncle suggested to get a job in a bank*. Additionally, the verb is not followed by an indirect object without a preposition. So, students forget about it and say: *Can you suggest a restaurant?*

B2 learners have difficulties with choosing the appropriate form of *–ing* after reporting verbs (Simple or Perfect): *He admitted (to)* **stealing/having stolen** *the plans*.

3.4   Problems with Pronunciation

If A2 learners are familiar with intonation in direct questions, they overgeneralize the rule and use the same intonation for indirect questions. For example: *He wanted to know if we would ∕come*.

Some reporting verbs are two syllable words and B1-B2 students know that many English two-syllable words are stressed on the first syllable. So, they can say: admit /'ədmɪt/, convince /'kənvɪn(t)s/, deny /'dɪnaɪ/. It is difficult for students to distinguish between /s/ and /z/, they pronounce accuse /ə'kjuːs/, advise /əd'vaɪs/ and refuse /rɪ'fjuːs/. Mispronunciation can cause misunderstanding.

4   Suggestions for Teaching

4.1   Activities for Practicing the Rules of Reported Speech

Reported speech is usually practiced with the help of a big number of exercises which can be easily found in different reference grammar books and course books. They all are aimed at re-writing the dialogues into the reported speech (or vice versa) and checking the answers with other learners or with teacher. It is useful to some extent as such activities focus on students' writing skills, precisely, re-writing. However, it isn unlikely to motivate students. Most of the exercises have separate sentences and it is difficult for students to understand the context and to choose the appropriate tense and "now and then" expression. Personally, it will be more helpful for students to work with the text and realize the context of it. Moreover, such an activity creates a whole picture which helps learners (especially visuals, and most of adult learners are visuals) to convey the context better and cope with the task.

To provide students with oral practice, they can play a game. It is necessary to prepare two-side cards and let learners work in pairs or small groups to practice reported speech. Students are to read one side of the card and change the sentence into reported speech or vice versa. This activity can be useful for A2-B1 learners. It can be also used with B2 learners at the beginning of the lesson to diagnose what learners remember and what is needed to be repeated or taught. This activity focuses not only on practicing form and use of reported speech, but on development of learners' independence and their ability to work in pairs or groups.

4.2   Activities for Practicing Word Order

There are some exercises which are aimed at practicing word order in reported questions. The second one is the dialogue and conveys the context as well. However, we think, it is beneficial to provide learners with a set of cards with words which can be used in indirect questions and ask students to work in pairs or small groups to practice the direct word order with the cards. It contributes a lot to memorizing word order of indirect questions, especially for visual and kinesthetic learners. Additionally, this activity can be valuable for practicing intonation of indirect questions. However, the model should be discussed and practiced with students beforehand. It can help to develop their abilities of self – and peer-correction, so they can learn from each other.

4.3   Activities for Practicing Reporting Verbs

To practice the form and meaning of the reporting verbs the range of exercises can be offered. Thornbury says, it is difficult to design "natural and productive speaking activities for practicing reported speech, perhaps because reported speech forms are a feature of written rather than spoken language" [11, p.189]. Thus, for semi-controlled practice with B2 learners a game can be used. This activity is challenging version of pair work dictation as they dictate direct speech sentences to each other and then change them into reported speech individually. They are to use reporting verbs and enlarge their knowledge of different reporting verbs (not only say and tell). A game which is close to real-life communication can be offered to B1-B2 students. It motivates learners to use reported speech, refers to their background knowledge and activates schemata on the topic "Famous People". As they have to choose themselves the famous person, it helps to develop their responsibility and independence.

4.4   Practice of Pronunciation

Pronunciation of some reporting verbs is confusing as they have different stressed syllables. Sounds and stress both contribute to the acoustic identity of a word, so both need studying at the same time. There are two exercises which are aimed at practicing pronunciation of reporting verbs. First, B2 learners underline the stressed syllable and then check with the recording. Then they are to "read" transcription of the verbs and again practice them. As the activities are done by students first, they help to develop their ability of noticing stress. It also aids their ability to work with dictionary (to check the stress) and raises their learner autonomy.

5   Conclusions

Learning how to use reported speech in English can become a challenging activity for students which will help them to gain a better understanding of how language itself work. Learners learning how to turn direct speech into reported speech often tend to start from what they know using translation from their native language. Learners can benefit from aroused interest by teachers applying suggested exercises and activities. They will benefit from learning something new and from gaining a new perspective on the understanding of reported speech.

References

1. Batstone, R.: Grammar. Oxford University Press, Oxford (1994)
2. Carter, R., Hughes, R., McCarthy, M.: Exploring Grammar in Context. Grammar Reference and Practice Upper-Intermediate and Advanced. Cambridge University Press, Cambridge (2000)
3. Close, R.A.: Teachers' Grammar. Language Teaching Publications, Homson-Heinle (1994)
4. Huddleston, R.: Introduction to the Grammar of English. Cambridge University Press, Cambridge (1995)
5. Leech, G.N.: Meaning and the English Verb. Longman, London (1997)
6. Leech, G., Svartvik, J.A.: Communicative Grammar of English. Longman, London (1994)
7. Lewis, M.: The English Verb. LTP Teacher Training, London (1986)
8. Parrott, M.: Grammar for English Language Teachers. Cambridge University Press, Cambridge (2000)
9. Swan, M.: Practical English Usage. Oxford University Press, Oxford (2005)
10. Thornbury, S.: About Language. Cambridge University Press, Cambridge (1997)
11. Thornbury, S.: Uncovering Grammar. Macmillan Heinemann English Language Teaching, London (2001)

Pedagogical Preconditions for the Development of Productive Communication Skills within the Special Course "Computers and Banking"

Gavriil A. Nizkodubov[1]([✉]) [iD], Svetlana S. Kuklina[2] [iD],
Alexander V. Obskov[1] [iD], Maria A. Vikulina[3] [iD],
and Nikolay A. Kachalov[1,4] [iD]

[1] National Research Tomsk Polytechnic University,
Tomsk 634050, Russian Federation
mnusa@yandex.ru, alexanderobskov@hotmail.com,
kachalov@tpu.ru
[2] Vyatka State University, Kirov 610000, Russian Federation
kss@ssk.kirov.ru
[3] Linguistics University of Nizhny Novgorod, Nizhny Novgorod 603005,
Russian Federation
centr_nid@lunn.ru
[4] The Russian Presidential Academy of National Economy and Public
Administration, Tomsk 634050, Russian Federation

Abstract. Undergraduate students seeking a degree in Management Studies at non-linguistic universities reveal insufficient level of the development of oral and written communication skills in professional and business communication in English. This underpins the need for more intensive training in productive types of speech by referring to a special textbook (a special course) that takes into account the specifics of professionally oriented training for students.

Computers and Banking is one of the possible foundations for building professionally oriented teaching of a foreign language in a non-linguistic university. In this regard, it seems necessary to determine the role and place of our "Computers and Banking" discipline both in the system of professionally oriented teaching of foreign languages in a non-linguistic university in general, and in a number of similar special courses in particular.

In this article, a concept of a "speech skill" and the pedagogical preconditions for the development of productive speech skills within the framework of a special course "Computers and Banking" are analyzed, the mechanisms for producing oral and written statements in a foreign language for pedagogical purposes are studied.

Based on the analysis of domestic and foreign literature on foreign language teaching methodology and psycholinguistics, textbooks and manuals on economics (banking) and the basics of computer science, special manuals for teaching professional-business communication in English, we established that the discipline "Computers and Banking" possesses a methodological potential for developing productive speech skills, and determined the structure of professional and business foreign language communication of undergraduate students in various fields of their activity.

© Springer Nature Switzerland AG 2019
Z. Anikina (Ed.): GGSSH 2019, AISC 907, pp. 173–182, 2019.
https://doi.org/10.1007/978-3-030-11473-2_20

Keywords: Undergraduate students ·
Professionally-oriented language training · Non-linguistic university ·
Oral and written speech · Communication skills

1 Introduction

One of the areas most susceptible to changes in the social environment is university education. In the light of serious changes in the life of Russian society in recent years, this thesis confirms its validity. At the moment, the Russian University has embarked on the path of integration into the international educational space.

In order to make foreign language learning in a non-linguistic university more professional and socially oriented, and to help domestic university graduates fulfill their potential in professional sphere, it is necessary to organize training with an account of the modern requirements of society and the new conditions in which a graduate will have to look for the use of professional knowledge and skills, even at the first steps after graduation. Thus, the essence of the social order of today consists not only in knowing a language in general but also in its purposeful training in the framework of various special courses presented in a foreign language and incorporating the specifics of the professional activity of future bachelor managers.

The special course "Computers and Banking" is one of the possible bases for arranging professionally oriented English language teaching at a non-linguistic university. Undergraduate students (bachelors) seeking a degree in Management Studies have been chosen as target learners (hereafter referred to as BMS students). In our opinion, the special course "Computers and Banking" can be organically integrated into the system of their professional training. It should be noted, that, in this context, by professional training we mean a specially organized educational process, which content involves the transmission of professional and business information providing the formation of an integral system of knowledge in a given field of activity [5].

As can be understood from the title of the special course "Computer and Banking", banking and computer science is the practical basis for teaching English. By "learning content" we, above all, imply the process of foreign language teaching, namely, the development of speaking and writing skills. However, this in no way means that the information component of training (fundamentals of banking and computer science) is only the thematic vector that corresponds to the essence of students professional training – study of economics. The fact that students acquire knowledge by learning new factual material of the proposed special course is the same priority as the development of their oral and written communication skills.

This makes a serious difference from many special courses for non-linguistic universities that aim at either teaching only a language based on the proposed topic related to students' specialization, or, on the contrary, predominantly teaching the specialization to the detriment of the language component. In our case, the factual and linguistic material of the special course forms a complementary unity. On the one hand, the factual material constitutes the language minimum, i.e. it is the basis for teaching speaking and writing; on the other hand, a foreign (English) language is a tool for mastering the factual (information) component of the special course.

We believe that a methodical system developed in line with the above-mentioned principle can become a full-fledged basis for the professionally oriented language training of BMS students at a non-linguistic university. As we see it, the following definition most closely corresponds to the term. By professionally oriented language training we mean the training of BMS students in the field of professional-business cooperation. Training is performed in a foreign language based on a holistic concept that includes an integral set of linguistic and cognitive knowledge [10].

2   Theoretical Framework

A line of reasoning for key provisions considered within this article is based on the best practices of Soviet, Russian, and foreign scientific researchers.

We consider professionally-oriented teaching of foreign languages as a pedagogical phenomenon on the basis of fundamental studies conducted by Darling and Dannels [1], Kakepoto [5], Kersiene and Savanevicien [7], Nakatani [9], Nizkodubov and Kachalov [10], Pimsleur [12], Rahman [13].

The State Educational Standard for university vocational education requires the account of professional specificity when learning a foreign language; it focuses on the implementation of the objectives of the future professional activity of graduates. Of particular relevance is the professionally-oriented approach to learning a foreign language at non-linguistic faculties of universities, which envisages the development of students' ability to speak foreign languages in specific professional, business, and academic fields and situations, taking into account the peculiarities of professional thinking, while organizing motivational and orientation-research activities.

"Professionally-oriented" refers to learning based on the needs of students in learning a foreign language, dictated by the features of a future profession or specialization. It involves a combination of mastering a professionally-oriented foreign language and the development of students' personal qualities, knowledge of the culture of the target language country, acquisition of special skills based on professional and linguistic knowledge.

In the paper, consideration of speech skills as a pedagogical phenomenon, including its psycholinguistic characteristics, is based on the findings of such researchers as Zimnyaya [3], Passov [11], and others. Psycholinguistic characteristics and methodological foundations of teaching speaking and writing prove the validity and possibility of constructing a special course focused on the development of skills in these types of speech activity, taking into account the characteristics of professional training of engineering and economics specialists: characterization of knowledge, skills and abilities in the structure of communicative competence; speech and language skills, their types and stages of development; speech activity, its types and stages of its implementation; speech skills and the stages of its development; the role of knowledge in learning a foreign language.

3   Methods

Based on the analysis of Russian and foreign literature on foreign language teaching methodology, psycholinguistics, textbooks for banking, the basics of computer science, and special textbooks for teaching professional and business communication in English, it was found out that the special course "Computers and Banking" incorporates a methodical potential for training BMS students, i.e. development of their productive communication skills for professional and business foreign language communication in various areas of their activities. A communicative approach to learning foreign languages is the implementation of such a way of learning, in which learning is a foreign language as a means of communication.

The communicative method involves the complete and optimal systematization of the relationship between the components of the learning content. These include: the system of general activity, the system of speech activity, the system of speech communication, the system of the foreign language being studied, the system of speech mechanisms (speech generation, speech perception, speech interaction), the system of structural speech formations (dialogue, monologue, different types of speech statements), the process of mastering a foreign language.

The communicative approach involves:

- communicative orientation of learning, i.e. a language serves as a means of communication in real life situations that require communication;
- interconnected learning of all forms of oral and written communication;
- authentic nature of training materials (even at early stages, original texts are selected without simplifications and adaptations);
- situational principle (In the classroom, the teacher recreates situations that students may encounter in real life, for example, talking in stores, on the street, discussing current topics, everyday situations at work or at school, etc. Modeling situations changes from day to day, creating new communication tasks for students);
- personality-oriented learning. During conversations, students express their opinions, talk about themselves, discuss relevant issues and topics. Working in a pair or in a group allows them to demonstrate creativity and actively interact in class.

Thus, communicative learning is arranged by us in such a way that all its content and organization is permeated with novelty. Novelty prescribes the use of texts and exercises containing something new for students, the variability of texts of different content, but based on the same material, the constant variability of the components of recreated situations, learning conditions, forms of speech utterances, tasks and methods for their implementation. Novelty provides for the rejection of arbitrary memorizing, develops speech production, heuristics and productivity of speech skills of bachelor managers, and arouses interest in educational and cognitive activity.

4 Ways to Develop Productive Communication Skills

Before considering the principles and methods for developing productive communication skills on the basis of our special course, it seems reasonable to determine what we mean by a communication skill as a pedagogical phenomenon. Besides, it is necessary to analyze the very possibility of developing speaking and writing skills on the basis of the special course material, and also to determine the predictable level of students' communication competence in these types of speech activity upon the completion of training.

As many researchers note [3, 8, 11], it is quite difficult to unambiguously define the concept of a "communication skill". There is a wide variety of approaches to this issue. Considering different views on the essence of communication skills, one can attempt to give definitions for each of them. In the physiological sense, the communication skill is an emergency closure of connections carried out on the basis of a system of speech dynamic stereotypes in order to solve communication problems. In the active sense, the communication skill is a conscious activity based on a system of subconsciously functioning actions aimed at solving communication tasks. Psychologically, the communication skill is a new qualitative formation that provides the ability to manage speech activity in the process of solving communication problems [11].

It seems to us that each of the abovementioned definitions is legitimate in its own way. After all that has been mentioned above, it is very difficult to provide a unique definition for the term of "communication skill", since communication skills, as we have found out, are extremely multifaceted, multilevel, and multidimensional. Thus, the communication skill can be characterized as the ability to mobilize experience and integrate it, an active synthesis of aspect skills, and solution to a new task, owning a system of mental and practical actions, regulating actions, a new qualitative stage, something that is self-programmed by a speaker.

Obviously, definitions based on a single quality, as well as definitions where all the qualities of communication skills are enlisted, cannot satisfy the definition rules. The reason for this will be, on the one hand, in the possible limitations of such a definition, and on the other, in the fact that a definition should express the essence of a concept rather than enlist its qualities.

Passov [11], emphasizing the presence of an intellectual component in the skill, makes an effort to define the skill as "a thought-based activity of a certain level based on actions of a lower level (aspect skills)". Moreover, the process of acquirement is seen in psychology as the process of knowledge accumulation (in this case, aspect skills and only partly knowledge) and mastering the way they operate. The latter statement is closely connected with thinking. Hence, it follows that the communication skill is the aspect skills as well as the ways of mastering them.

However, it should be noted that this approach covers only two components of the communication skill (aspect skills plus their acquirement). Still, this definition fails to reflect the main essence of the communication skill as an activity.

The definition, ultimately proposed by Passov [11], states that the communication skill is the ability to manage speech activity in the conditions of solving communication problems. Hereafter, we are going to refer to this definition as a fundamental one. With regard to the teaching of the "Computers and Banking" special course, this definition of the communication skill, namely its part concerning the conditions for solving communication tasks, will find its confirmation in the arrangement of the learning process based on a number of communication situations.

5   Compositional Analysis of Communication Skills

It is of paramount importance to mention a few words about the components of the communication skill; otherwise, it will not be possible to adequately build the practical basis for the development of communication skills (a set of exercises as per pedagogical specialization) without realizing the essence of the teaching object.

The analysis of various approaches to this issue reveals no consensus. In general, various researchers outline different components of the communication skill [2, 7, 9, 12, 14]. Yet, as the most consistent effort to identify the components of the communication skill can be considered an attempt made by Passov [11]. He states that it is important to distinguish the components of two levels: aspect skills and communication skills. By all means, this classification seems to be rather idealized; nevertheless, from a pedagogical point of view, it seems significant since it puts more or less clear tasks for the stages of mastering the material and for developing a set of exercises on this basis.

Passov [11] also notes that, at the first level, three skill groups (subsystems) should be singled out in a speech utterance: lexical, grammar and pronunciation when teaching speaking, and lexical, grammar and graphical (writing mechanism) when teaching writing.

At the level of the communication skill itself, for instance, speaking, it is necessary to distinguish skills of the operational nature (for example, a skill to combine words within a word combination, statements, several topics, etc.) and thinking skills (a skill to program statements, to evaluate adequacy of linguistic material in compliance with the goal and situation, etc.) [1, 5, 6, 13].

Therefore, each skill should be developed (mainly) through exercises [4] aimed at developing a specific type of speech activity. This idea can be realized in the proposed special course where the main part of the system of tasks and exercises for mastering the material is precisely the exercises for the development of speaking and writing communication skills as the ultimate goal. The abovementioned, however, does not exclude the tasks for listening and reading (the special course provides listening material and a number of additional texts for reading) but their number in the overall system of tasks is not large. Thus, the course incorporates the principle of simultaneous learning of different types of speech activities, with the leading role of one of them (in our case, speaking and writing).

6 Mechanisms for Production of Oral and Written Statements in Foreign Language Classroom

Having found out what is meant by the communication skill as a pedagogical phenomenon, it is essential to explain why it is the teaching of foreign speaking and writing speech (i.e. productive types of speech) that is combined in the framework of the developed special course.

In our viewpoint, the relevance of the choice of speaking and writing communication skills as the objects of development within the training of BMS students in English on the basis of the "Computers and Banking" special course can be justified by a number of psycholinguistic factors that prove the similarity of the mechanisms for producing oral and written statements in a foreign language.

With regard to speaking and writing, these mechanisms are as follows. Both oral and written statement is the result of the formation of a motive or "speech intention". Based on the motive and other factors, the grammatical-semantic side of an utterance is programmed including the grammatical realization of an utterance and the choice of words, the motor programming of the components of an utterance (syntagm), the choice of sounds and the "output" [8]. With regard to written message production, the mechanism consists in translating a sound code into a graphic code. Thus, in all the cases of using oral or written speech in a foreign language, there are two stages of using the code of this language: (1) encoding or decoding, which results in the production of a sound code, and (2) encoding using a graphic code [3].

When expressing thoughts in writing, this activity is supplemented by the association of the elements of sound communication and those or other graphemes, by motor activity, accompanied by pronunciation in internal speech. Pronunciation in the process of anticipating a written statement can differ in intensity depending on the complexity of written communication and the level of foreign language proficiency [3].

Anticipation of a written statement in internal speech is evaluated differently by the researchers on the issue [3]. There is an opinion that anticipation in inner speech is not an obligatory condition for the implementation of written speech, but only a habit brought in from the native language. However, most researchers agree that, regardless of whether an internal pronunciation is a component of written speech or only a habit brought in from the native language, the fact of oral anticipation is of great pedagogical importance when teaching speaking and writing [3, 8, 11].

The two stages of written speech formation (encoding or decoding and coding in graphics) and the presence of oral anticipation determine the relationship between written and oral speech. The difference between oral and written utterances boils down to the fact that in the first case the formation process ends with sound production and requires a high degree of automation, while for writing it is only the first stage, and in the process of writing, changes, additions, and corrections are possible. Thus, in the methodical sense, written speech, in its turn, can act as an important means of promoting the development of oral speech [11].

Expediency and relevance of the development of communication skills of speaking and writing in the framework of the special course confirm the characterization of these types of speech activity proposed by Zimnyaya [3]. This comparative analysis, which includes both psycholinguistic and psycho-physiological characteristics, proves the possibility and pedagogical expediency of the development of productive communication skills within the special course "Computers and Banking":

- speaking and writing are the initial communication processes stimulating listening and reading;
- both speaking and writing refer to the productive types of speech activity, i.e. a person produces a voice message by means of speaking and writing. In addition, from a psychophysiological viewpoint, in both types of speech activity the so-called speech-sound motor analyzer is involved. Writing, as a kind of activity, is realized on the basis of the connections of motor, visual-graphical and speech-sound images of linguistic phenomena [14]. These connections are realized in graphic and calligraphic skills – the material basis of written speech;
- as per the way of forming and formulating thoughts, speaking is an expression of the external oral way of forming and formulating thoughts in the oral form of communication. Writing fixes written and sometimes oral ways of forming and formulating thoughts;
- as per the nature of feedback controlling the processes of speech activity, in both productive types of speech activity (speaking and writing), the muscular (kinesthetic, proprioceptive) feedback from a doer's organ (articulatory apparatus, writing hand) to the brain arranging the performed action is involved. This muscular feedback serves as an internal control function. At the same time, both forms of this muscular, kinesthetic control from the organs of articulation and hands participate in the regulation of writing, especially at the initial stages of its formation. Along with the muscular "internal" feedback, productive types of speech skills are regulated by "external" audio feedback in speaking and "external" visual feedback in writing. In this case, noting the advantages of spatial visual perception in comparison with irreversible audio perception in time, Zimnyaya [3] emphasizes the greater arbitrariness and controllability of writing by the subject of activity as compared to speaking.

Apart from the above, it is also obvious that there is an inextricable link between all communication skills. Speaking and writing skills are similar and interrelated. A characteristic feature of the relationship between oral and written speech is that the former becomes the basis for the latter (the principle of the primacy of the audio-motor connections). Written speech requires reliance on oral speech, since before learning to write, a person must first master his oral speech. The relationship between oral and written speech is not one-sided, but mutual. According to the tradition of the national methodical school, written speech not only arises under the influence of oral speech, but also affects it. At the same time, the development of communication skills in written speech is itself capable of influencing the development of other types of speech activity.

Thus, taking into account the mechanisms for producing oral and written utterances, simultaneous development of communication skills in these types of speech activity in the framework of one academic discipline, in this case, the special course "Computers and Banking", is quite justified.

This clearly confirms the need for a comprehensive development of all communication skills, with the leading role of that communication skill, which is set as a goal at a particular stage of training. In our case, these are speaking and writing skills.

7 Conclusion

With an eye to make foreign language teaching at a non-linguistic university more professional and socially oriented and to help university graduates fulfill their potential in the professional sphere, it is recommended to arrange training with respect to modern requirements of the society and new conditions in which a graduate, starting with the first steps after graduation, will have to seek implementation of the acquired professionally oriented knowledge, aspect skills and communication skills. Thus, the essence of today's social order is not only a good command of a foreign language in general, but also the focused foreign language training with regard to the specific character of the professional activity of future specialists.

Based on the analysis of pedagogical and psycholinguistic features of teaching speaking and writing skills, as well as the specifics of professional training of BMS students at a non-linguistic university, it has become possible to determine ways of selecting and organizing materials for the "Computers and Banking" special course. Methodical organization and structuring allow it to be used equally effectively for training oral and written skills in professional and business communication, regardless of the level of prior training in the subject and specialization of BMS students. The analysis of the typology of areas, situations and roles of professional and business communication of bachelor managers will determine the criteria for the selection of language and factual material, its methodological organization, organization of the exercise system and the method for controlling the level of the development of productive speech skills within the "Computers and Banking" special course.

References

1. Darling, A., Dannels, D.: Practicing engineers talk about the importance of talk: a report on the role of oral communication in the workplace. Commun. Educ. **52**(1), 1–16 (2003)
2. Gurvich, P.B.: Obuchenie nepodgotovlennoi rechi [Teaching unprepared speech]. Inostrannye yasyki v shkole **8**(6), 11–17 (2012). (In Russian)
3. Zimnyaya, I.A.: Psihologicheskie aspekty obucheniya govoreniyu na inostrannom yasyke [Psychological aspects of learning to speak a foreign language]. Prosveschenie, Moscow (2012). (In Russian)
4. Kachalov, N.A.: Metodicheskaya organizatsiya kompleksa upradgnenij po razvitiyu umenij govoreniya i pismennoi rechi [Methodical organization of a set of exercises on development of speaking and writing skills]. Vestnik Tomskogo pedagogicheskogo universiteta **7**(135), 143–145 (2013). (In Russian)
5. Kakepoto, I.: Perspectives on oral communication skills for engineers in engineering profession in Pakistan. Int. J. Appl. Linguist. Engl. Lit. **1**(5), 176–183 (2012)

6. Kasapoğlu-Akyol, P.: Using educational technology tools to improve language and communication skills of ESL students. Novitas-ROYAL (Research on Youth and Language) **4**(2), 225–241 (2010)

7. Kersiene, K., Savaneviciene, A.: The formation and management of organizational competence based on cross-cultural perspective. Inz. Ekon.-Eng. Econ. **5**, 56–65 (2009)

8. Leontiev, A.A.: Osnovy psiholingvistiki [Fundamentals of psycholinguistics]. Smysl Academy, Moscow (1999). (In Russian)

9. Nakatani, Y.: Developing an oral communication strategy inventory. Mod. Lang. J. **90**, 151–168 (2006)

10. Nizkodubov, G.A., Kachalov, N.A.: Prepodavanie yasyka spetsialnosti [Teaching language of the specialty]. Vestnik Kemerovskogo gosudarstvennogo universiteta **7**, 24–27 (2008). (In Russian)

11. Passov, E.I.: Metodologiya metodiki: teoriya i opyt primeneniya [Methodology of teaching methods: theory and application experience]. Publishing house of LSPU, Lipetsk (2002). (In Russian)

12. Pimsleur, P.: Testing foreign language learning. In: Valdman, A. (ed.) Trends in Language Teaching, pp. 175–214. McGraw-Hill, New York (1966)

13. Rahman, M.M.: Teaching oral communication skills. A task-based approach. ESP World 1 (9). http://www.esp-world.info/Articles_27/Paper.pdf (2010). Last accessed 08 Aug 2018

14. Shatilov, S.F.: Metodika obucheniya nemetskomu yasyky v srednej shkole [Methods of teaching German at high schools]. Prosveschenie, Moscow (1986). (In Russian)

Modes of Wording Direct into Indirect Speech in Intercultural Communication

Olga A. Obdalova[1,2]([⊠]) [iD], Ludmila Yu. Minakova[1] [iD],
Aleksandra V. Soboleva[1] [iD], and Evgeniya V. Tikhonova[1] [iD]

[1] Tomsk State University, Lenin Avenue 36, 634050 Tomsk, Russian Federation
{o.obdalova, ludmila_jurievna}@mail.ru,
alex_art@sibmail.com, evkulmanakova@gmail.com
[2] Tomsk Scientific Center SB of RAS, Avenue Akademichesky 10/4,
634055 Tomsk, Russian Federation

Abstract. This paper aims to determine the modes of wording indirect reporting of authentic direct utterances by Russian learners of English. We claim that since the process of transferring someone else's speech involves implication and inferences of the speaker and the hearer correspondingly, when conveying the meaning of the speaker's authentic message in the form of indirect speech the personal context of the utterance plays a vital role. The experiment to check the hypothesis that direct speech requires not only grammatical and lexical transformations but also a complex pragmatic enrichment was organized. The reporting verbs used by the participants of the experiment to convey the speaker's intention and the presentation of the speaker's identity were analyzed. The study proved that when conveying the speaker's authentic speech meaning in the form of indirect speech the listeners need to shift from the reporting speaker's perspective to the reported speaker's perspective to comply with an actual communicative meaning of the utterance. Thus, a foreign language context of communication imposes additional linguistic, extra linguistic, and pragmatic difficulties on the process of English language learners' interpreting of the utterance which is cognitively demanding and needs to be persistently developed.

Keywords: Direct speech · Reported speech · Cognitive processes ·
Person identification · Reporting verbs

1 Introduction

1.1 A Theoretical Underpinning of the Reported Speech Research

In teaching English as a foreign language, it is important to take into consideration the difference between the communicants belonging to different cultural backgrounds as well as the discursive factors in order to achieve adequacy in communication [1–11]. All the authors pay special attention to the pragmatic factors which are to be developed when teaching a foreign language as a means of intercultural communication and interaction.

© Springer Nature Switzerland AG 2019
Z. Anikina (Ed.): GGSSH 2019, AISC 907, pp. 183–194, 2019.
https://doi.org/10.1007/978-3-030-11473-2_21

Learners should be able to use different modes of wording the indirect reports which have to comply with the language norms and pragmatic factors of the referential communicative situation. The spotlight of this study is determining the modes of wording indirect reporting by Russian learners of English of authentic direct utterances delivered by American native speakers. Reported speech has been a topic of interest for many researchers of linguistics, sociology, philosophy and pragmatics in past years. There has been a growing interest in indirect reporting in bilingual education [12–15]. A lot of publications deal with the issue of indirect speech acts with English as a medium of communication [16–19], while other studies examine indirect reporting using a foreign language [20, 21]. Some research of indirect reports focuses on the contrasting effects of direct and indirect speech on language comprehension [22–24].

It has been established that while direct speech, as a rule, conveys the expression of another person, preserving the lexical composition, grammatical structure and stylistic features, indirect speech usually reproduces only the content of the statement, changing its structure under the influence of the author's position. Therefore, in addition to the fact that the process of transferring someone else's speech is based on the generality of the language and the rules for its use, they highlight the implicit information that the speaker puts in the message, and the listener's inferences which are based on what the listener extracts.

When conveying the meaning of the speaker's authentic speech in the form of indirect speech, the personal context of the utterance, which is encoded in lexical units and framed in the utterance, assumes great importance and is pronounced by the communicant in the actual situational context. The result of this process is a statement having an actual communicative meaning. The act of indirect speech has a number of characteristics:

- ability to influence judgments and actions of communicants;
- cooperative speech activity of two communicants – a speaker and a listener - that affects the linguistic choice reflected in the transformations of the original utterance;
- representativeness.

While direct speech, as a rule, conveys one person's utterance and communicative intent, preserving the lexical composition, grammatical structure, and stylistic features of the addressee's speech, indirect speech usually reproduces the content of the utterance, changing its structure under the influence of the personal context of the addressee. By the personal context we mean the communicant's ability to adequately perceive and interpret a foreign language utterance in accordance with the level of his/her foreign language competence and cognitive abilities. In this case, the implicit information that the addressee puts, and the inferences – the information that the addressee retrieves and transmits – become instrumental in indirect communication.

The speaker, creating an utterance, exercises control over what he/she says and how he/she shapes his thoughts. The listener interprets the speaker's message, and this interpretation may not coincide with the content implicit in the given utterance by the speaker, which will affect the content of the statement when it is communicated to a third party. The foreign language context of communication imposes additional linguistic, extra linguistic and pragmatic difficulties on the process of interpreting the utterance. The study of the process of conveying foreign speech, in particular indirect speech-making, aims at studying a wide range of phenomena related both to their grammatical nature [25] and to the pragmatic load [26].

Based on the above-mentioned considerations, the hypothesis of this research was the assumption that the process of conversion of foreign direct speech and the implicit nucleus embedded in the utterance is a complex cognitive task, especially in the situation of intercultural communication, since it depends on the receiver's inference and his/her ability to process input information in accordance with the situational context and realities of the communicative situation. The task of this study is to determine by what language means the semantic core of the original foreign language utterance is conveyed by a foreign language learner and what reasons may lie behind failures in intercultural communication.

1.2   Cognitive and Linguistic Factors in Conversion of Direct Speech into Indirect Speech

As we know, perception and comprehension in the conditions of intercultural communication take place as a result of the functioning of the perceptual mechanism which processes the input information at a multi-level cognitive activity [5]. Due to various contextual factors, the recipient of the discourse needs additional knowledge about the specific communicative situation in which the entire complex of incoming information is encoded implicitly [27–29].

A core part of the reporting clause is a reporting verb (RV). Reporting verbs are the most important features of a reporting clause and occur in most reporting sentences [30, p. 2]. As far as English grammar is concerned, a reporting verb is used to indicate that discourse is being quoted or paraphrased. It is also called a communication verb.

We agree with Thomas and Hawes [31] and Hyland [32] who identify types of content based on the choice of verbs of communication which are used to convey the speaker's intention. There are three basic types of utterances distinguished according to discourse functions: (1) *statements,* (2) *questions (general and special),* (3) *requests and commands.*

Thompson and Ye (1991) argue that reporting verbs permit a hearer to express his/her own judgment of what is being reported [33]. From this point of view, the speaker who conveys someone else's utterance reveals some attitude or value to what is reported, being tentative (without being absolutely certain, e.g., *imply, propose, recommend, suggest*). There can be also distinguished a group of verbs called the verbs of speech, which describe a speaker's intention through the way of saying: *say, ask, answer, suggest.* Table 1 represents the list of the most common reporting verbs (RV) assorted into groups.

Bell [34, p. 206] asserts that '*to say*' is the most frequent reporting verb. Thompson [35, p. 34] calls this verb "a neutral reporting verb" that can be used when reporting any type of language event, no matter if it is oral speech or writing.

Table 1. The list of reporting verbs used in indirect speech.

Communicative types of utterances	Most frequently used and registered variation of reporting verbs in naturally occurring reported speech
Statements	say, tell – verbs that behave like *say* verbs that do not require an indirect object; e.g. admit, announce, comment, complain, confess, explain, indicate, mention, point out, remark, reply, report, shout, state, swear, whisper; – verbs that behave like *tell* (i.e. verbs that do require an indirect object; e.g. assure, convince, inform, notify, persuade, remind), – other reporting verbs for which no 'say vs. tell behavior' distinction is offered (e.g. advise, answer, demand, insist, promise, propose, recommend, require, suggest, want to know, order, request) believe, reply, respond, admit, explain, emphasize
Questions	ask, inquire, wonder, want to know
Requests/commands	ask, beg, order, tell

A reporting verb can show the speaker's purpose (e.g. *admit, explain, emphasize*) or manner of speaking (e.g. *whisper, scream, mutter*) [35, p. 71–77]. A reporter can also display a neutral, positive or negative personal attitude toward the reported message, which is achieved by using reporting verbs that have positive or negative connotations.

Another canonical component of indirect speech is the person distinction. Pragmatic factors such as the communicative situation influence the processing of speech reports. In direct speech pronouns have to be evaluated with respect to the reported speaker's perspective and in indirect speech – with respect to the reporting speaker's perspective. Pronouns such as *I, you* and *she/he* are context-dependent. This means that listeners need to have knowledge of the speech context – in particular the distribution of speech-act roles – in order to determine their meaning. The actual speaker constitutes first-person *I* and second-person *you* refer to the primary participant of an interaction; speaker and addressee. Third-person *he* and *she* refer to a male or female person other than speaker and addressee [36]. Since both speaker and addressee are aware of their communicative roles, the referents of the first-person and second-person pronouns are automatically salient in the discourse [37]. In Kaplan's [38] framework, for example, first- and second-person pronouns are identified as pure indexicals, directly getting their reference from the context parameters. Presentation of the identity of the reporting speaker involves the presentation of the agent by such linguistic means as names. The studies on comprehension proved that gender-marking is a very salient feature. Arnold et al. [39] results indicate that in English gender easily marked pronouns can be interpreted even at an early age. While pronouns in indirect speech have to be interpreted with respect to the actual speech context, pronouns in direct speech are anchored in the reported speech context. This means that listeners need to shift from the reporting speaker's perspective to the reported speaker's perspective. This perspective shift could be cognitively demanding for learners of English.

2   Methodology

2.1   Subjects

38 subjects who represented Russian learners of English were engaged in the experiment, including male and female learners aged from 18 to 26 (Table 2). The focus group included first and second year non-linguistic students from Tomsk State University, majoring in Science; and students of the Faculty of Foreign Languages, majoring in Linguistics and Translation. The average level of language proficiency in non-linguistic students was intermediate while the average level of language competency in linguistic students was pre-advanced.

Table 2. Participants.

Number	Mean age	Gender (f/m)
38	20 (from 18 to 26)	28/10

2.2   Research Procedure

This paper aims to investigate the ways Russian learners of English convey speech utterances presented by American English native speakers to third parties in written form.

The study was carried out in the framework of the socio-cognitive approach proposed and developed by Kecskes [40–42] and Kecskes and Zhang [43, 44]. We assume that when conveying speech utterances of English native speakers, Russian learners of English rely on the semantic content of utterances rather than the pragmatically enhanced message. We also seek to explore if Russian learners of English are influenced by their socio-cultural background when conveying speech utterances.

The reporting material included 12 utterances of three communicative types (requests/commands statements, questions):

Request/Command

(1)   John: Don't open the window, please. It is chilly here.
(2)   Mary: Don't even think about lying to me.
(3)   John: You should meet with the professor on Friday.
(4)   Mary: They must be more careful with what music they select.

Statements

(1)   John: I think I will need your help in an important matter.
(2)   Mary: I am tired of answering your silly questions.
(3)   John: I do not want to tell you what I think about Tom.
(4)   Mary: Mary knows what Jim is hiding from us.

Questions

(1)   John: I wonder why you look so happy.
(2)   Mary: Do you know when the accident happened?
(3)   John: Where do you think Jill has put the book?
(4)   Mary: How much money can I spend on the trip?

The video was recorded so that each utterance was repeated twice with 10-s breaks between the stimuli. The subjects were instructed to watch and listen carefully to the speakers, John and Mary by names, and report in writing what they understood. The way the subjects would shape their reports was the most important issue for the analysis.

One of the interesting reporting devices to analyze is the reporting verbs which have the main function of reporting other people's utterances of 3 types: statement, question, and command/request. Another instrument of reporting speech analysis is the use of pronouns and proper names when transforming the speaker's utterance into indirect speech. We used discourse and error analyses in our study. The study will be guided by the following research questions:

1. Which reporting verbs are used in the reported speech by Russian learners of English?
2. What linguistic means are used by Russian learners of English when referring to the reporting speaker?

2.3 Data Collection

The data set comprises indirect reports produced by Russian EFL learners. We assume that in reporting in a foreign language what others say in their own language the listener involves cognitive processes of perception and interpretation of the original utterance based on his or her prior linguistic and communication experience as well as socio-cultural background.

More specifically our interest concerns the reporting verbs which constitute an important factor in the speaker's intention presentation. Table 3 illustrates the distribution of reporting verbs (RV) used by the subjects in their reports. The minimum is represented by a pair 2–3 commonly used RV (*say, tell, ask*), the other groups constitute a combination of these with other RVs represented in a separate column.

Table 3. Reporting verbs used in the reports.

Number of RVs	Number of subjects Linguists/Non-linguists	RV and frequency of use (%)
2–3	4/1	say, tell, ask (65.8%)
4–5	7/1	advise, affirm, allow, beg, claim, confirm, declare, insist,
6–7	5/11	mention, prevent, prohibit, refuse, suggest, suppose, think,
8–9	1/5	threaten, utter, want, warn, wonder (34.2%)
10–11	1/1	

As Table 3 shows, this study revealed that most participants used common verbs such as *say, tell,* and *ask,* representing the speaker's action, as well as other reporting verbs comprising a large number including 18 various units. The use of three verbs, *say, tell, ask* reaches 65.8% which complies with the rule of using reporting verbs in the indirect speech.

Another focus is on how person distinction is done in the reported speech in connection with the reporting speaker and the speaker's gender. In Table 4 one can find the distribution of errors made by the subjects belonging to different categories.

Table 4. Distribution of errors in the speaker's identity representation.

Speaker's name	Number of errors	Male/female	L/NL
John	6	2/13	11/4
	5	2/2	4/0
	4	0/3	0/3
	3	1/4	2/3
	2	1/4	1/4
	1	0/1	0/1
	0	4/1	1/4
Mary	6	3/13	12/4
	5	1/2	1/2
	4	0/5	3/2
	3	2/5	2/5
	2	0/2	0/2
	1	1/0	1/0
	0	2/1	0/3

The maximum number of errors in each block of utterances (by *John* and *Mary*) is equal to 6 according to the number of utterances and the minimum number (0) corresponds to the absence of mistakes in the reports. By an error we mean substitution of the proper name of the speaker by personal pronouns *he* or *she*. We also distinguished errors made by female and male participants as well as by learners belonging to either the linguistic (L) or non-linguistic (NL) language group.

2.4   Data Interpretation and Discussion

Based on the data of Table 3, we can say that Russian learners of English as a foreign language used 21 various reporting verbs when conveying the speakers' intention in the indirect speech. This variety testifies to the work of cognitive processes in the learners in the search for adequate language facilities for conveying the implicit nucleus of utterance.

The data of discourse analysis showed that the subjects had chosen various ways of conveying the speakers' intention [45]. In Table 5, we provide some samples of interpretation of the original utterances belonging to three types of communicative acts:

Table 5. Samples of reported utterances.

Original utterance	Communicative type	Reported utterances
– I wonder why you look so happy	Statement	– He **said** that he looked so happy – He **wondered** why I was so happy – He **wonders** why you look so happy – He **told** that he wants to know why he looks so happy – John **wondered** why I was looking so happy – John **asked** me why I looked so happy
– Don't open the window, please. It is chilly here	Request	– He **asked** me not to open the window – He **asks** not to open the window because it's chilly – He **didn't allow** me to open the window because it was chilly – John **prohibited** me to open the window as it was chilly – John **begs** me not to open the window – He **claims**, it's chilly in the room and prevents me from opening the window – He **told** me not to open the window because it's chilly
– How much money can I spend on the trip?	Question	– She **wanted** to know how much she could spend on her trip – She **wondered** what amount of money she could spend on the trip – She **asked** me how much money the trip cost – Mary **thinks** how much money she can spend on her trip – Mary **asks** how much money she can spend on the trip

These examples indicate that the participants tried to convey the original utterances by using not only commonly used RVs, but also by preserving some lexical items in their indirect reports, e.g. *so happy, open the window, it is chilly, how much, spend on the trip*. The variety of verbs used indicates the difference in the subjects' perception of the original message and the ability to represent the implicit core of the speaker's speech.

It should be noted that in the interpretation of the same utterances by native speakers who were engaged in the experiment as experts the same commonly used RVs were found. This fact demonstrates that Russian learners of English use a wide variety of RVs in order to convey the implicit core of the referential utterance.

A closer look at the participants' errors in indirect speech focusing on the person's identity reveals that the masculine pronoun '*he*' was used instead of '*John*' in the majority of cases and the feminine pronoun '*she*' was used for the substitution of '*Mary*'. We analyzed participants' production of speech reports and found out that both male and female participants in linguistic and non-linguistic subgroups made mistakes in interpretation of the speakers' identities in comparison with native speakers (experts) who preferred labeling the speakers' identities by their names (*John* and *Mary*). We assume that this situation can be explained by the socio-cultural background of the Russian learners. As we know, American and Russian cultures belong to different types. In American culture, individualism and personal value are typical while in Russian culture collective relationships are more valuable. This complies with the assumption of Larina that 'the value of privacy in American culture and the lack of it in Russian explain a lot of characteristics peculiar to both politeness systems, as well as to their communicative styles' [46, p. 3]. However, it is important to mention that none of the participants made errors in identification of a male or a female speaker.

The mistakes made by the subjects in the reports were also subjected to statistical analysis.

Table 6 shows the values of the pairwise Pearson correlation coefficients

$$r_{xy} = \frac{\sum\limits_{i=1}^{n}(X_i - \bar{X})(Y_i - \bar{Y})}{\sqrt{\sum\limits_{i=1}^{n}(X_i - \bar{X})^2 \sum\limits_{j=1}^{n}(Y_j - \bar{Y})^2}}, \tag{1}$$

where n is the number of observations for each variable, $\bar{X}, \bar{Y}$ – sample mean values for the variables x and y, respectively. We should note that there is a strong direct linear relationship between errors when dealing with John and Mary – those who transferred John's name from original utterance to reported also successfully coped with the presentation of Mary's identity, and vice versa. Moreover, for linguists this dependence is the strongest.

Table 6. The values of the Pearson correlation coefficients of student errors in the presentation of the speaker's identity.

	Non-Linguists		Linguists	
	Errors (John)	Errors (Mary)	Errors (John)	Errors (Mary)
Error (John)	1		1	
Error (Mary)	0.948066	1	0.87675	1

3 Conclusion

The analysis of the empirical results obtained during the study suggests that indirect speech is not only a syntactically organized form of the transmission of someone else's speech, which requires certain transformations when conveying an utterance from direct speech to indirect speech, but also a complex pragmatic expression showing how the reporter interpreted the original message, i.e. of the speaker, rendering its content (statement, question, request/command) as well as the speaker's assertion in a particular context. The experiment confirmed our hypothesis that the implicit core of the direct utterance is represented in two crucial elements: reporting verbs used to convey the speaker's intention and the language means applied to represent the speaker's identity. At the same time, the original authentic utterance undergoes specific transformations in connection with the socio-cultural background, language norms and cognitive mechanism functioning in representatives of various cultures involved in intercultural communication. When teaching English as a foreign language to Russian learners it is important to take into consideration the difference between the communicants belonging to different cultural backgrounds in order to achieve adequacy in indirect reporting. The learners should be able to use different modes of wording the indirect reports which have to comply with the language norms and pragmatic factors of the referential communicative situation.

Acknowledgements. We would like to thank Istvan Kecskes, Professor of Linguistics and Education at the State University of New York, Albany, the President of the American Pragmatics Association and Editor-in-Chief of the journals Intercultural Pragmatics (De Gruyter) and the Mouton Series in Pragmatics for his supervision, advice and guidance from a very early stage of this research.

References

1. Gural, S.K., Mitchell, L.A.: Model' formirovaniya grammatiko-diskursivnyh navykov u studentov neyazykovogo vuza na osnove kognitivnogo podhoda [Grammar-discursive skills formation model in students of English (EFL) at a high school on the basis of the cognitive approach]. Lang. Cult. **3**(35), 146–154 (2016). (in Russian)
2. Gural, S.K., Smokotin, V.M.: Mezhyazykovaya i mezhkul'turnaya kommunikaciya v period globalizacii. [Interlingual and cross-cultural communication during the period of globalization]. Lang. Cult. **4**(24), 14–23 (2013). (in Russian)
3. Sysoyev, P.V., Ezhikov, D.A.: Teaching students verbal communication on the bases of synchronous video-internet-technologies. Lang. Cult. **2**(6), 58–68 (2015)
4. Sysoyev, P.V., Evstigneeva, I.A., Evstigneev, M.N.: The development of students' discourse skills via modern information and communication technologies. Procedia Soc. Behav. Sci. **200**, 114–121 (2015)
5. Kecskes, I., Obdalova, O.A., Minakova, LYu., Soboleva, A.V.: A study of the perception of situation-bound utterances as culture-specific pragmatic units by Russian learners of English. System **76**, 219–232 (2018)
6. Bezukladnikov, K.E., Zhigalev, B.A., Kruze, B.A., Novoselov, M.N., Vikulina, M.A., Mosina, M.A., Dmitrieva, E.N., Novoselova, S.N., Oskolkova, V.R.: ESP Teaching, Learning, Assessment: Modern Tools, Strategies, Practices, 2nd edn. Perm State Humanitarian Pedagogical University; Linguistic University of Nizhniy Novgorod, Perm (2018)
7. Zhigalev, B.A., Bezukladnikov, K.E.: Writing as the aim and means in teaching a foreign language: problems of assessment. Life Sci. J. **11**, 685–689 (2014)
8. Bezukladnikov, K.E., Novoselov, M.N., Kruze, B.A.: The international teachers foreign language professional communicative competency development. Procedia Soc. Behav. Sci. **154**, 329–332 (2014)

9. Obdalova, O.A., Minakova, LYu., Tikhonova, E.V., Soboleva, A.V.: Insights into receptive processing of authentic foreign discourse by EFL learners. In: Filchenko, A., Anikina, Zh (eds.) Linguistic and Cultural Studies: Traditions and Innovations 2017, LKTI, pp. 231–242. Springer, Heidelberg (2017)

10. Obdalova, O.A.: Modelling conditions for students' communication skills development by means of modern educational environment. In: Al-Mahrooqi, R., Denman, C. (eds.) Bridging the Gap Between Education and Employment: English Language Instruction in EFL Contexts, pp. 73–91. Peter Lang International Academic Publishers, Bern (2015)

11. Obdalova, O.A., Minakova, LYu., Soboleva, A.V.: The study of the role of context in sociocultural discourse interpretation through the discursive-cognitive approach. Vestnik Tomskogo gosudarstvennogo universiteta **413**, 38–45 (2016)

12. Kecskes, I.: Indirect reporting in bilingual language production. In: Capone, A., Kiefer, F., Lo Piparo, F. (eds.) Indirect Reports and Pragmatics. Perspectives in Pragmatics, Philosophy & Psychology, vol. 5, pp. 9–29. Springer, Cham (2016)

13. Kperogi, F.A.: Common errors of reported speech in Nigerian English. http://www.farooqkperogi.com. Accessed 05 June 2018

14. Oluwakemi, T.O.: I was like as a quotative device: implications for indirect or reported speech in Nigerian English usage. Asia Pac. J. Multidiscip. Res. **5**(2), 94–103 (2017)

15. Köder, F.M.: Between Direct and Indirect Speech: The Acquisition of Pronouns in Reported Speech. University of Groningen, Groningen (2016)

16. Coulmas, F.: Reported speech: Some general issues. In: Coulmas, F. (ed.) Direct and Indirect Speech, pp. 1–28. Mouton de Gruyter, Berlin (1986)

17. Groefsema, M.: Processing for relevance. A pragmatically based account of how we process natural language. Ph.D. thesis, University of London (1992)

18. Biber, D., Johansson, S., Leech, G., Conrad, S., Finegan, E.: Longman Grammar of Spoken and Written English. Pearson Education, Harlow (1999)

19. Asher, N., Lascarides, A.: Indirect speech acts. Synthese **128**, 183–228 (2001)

20. Marinchenko, D.B.: Sposoby peredachi chuzhoj rechi v rechi mladshih shkol'nikov [The ways of speech reporting in grade school students]. Ph.D. thesis, Taganrog state pedagogical institute (2006). (in Russian)

21. Latysheva, S.V.: Modusnaya obuslovlennost' aspektual'noj formy predikata v pridatochnom predlozhenii vyskazyvaniya s kosvennoj rech'yu [Mode dependence of predicate aspectual form in reported clause]. Ph.D. thesis, Irkutsk State Linguistic University (2008). (in Russian)

22. Bohan, J., Sanford, A.J., Cochrane, S., Sanford, A.J.S.: Direct and indirect speech modulates depth of processing. In: The 14th Annual Conference on Architectures and Mechanisms for Language Processing (AMLaP), pp. 287–307. Springer, Cham/Cambridge (2008)

23. Yao, B., Scheepers, C.: Contextual modulation of reading rate for direct versus indirect speech quotations. Cognition **121**, 447–453 (2011)

24. Eerland, A., Engelen, J.A.A., Zwaan, R.A.: The influence of direct and indirect speech on mental representations. PLoS One **8**(6), 1–9 (2013)

25. Savel'eva, G.K.: Algoritm obrazovaniya standarta dlya zameny pryamoj rechi kosvennoj [Standard formation algorithm for substituting direct speech for indirect]. Philol. Sci. Issues Theory Pract. **3**(10), 141–144 (2011). (in Russian)

26. Anderson, L.: When reporting others backfires. In: Capone, A., Kiefer, F., Lo Piparo, F. (eds.) Indirect Reports and Pragmatics. Perspectives in Pragmatics, Philosophy & Psychology, pp. 253–264. Springer, Cham (2016)

27. Obdalova, O.A.: Exploring the possibilities of the cognitive approach for non-linguistic EFL students teaching. Procedia Soc. Behav. Sci. **154**, 64–71 (2014)

28. Soboleva, A.V., Obdalova, O.A.: Kognitivnaja gotovnost' k mezhkul'turnomu obshheniju kak neobhodimyj komponent mezhkul'turnoj kompetencii. [Cognitive readiness for intercultural communication as an essential component of intercultural competence]. Lang. Cult. **1**(29), 146–155 (2015). (in Russian)

29. Soboleva, A.V., Obdalova, O.A.: Organizaciya processa formirovaniya mezhkul'turnoj kompetencii studentov s uchetom kognitivnyh stilej obuchayushchihsya [Organization of students' intercultural competence formation in view of their cognitive styles]. Vestnik Tomskogo gosudarstnevvogo universiteta **392**, 191–198 (2015). (in Russian)

30. Nkansah, N.B.: Reporting verbs and stance in front page stories of Ghanaian newspapers. Engl. Specif. Purp. World **40**(14), 1–22 (2013)

31. Thomas, S., Hawes, T.: Reporting verbs in medical journal articles. Engl. Specif. Purp. **13** (2), 129–148 (1994)

32. Hyland, K.: Disciplinary Discourses: Social Interactions in Academic Writing. Longman, London (2000)

33. Thompson, G., Ye, Y.: Evaluation in the reporting verbs used in academic papers. Appl. Linguist. **12**(4), 365–382 (1991)

34. Bell, A.: The Language of News Media. Blackwell, Oxford (1991)

35. Thomson, G.: Collins Cobuild English Guides: 5. Reporting. HarperCollins Publishers, London (1994)

36. Lyons, J.: Semantics, vol. 1–2. Cambridge University Press, Cambridge (1977)

37. Diessel, H.: Deixis and demonstratives. In: Maienborn, C., von Heusinger, K., Portner, P. (eds.) Semantics. An International Handbook of Natural Language Meaning, pp. 2407–2431. Mouton de Gruyter, Berlin (2012)

38. Kaplan, D.: Demonstratives. In: Almog, J., Perry, J., Wettstein, H. (eds.) Themes from Kaplan, pp. 481–563. Oxford University Press, Oxford (1989)

39. Arnold, J., Brown-Schmidt, S., Trueswell, J.: Children's use of gender and order of mention in pronoun processing. Lang. Cogn. Process. **22**, 527–565 (2007)

40. Kecskes, I.: The paradox of communication: a socio-cognitive approach. Pragmat. Soc. **1**(1), 50–73 (2010)

41. Kecskes, I.: Intercultural Pragmatics. Oxford University Press, Oxford (2013)

42. Kecskes, I.: Explorations into Chinese as a Second Language. Springer, Cham (2017)

43. Kecskes, I., Zhang, F.: Activating, seeking and creating common ground: a sociocognitive approach. Pragmat. Cogn. **17**(2), 331–355 (2009)

44. Kecskes, I., Zhang, F.: On the dynamic relations between common ground and presupposition. In: Capone, A., Lo Piparo, F. (eds.) Perspectives on Pragmatics and Philosophy, pp. 375–395. Springer, Dordrecht (2013)

45. Obdalova, O.A., Minakova, LYu., Soboleva, A.V.: Issledovanie roli konteksta v interpretacii sociokul'turno-markirovannogo diskursa na osnove kognitivno-diskursivnogo podhoda [A study of the role of the context in the adequacy of the representation of foreign speech in indirect communication]. Vestnik Tomskogo gosudarstnevvogo universiteta **413**, 38–45 (2016). (in Russian)

46. Larina, T.: Directness vs. Indirectness in Russian and English Communicative Cultures. LAUD, Essen (2006)

Developing Foreign Language Regional Competence of Future Foreign Language Teachers: Modeling of the Process

Olga Oberemko⬝, Elena Glumova$^{(\boxtimes)}$⬝, and Alexey Shimichev⬝

Linguistics University of Nizhny Novgorod, Minina Street 31a,
603155 Nizhny Novgorod, Russian Federation
dolober@rambler.ru, el.glumova2010@yandex.ru,
alexshim@list.ru

Abstract. The article considers modeling of the process aimed at developing the future foreign language teacher regional competence. It is noted that modeling is one of the practice-oriented methods for description and generalization of research results. The process of model development has revealed a number of contradictions in the systems of secondary and higher education. The structure of the regional competence development model comprises the following linked components: theoretical-methodological, purpose, contentful, process-operational, and evaluative-result. Methodology of the obtained model is based on the intercultural, regional, competence, professionally-oriented, and personal-activity approaches. The process of the experimental verification of the developed model has revealed a complex of professional knowledge and pedagogical skills that have a direct link with the foreign language learning process in the context of intercultural communication at the regional level. The experiment is conducted by employing experimental and control groups of students, as well as invariant and varied learning conditions. The result of the study is the assessment of the regional competence of future foreign language teachers in terms of six criteria. The research outcomes prove the purposefulness of incorporating materials with a regional component of native and foreign-speaking society in the process of foreign-language training of future foreign language teachers.

Keywords: Intercultural approach · Regional competence ·
Teaching foreign languages · Future foreign language teachers training

1 Introduction

The purpose of the article is to analyze modeling as a practice-oriented process for optimizing the development of foreign-language regional competence of students in linguistic areas in the framework of profiles that provide the right to teach foreign languages.

The role of a foreign language teacher as an organizer of students' intercultural communication with the representatives of different cultures is of high importance [4, 5, 10]. When organizing the international activity of an educational institution, a teacher is obliged both personally and via his students to know and be able to present the

© Springer Nature Switzerland AG 2019
Z. Anikina (Ed.): GGSSH 2019, AISC 907, pp. 195–209, 2019.
https://doi.org/10.1007/978-3-030-11473-2_22

region of their residence in various aspects, i.e. to obtain a foreign language regional competence, manifesting itself in the mentality and ethno-sociocultural characteristics of representatives of other cultures, since the viewpoints on the same phenomenon often do not coincide in different socio-cultures [9, 11, 12]. Thus, modeling of the foreign language teaching process at university largely systematizes, organizes, structures the work of a teacher. It contributes to the effectiveness of regional competence development of future foreign language teachers.

2   Modeling as a Basis of Didactic Studies

In recent years, modeling as a method of indirect study has been widely used in various fields of science and practice. There is a great number of works focused on the issues of modeling. The possibilities of applying models in particular sciences are described. Modeling helps to systematize knowledge about a phenomenon or a process under study, determines the ways of their more holistic description, outlines the more complete links between the components, opens up the possibilities for extracting the most significant units of information with a reference to the studied object. Arkhangelsky supposes that **modeling is a scientific method of research of all kinds of objects, processes, etc. on the construction of their models, which keep the basic, distinguished features of the object of study** [1].

A model serves as a generalized reflection of the phenomenon, a result of the generalization of experience and the correlation of theoretical concepts and empirical knowledge about the object.

Two types of models are generally distinguished. The first type treats a model as an example, a standard, a cultural standard that comprises a system of attributes and links proper to the ideal pattern. The second type is a model presented as a structure, an action, a mechanism that ensures functioning and development of the created process.

A model as a replacement of the original object has a practical significance consisting in storing and expanding knowledge about the original, its construction, transformation and management. According to Nikiforov, **a model is an abstract hypothesis, but not the immediate result of an experiment** [8]. Therefore, it should be borne in mind, that by developing a model for studying a particular process, we simplify the actual process, accentuate the most significant and disregard the insignificant features.

In our research, a model is considered as a holistic description of a phenomenon, of a cultural standard with regard to the educational process; a system that transmits and develops knowledge, skills and abilities to realize intercultural communication based on the regional functions of communicators, thus, enabling the development of the theoretical and practical readiness of a future foreign language teacher to act as a mediator of cultures in professional activity.

3 Components of the Foreign Language Regional Competence Model

The performed analysis of the problems is the basis of the author's model - an analog of the process aimed at developing the regional competence of Linguistics students in a foreign language classroom.

The necessity for the model development rises from the following contradictions in the system of secondary and higher education:

- contradiction between the low level of the intercultural competence of school students and university students in *Linguistics* and *Pedagogy* and the requirements of new FSES (Federal State Educational Standard) for foreign language teaching in secondary and higher education;
- contradiction between the particular research of the regional component in foreign language teaching and the objective need for a scientifically grounded approach presented in the form of a methodological system that realizes the principles of an interculturally oriented process in foreign language teaching;
- contradiction between the complex of scientific data on the regional component, understood as local lore and reflecting only the students' native culture, and the broader meaning of the 'region' concept, which can be considered at several levels;
- contradiction between the requirements of the secondary and higher education FSES for the education of a Russian citizen and the lack of materials in foreign language teaching that contribute to the formation of self-identity required in the process of intercultural communication;
- contradiction between the actual reality in which a foreign language functions and fragmentary information about a culture, as a rule, of only one country of a studied language, represented in the educational content.

The experimental base for the research was Nizhny Novgorod State Linguistic University named after Dobrolyubov.

The developed model includes several logically linked components: theoretical and methodological, purpose, contentful, process-operational, evaluative-result.

3.1 The Theoretical and Methodological Component of the Model

Theoretical and methodological bases for the model of regional competence development are modern approaches to a foreign language teaching process: intercultural approach, regional approach, competence approach, professionally oriented approach, and personal-activity approach. Let us briefly consider their essence and content with regard to the subject of our research.

The intercultural approach defines the intercultural competence of a foreign language teacher as a strategic goal of foreign language training. Development of the ability to use a foreign language in various communicative situations based on the awareness of native and foreign cultures presupposes a system of special strategies, methods, forms of training, as well as exercises aimed at comparison and analysis of facts in both native and foreign cultures. **The use of a way of life as a form of**

structuring educational information prepares students for intercultural communication, prompting them to realize their own life experience in their country from the perspective of a foreign way of life and culture of another people and another country [6].

The regional approach concretizes and clarifies the content and technology for the development of students' intercultural communication skills. This approach is based on the necessity to take the account of the cultural characteristics of region representatives and the functioning of a foreign language based on the regional knowledge about the native society.

The competence approach is one of the innovations in methodological science and is defined in FSES as a fundamental approach to students training. Social needs in training foreign language teachers are not limited only to language system knowledge and professional and pedagogical skills. The modern process of training students for intercultural communication requires the use of their personal potential, creativity, mobility and the ability to ensure their economic, cultural, political and self-development. Regional competence, being an integral part of the professional-pedagogical intercultural competence of a foreign language teacher, ensures the educator's ability and readiness to teach students how to realize a full-fledged process of intercultural communication based on the knowledge of the significant regional characteristics of communication partners and the awareness of their own region. When training regional students in a foreign language classroom, it is necessary to develop their ability to compare the studied lingua-cultural regional facts, to correctly interpret regional differences in intercultural communication and adequately act in situations of intercultural misunderstanding.

The focus of the professionally oriented approach in foreign language training is on mastering the students' regional competence as a part of the pedagogical competence of a foreign language teacher.

The personal-activity approach is based on the leading role of various forms of social activity in educational process. It is caused by the necessity to perfect the professionalism of a future foreign language teacher in the process of joint emotional and intellectual activity, which contributes to students' personal development. The essence of this approach is the communicative activity of the learning process, the maximal consideration of the students' speech needs, their individual psychological, age, national and regional characteristics. With the students' interests in focus, a foreign language teacher formulates the objectives of a lesson, constructs the educational process according to the above approaches and the formulated educational principles. All this results in the development of a personal position, self-identification, regional self-consciousness of a future foreign language teacher.

3.2 The Purpose Component of the Model

The purpose component of the model determines the direction of the foreign language educational process at university. The main, strategic goal of the obtained model is to develop the regional foreign language competence of a future teacher as the ability to perform intercultural communication based on the regional characteristics of communication partners and to teach students.

When determining the practical goal of teaching in the context of the dialogue of cultures, we need to consider a system of knowledge and skills that define the ability for intercultural communication at the regional level. This system consists of specific objectives of foreign language teaching in the types of communicative activity.

The educational goal can be as follows: identification of value orientations in native culture; acquisition of regional consciousness including national and civic consciousness; development of a value-based positive attitude to other cultures; cultivation of tolerance and respectful attitude towards representatives of other cultures; development of interest, openness to communication with foreigners.

The general educational goal is achieved through the in-depth study of the native region culture, which presumes the assimilation of knowledge about the history and modern life of the native region (regional information at the microlevel) and allows the development of ethnic and national consciousness; study of the culture of other Russian regions, their interconnection, interdependence (mesolevel of regional information); creation of the modern image of foreign-speaking countries and regions; creation of the modern image of Russia and the native region; study of the regional features of a foreign-speaking society at different levels; comparison and analysis of regional problems on a global scale; familiarization with global problems and conflicts; study of the ways to avoid conflicts.

Achievement of the developmental goal with regard to the dialogue of cultures implies: development of the ability to compare facts of the native region culture and regional culture of target-language countries; awareness of the involvement in the global problems of the mankind, responsibility for personal acts; development of the ability to highlight differences in the studied cultures and societies and the ability to appreciate contribution of different cultures and societies to human creativity progress; development of the ability to interpret regional knowledge and apply it taking into account the range of foreign communication partners [2].

3.3   Principles of Foreign Language Regional Competence Formation

The system of principles underlies the process of regional competence development and determines the direction and content of educational and pedagogical activities.

Dedicated and grounded principles of regional competence development of Linguistics students in a foreign language classroom determine the content component of the model. The content of the regional component includes three aspects: linguistic, psychological and methodological.

The interdisciplinarity principle of the development of regional competence makes it possible to integrate these components into the content of various disciplines where the subject is regional knowledge. This integration should provide students with a holistic view of the regional culture of the target-language countries and the native linguistic environment. A parallel study of certain regional topics will enable the implementation of the principle of proportioning of the regional information [13].

A system of methodological approaches, a system of principles that determines the content of foreign language teaching aimed at developing students' regional competence reveal the patterns for designing techniques for teaching Linguistics students in a foreign language classroom. The technique component of the model is justified by the necessity to consider the process of regional competence development as a techno-logical one yielding the guaranteed result.

The technique component of the model is a synthesis of methods, techniques, strategies, forms and exercises aimed at developing the regional competence of future foreign language teachers.

3.4 The Evaluative-Result Component of the Model

The evaluative-result component is aimed at revealing the level of the regional com-petence of Linguistics students, analyzing the effectiveness of the selected technology and taking the corrective measures that will allow intensive development of the system of regional knowledge, skills and abilities for intercultural communication based on the regional characteristics of communicants, and a complex of professionally significant personal qualities that contribute to the achievement of the results in preparing students for the real dialogue of cultures.

The experimental verification of the proposed thesis was performed in the Nizhny Novgorod State Linguistic University and lasted 10 years, from 2007 to 2017. The first part of the experiment (2007–2010) was of a reconnaissance character. 208 students of the Romance and Germanic Faculty participated at that stage. The results of the first stage of the experiment convinced us of the truth of our propositions and made it possible to select the main content of foreign language teaching with an account of the development of the regional competence components.

3.5 The Psychological Aspect of Foreign Language Regional Competence

Professional knowledge and skills required for intercultural communication between Russian students in a specific region and representatives of the target-language regions, as well as the development of students' ability for effective intercultural communica-tion are based on the information that is borrowed from pedagogy, psychology, didactics, foreign language teaching methodology, ethics, linguistics, sociology, local history, regional studies, and contributes to the recognition of the importance of intercultural contacts in modern society.

In order to successfully facilitate the adaptation of foreign guests in the Russian regions or Russian students in the target-language country, as well as to develop the students' ability for intercultural exchanges, a foreign language teacher must possess a system of specific knowledge, namely:

1. knowledge of the objectives of teaching the regional component within the content of foreign language teaching at each stage at school;
2. knowledge of the basic concepts of teaching the regional component within the content of foreign language teaching;
3. knowledge of the different techniques for implementing the regional component in foreign language teaching;
4. awareness of the interests of students in a class, which can help to predict typical difficulties while translating general objectives into personally significant objectives in the process of studying of the regional component and training students for intercultural communication;

5. knowledge of the history of the native region and its modern political, economic, and social life;
6. knowledge of the particular features of the Russia's region cooperation at the meso- and macrolevels;
7. knowledge of the regional peculiarities of a foreign-speaking society in order to establish personal and regional similarities and differences;
8. knowledge and consideration of a complex of objectives of foreign language teaching with regard to regional component.

Professional pedagogical skills that make up the regional competence of a foreign language teacher comprise the teacher's actions in organizing intercultural communication, performed at a high level and aimed at developing students' skills to carry out this type of activity.

The analysis of various classifications of professional and pedagogical skills made it possible to identify five areas of an activity of a foreign language teacher as an organizer of intercultural communication in the region:

1. Organizational activity includes the ability to introduce and semantize a lexical unit with a regional component of meaning; the ability to distinguish lexical units that reflect the regional culture in the reading or listening texts and to structurize them; the ability to select lexical means with a regional component in order to perform the speech task; the ability to select and prepare exercises for teaching language skills based on the regional material; the ability to select texts on regional themes (from press materials, pragmatic materials, classic novels); to adapt the information about the region for students of different levels; the ability to organize intercultural exchanges: to establish contact with foreign partners, regulate problems connected with accommodation, food, transfer, etc.; the ability to organize communication of Russians and foreigners at the regional level.
2. Design activity is related to the realization of the following skills: the ability to draw up lesson plans taking into account the presentation specifics of the lexis with the regional component of meaning; the ability to determine the possible difficulties of the regional component assimilation by students of the 5th - 10th grades; the ability to plan intercultural exchanges: to come up with and perform interesting projects and reject the inexpedient ones; the ability to determine the objectives of the regional component teaching at each stage of training in modern conditions; the ability to determine the conditions for the reduction of cultural shock; the ability to create conditions for regional acculturation.
3. Research activity implies the ability to explore and analyze the influence of the native language on a foreign language in order to prevent possible mistakes in pupils' speech when rendering regional realities by means of a foreign language; the ability to analyze not only the educational, but also the social value of mastering the regional component; to study the conditions of communication, interaction of cultures; to study the society of a target-language country in order to take into account its mentality in the communication process, to identify different groups of visitors and establish their interests as well as the purpose of traveling to Russia; to

study regional culture and ways of transferring knowledge about the native region with an account of differences in mentality; to transform general pedagogical and methodological objectives of regional component teaching to the objectives personally important for each individual student.

4. Communicative activity is the ability to communicate in and outside of the classroom, to easily get in touch with the representatives of other cultures; to represent native culture through a foreign language; to construct the communication process based on striking similarities and differences in the components of the regional cultures, taking into account the interests of foreign visitors as well as their age, gender, ethnic, professional and sociocultural characteristics; to represent regional information according to the visitors' needs.

5. Supervising activity requires the ability to evaluate actions for organizing intercultural communication in a specific region; the ability to estimate the practical value of your own actions when teaching the regional component and prepare students for intercultural exchanges.

3.6   The Affective Aspect of Foreign Language Regional Competence

Along with cognitive (knowledge) and strategic (skills) components, it seems important to point out the affective component of the regional competence. The German scientist Knapp-Potthoff, highlighting the affective aspect of intercultural communication, believes that this aspect is related to: (a) the cultivation of tolerance towards different phenomena in a studied culture; (b) the ability to perceive, evaluate and accept differences in the phenomena of a studied culture; (c) the development of the empathic abilities of communication partners [7].

The affective component of the regional competence of a foreign language teacher presupposes a high level of motivation for the realization of intercultural communication at the regional level. The basis of the affective component is the ability of partners in intercultural communication to evaluate and accept the differences in the phenomena of regional culture through the prism of their own cultures. The affective component includes:

- high need for communication;
- ability to communicate with people and establish good relationships with them;
- empathy;
- comprehension of self-identity;
- role-playing distance;
- tolerance;
- reflection;
- mobility;
- erudition;
- kindness;
- tact, endurance;
- effort for self-perfection;
- optimism;

- variety of interests;
- creative initiative;
- ability to understand others;
- ability for self- reflection.

Foreign language teachers should develop their psychological readiness for intercultural communication manifested in such personality traits that form the objectives, the condition and the result of intercultural communication. These personal qualities include:

- role distance: a person's ability to abstract from his/her own position, to look at it from the other side, understanding that there are other viewpoints;
- tolerance: a person's ability not to be afraid of meeting "strangers", not to avoid them, but vice versa, to establish a contact, to support the most contradictory requests;
- empathy: a person's ability to analyze a situation of the other sociocultural community and try to understand its commonness.

Empathy development is a one of the main conditions for the formation of the personality of a foreign language teacher. Furmanova defines empathy as a person's integrative quality that possesses certain ethno-sociocultural statuses, linguistic and cultural information in the form of levels of cultural and linguistic personal development, and the ability to interpret linguistic and cultural phenomena [3].

Students have more chance to meet a "stranger" when learning a foreign language. It allows them to discuss their perceptions of another culture, to analyze the foreigners' perception of the learners' reality.

One of the forms of group community perception that arises on the basis of a sense of belonging to a territorial unit is regional consciousness. Regional consciousness is often associated with manifestation of nationalism as a product of feelings, thoughts and actions of the people's territorial community, especially if the sovereign state is divided into regions on a national basis. This indicates a pathology of public consciousness, a philosophical aberration. Such nationalism is associated with aggressiveness towards neighboring ethnic groups, with ignoring, infringing other national feelings or their misinterpretation, which leads to interethnic confrontation. Yet, it is not always that regional consciousness in polyethnic states causes enmity between separate ethnic groups (which is confirmed by the example of the Switzerland cantons).

Regional self-consciousness can be initiated by a variety of reasons: socio-economic inequality of regions, a sense of danger coming from a neighboring state, existence of some social and economic functions in the territory, lifestyle and cultural traditions, consequences of an environmental catastrophe, etc.

The process of the development of regional consciousness is closely related to the factors that provoke outbreaks of ethno-nationalism:

- adherence to the principle of state identity and ethnic boundaries by the population of the region;
- tendency of ethnic groups to self-identity;
- tendency of ethnic groups to the formation of super-ethnoses;

- economic fight for territory, natural resources, etc.;
- uncontrolled demographic development in particular countries and regions;
- assimilation processes and depopulation of ethnic minorities;
- the nations' "senescence" of the countries with developed market economies and a sense of the fear of "ebb";
- environmental problems, etc.

4   Process and Results of the Experiment

The second part of the experiment proceeded from 2010 to 2015. During that period, the model of the regional competence development was created, and the content of the technology component was formed.

In 2010, the task was to form the experimental and control groups in order to verify the hypothesis and evaluate the results of the regional competence development. To evaluate the obtained results, we distinguished the invariant and variable conditions. The main stage of the experiment was performed in 2015–2017 in the Romance and Germanic Faculty of the Nizhny Novgorod Linguistics University. The participants were 60 3rd- and 4th-year students of the German language department. In this article, we present the data of the 4th-year students. Experimental and control groups were selected according to the following conditions:

Invariant conditions:

1. 4th-year students with the same level of language acquisition, both in the experimental and in the control group.
2. Identical duration of the experimental training - 4 months in one academic year.
3. Identical gender content of the groups. Both in the experimental and in the control group, there were three males and seventeen females.
4. Identical level of the professional competence of teachers engaged in the experimental and control groups. Both teachers had the same pedagogical experience (24 years, 22 years). The age and temper of the teachers were also similar.
5. Identical level of the teachers' and students' motivation in the experimental and control groups.
6. Identical linguistic material to be learned in accordance with the requirements of the work program *Workshop on the Culture of Speech Communication*.

Variable conditions:

1. In the control group, teaching was based on traditional didactics. In the experimental group, the author's model was implemented.
2. The experimental group employed the methodology that included an additional aspect of the regional competence development that modified the content of foreign language teaching.

As the criteria for the evaluation of the regional competence development, the most significant positions were identified from the cognitive, strategic and affective blocks:

1. knowledge of the native region history and its current political, economic, and social life;
2. knowledge of the regional characteristics of a foreign society used to establish personal and regional similarities and differences;
3. knowledge of the ways and means of implementing the regional component in foreign language teaching;
4. ability to organize a proper presentation, training, and practical use of the language material that contains the regional component of meaning;
5. skills related to the development of all types of speech activity on the basis of the regional component of the Russian and foreign society;
6. regional self-identity related to lifestyle, presence of specific social, economic and environmental functions, a sense of belonging to a territory unit.

As a result of the experiment, the following indicators were obtained:

After modification of the course content by introducing the materials that reveal historical, political, economic and social characteristics of the native region, the results (see Fig. 1) were higher in the experimental group. The system of tests was used to check the results.

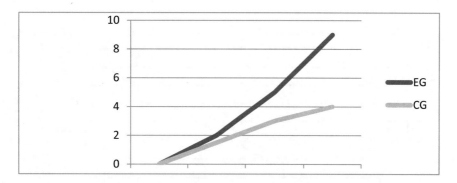

Fig. 1. Knowledge of the native region history and its current political, economic and social life.

In the course of the experiment, attention was paid to the materials that reflect mental characteristics, traditions, customs of the specific regions of the target-language countries, accompanied by a combination of regional characteristics both within the country of the studied language and the native culture region. The obtained learning outcomes in the control and experimental groups are presented in Fig. 2. To check the results, we used a system of tests and created the situations of intercultural communication in a foreign language classroom.

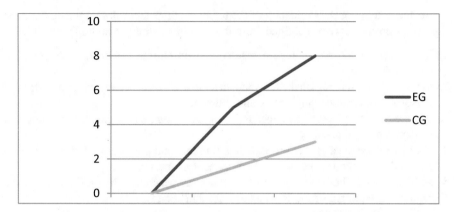

Fig. 2. Knowledge of the regional characteristics of a foreign society used to establish personal and regional similarities and differences.

Knowledge of the ways and means of implementing the regional component in foreign language teaching was presented to the students in the theoretical course on Methodology of foreign language teaching. The continuous assessment was performed by using a system of tests, final control – by means of an exam. The results are given in Fig. 3.

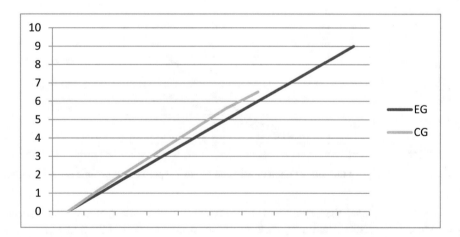

Fig. 3. Knowledge of the ways and means of implementing the regional component in foreign language teaching.

The ability to organize, present, and practise the language material with a regional meaning was developed and evaluated within the pedagogical (teaching) practice, when students worked as foreign language teachers. In their classroom, they employed the materials that reflected the regional component of the meaning (Fig. 4, 5 and 6).

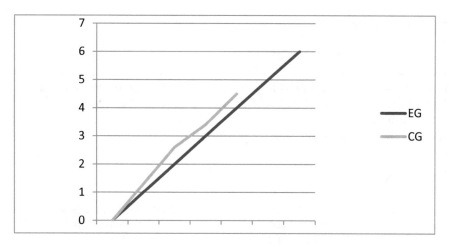

Fig. 4. Ability to organize presentation, training, and practical use of the language material that contain the regional component of the meaning.

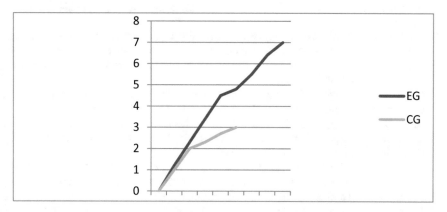

Fig. 5. Skills related to the development of all types of speech activity on the basis of the regional component of the Russian and foreign society.

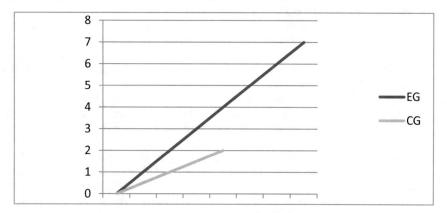

Fig. 6. Regional self-identity related to lifestyle, presence of specific social, economic and environmental functions, a sense of belonging to a territory unit.

5 Conclusion

To sum up, in the post-globalization society, the criteria for self-identification are blurred or seen by each individual as fragmentary and chaotic. The presented research outcomes clearly demonstrate an increase in the level of the development of the foreign language competence of a future foreign language teacher. The obtained results prove the necessity to introduce materials with the regional component of native and foreign-speaking societies in foreign language teaching.

References

1. Arkhangel'skiy, S.I.: Uchebnyy protsess v vysshey shkole, yego zakonomernosti, osnovy i metody [The educational process in higher education, its patterns, basis and methods]. Higher School, Moscow (1980). (in Russian)
2. Dmitrieva, E.N., Oberemko, O.G.: Linguisticheskoye obrazovanie v kontexte issledovaniya fenomena mezhetnicheskoy kommunikatsii [Linguistic education in the context of studying the phenomenon of interethnic communication]. Lang. Cult. **41**, 241–254 (2018). (in Russian)
3. Furmanova, V.P.: Mezhkul'turnaya kommunikatsiya i kul'turno-yazykovaya pragmatika v teorii i praktike prepodavaniya inostrannykh yazykov [Intercultural communication and cultural-linguistic pragmatics in the theory and practice of teaching foreign languages]. Nauka, Moscow (1994). (in Russian)
4. Glumova, Y.P.: Regional'nyi aspect mezhkulturnoy kommunikatsyi [Regional aspect of intercultural communication]. Prob. Modern Educ. **6**, 92–104 (2017). (in Russian)
5. Glumova, E.P.: Regional'nyi podkhod v obuchenii inostrannym yazikam [Regional approach to foreign language teaching]. Vestnik of Nizhny Novgorod State Linguistics University **32**, 141–147 (2015). (in Russian)
6. Glumova, Y.P.: Tseli inoyazychnoy podgotovki budushchikh bakalavrov-lingvistov k mezhkul'turnoy kommunikatsii na regional'nom urovne [The purposes of foreign language training of future bachelor in linguistics for intercultural communication at the regional level] Vestnik of Nizhny Novgorod State Linguistics University, vol. 35, pp. 130–137 (2016). (in Russian)
7. Knapp-Potthoff, A.: Interkulturelle Kommunikationsfahigkeitals Lernziel. Aspekteintercul-turreller Kommunikationsfahigkeit, Munchen (1997)
8. Nikiforov, A.L.: Filosofiya nauki: Istoriya i metodologiya [Philosophy of Science: History and Methodology]. House of Intellectual Books, Moscow (1998). (in Russian)
9. Oberemko, O.G.: Aksiologicheskiye osnovy obucheniya inostrannomu yazyku [Axiological basis of foreign language teaching]. Nizhny Novgorod State Linguistics University, Nizhny Novgorod (2015). (in Russian)
10. Shimichev, A.S. Mezhkul'turnyy aspekt v soderzhanii obucheniya inoyazychnoy gram-matike studentov po napravleniyu podgotovki « Lingvistika » [Intercultural aspect in the content of teaching foreign grammar of students in Linguistics]. Vestnik of Nizhny Novgorod State Linguistics University, vol. 32, pp. 177–186 (2015). (in Russian)
11. Shitikova, I.B., Pleshkova, A.V.: Obrashcheniye k regionalnim etnokulturnym traditsyjam kak vazhnejshemu komponentu naridnogo iskusstva [Appeal to regional ethnocultural traditions as the most important component of folk art]. In: Actual Problems of Development of Science and Modern Education: International Science Conference, pp. 548–550. Belgorod State University, Belgorod (2017). (in Russian)

12. Ulrich, YuA, Sadykova, O.V.: Metodicheskiye podkhody v integratsyi profil'nogo obrazovaniya s regional'nymi soderzhatel'nymi componentami [Methodological approaches in integration of profile education with regional substantial componants]. Int. J. Adv. Stud. **2**, 154–158 (2017). (in Russian)

13. Yermakova, YuI: Sredstva formirovaniya professional'noy kommunikativnoy competentsii s integrirovannym professional'nym komponentom [Tools of professional communication competence formation with included regional component]. Eur. J. Soc. Sci. **8**, 300–306 (2017). (in Russian)

Teaching Writing Skills to Students via Blogs

Pavel V. Sysoyev[1,2](✉) [iD], Maxim N. Evstigneev[1] [iD],
and Ilona A. Evstigneeva[1] [iD]

[1] Derzhavin Tambov State University, Tambov 392000, Russian Federation
psysoyev@yandex.ru, maximevstigneev@yahoo.com,
ilona.frolkina@mail.ru
[2] Moscow Pedagogical State University, Moscow 119991, Russian Federation

Abstract. The global informatization of society is one of the dominant trends
in the development of mankind in the 21st century. Rapid development and
spread of modern information and communication technologies in Russia affects
the state policy in the field of education. Informatization of education has
become one of the priority directions in the Russian education system modernization, aimed at development of methodology, methodological systems,
technologies, methods and organizational forms of education to improve the
mechanisms of education system management in the modern information
society. Integration of Web 2.0 social services into the process of a foreign
language teaching also contributes to the development of foreign language
communicative competence and the achievement of the main learning objectives. In this paper authors consider using blogs in teaching the writing skills to
students: they explore various types of blogs and review their didactic features
and methodological function; suggest a list of writing skills developed via blogs;
and determine the stages of students' writing skills development via blogs.

Keywords: Writing skills · Blog technology ·
Modern information and communication technologies · ICTs

1 Introduction

The information and communication technologies (ICTs) are becoming indispensable
means of communication for more and more people. According to the World Centre for
Internet Statistics, there are about four billion people using ICTs, i.e. every second
inhabitant of the planet takes part in educational, professional and personal communication through ICTs every day.

These figures show that today it is almost impossible to carry out professional and
interpersonal activities without ICTs. As an example, we can consider the system of
general secondary education. All applications for national and international competitions, contests and grants are distributed, filled and submitted exclusively via the
Internet; communication with participants of national and international educational
programs takes place mainly by e-mail; the Internet search engines are the most
effective means for students to choose an educational institution for higher education;
the teachers are offered new distance learning courses which can be completed only
with ICTs, etc. The ICTs create a new global environment which allows students not

© Springer Nature Switzerland AG 2019
Z. Anikina (Ed.): GGSSH 2019, AISC 907, pp. 210–218, 2019.
https://doi.org/10.1007/978-3-030-11473-2_23

only to communicate, but also to build professional and personal relationships, to represent themselves and their interests. That is why nowadays the great importance is placed on the school system informatization, including language education [15].

One of the goals of foreign language teaching in secondary school is development of foreign language communicative competence and all its components (linguistic, discourse, socio-cultural, strategic, educational and cognitive) [13, 14]. However, the level of formation of students' foreign language communicative competence will be determined not only by the ability to use a foreign language for communication in real life, but also the ability to communicate through various social Internet services. Therefore, in this paper we are going to discuss the issue of teaching writing skills to students via blogs.

2 Students' Writing Skills Development

Writing is one of the four types of speech activities along with speaking, listening and reading. Unfortunately, the scholars pay less attention to it than to other speech activities for some reasons. This is primarily due to that exchange of information in communication is carried out mainly through oral speech: speaking (production) and listening (reception).

Considering the writing both as a tool and as a learning goal, it is allowed to use the modest language means for real communication at the initial stage of study (i.e. filling out a questionnaire) and complex language tools to achieve significant results (i.e. writing a review).

Written speech is formalized both in terms of language and in terms of the text [1]. This formalization has been increased many times, according to development of modern communication means. The ability to produce communicative messages in clear conventional forms facilitates the correct perception and understanding of information outside the communication situation. Mastering the formal structure of different types of written texts (language and text) in the target language is an important goal of learning written communication [7].

According to academic foreign language programs for secondary school [12], the following writing skills are developed in the 10th and 11th grades: writing a personal letter; filling out questionnaires; making a plan for abstracts of oral/written communication, including extracts from a text; inquiring about the news in a personal letter and reporting them; telling about facts and events, expressing personal judgment and feelings; describing the future plans. This content of writing skills teaching corresponds to B1level in the Common European Framework of Reference [4].

Taking into account the profile language education (level B2 or higher), the content of teaching writing can vary, depending on the stage of training. At the initial stage of training students should be able to create written statements on the proposed topic and based on the text read or listened. At the main stage of training students should be able to create written texts: essay, application, explanatory note, etc. At an advanced stage of training the content of student written language training should include: writing research papers, a thesis, a review of a scientific article, writing and filling out business documents, design of scientific texts in accordance with the necessary requirements (correct citation of sources, references and footnotes, bibliographic list, etc.) [23].

More successful content of teaching writing skills to students is given in the research paper by Markova who singled out the following skills, which students should master by the time of bachelor program completion: writing business letters, reports and messages, reviews, different types of essays, short articles on professional and social topics [11].

In our opinion, Evstigneeva suggests the most complete content of writing skills teaching to students by noting the difference between discourse and such concepts as "speech act" and "communication" [7]. The main difference between discursive skills and the speech skills is that the latter focuses on management of language skills to solve specific communicative tasks: to be able to speak, read and write. Opposite to it, discursive skills are considered slightly wider than speech skills due to extralinguistic factors: stylistic, genre, social, cultural, psychological and emotional [17, 18].

Analysis of modern requirements for the training of secondary school graduates that is defined in the Federal State Educational Standard [8] shows that the following writing skills can be developed: using a foreign language lexical resources to create a written text; using the stylistic and genre means to create a written text in a foreign language; organizing a sequence of statements in a foreign language so that they represent a coherent written text; building logical statements in a foreign language; formulating and expressing the point of view by means of a foreign language; describing and explaining the facts in a foreign language; panning a text in a foreign language; providing motives and goals of communication in a foreign language; forming a communicative behavior according to the main theme of communication; building the verbal behavior adequately to the socio-cultural specifics of the target language country [7, 17, 18].

3   Blogs in Teaching a Foreign Language

Blog is a user's personal site in the form of a diary or a journal. A blog is usually created and moderated by one person who can place text materials and photos, audio and video recordings, and links to other Internet resources on the page. Any visitor of the blog, after reading the content of the site, can react to the published text or viewed photos by posting comments. Blogs have a linear structure. This means that all messages are arranged chronologically one after another on the same page. Blog is considered one of the Internet social services (Web 2.0) as they create conditions for communication among people, united by common interests, but separated by distance [16]. The main didactic features and methodological functions of blogs in a foreign language teaching are given in Table 1.

These didactic features of blogs allow developing such types of speech activity as writing and reading. Also, it should be noted that the blog is one of the widely used Web 2.0 technologies in foreign language teaching, and nowadays it is the most described and discussed in the scientific methodological literature [2, 5, 6, 10, 19–22].

Table 1. Didactic features and methodological functions of blogs in foreign language teaching.

Didactic features of blogs	Methodological functions of blogs in a foreign language teaching
Publicity	The content of a particular blog may be available for any participant of the Internet project. This technology can be used for the networking organization among foreign language learners and organization of extracurricular classes to develop the language aspects (vocabulary and grammar), types of speech activity (reading and writing), as well as socio-cultural and intercultural competences
Linearity	The information is posted by the blog author (moderator) or blog visitors chronologically. The blog technology does not allow making additions and corrections to the previously published information in the blog. Only moderator of the blog can delete it. This didactic feature of the blog technology makes it possible to develop methods of students' speech skills through an individual work of each student (within the framework of the general group project) and organization of network communication between project participants
Authorship and moderation	The author is the moderator of the blog. The moderator defines the blog's purpose and thematic focus, coordinates posting of materials (text, graphical, audio and video) by other members of the network. In case the material does not meet some criteria, the moderator can delete it. Unlike the wiki technology, which is aimed at real life-organization of group projects, the blog technology allows hearing the "voice" of each participant of educational process. In a foreign language teaching the blog technology can be used to develop types of speech activities in individual and group forms of training
Multimedia	The blog technology allows using a wide range of materials: text, graphic, photo, video, audio. It provides the opportunity to enrich the language and socio-cultural material while writing essays and reviews

Many scholars consider blogs as an effective means for writing skills developing - as a type of speech activity, and also use blogs as a platform for intercultural telecommunication projects [3, 9, 16].

Description of blogs methodological potential shows that they allow creating unique conditions for foreign language writing and reading skills development. In Table 2 the range of skills developed via blogs is presented.

All these writing skills fully meet the requirements of modern educational standard for foreign language teaching [8] and are reflected in the state exam materials for the English language (writing section). As for a sample task, each student can post their letter in a personal blog. The other students can be given the assignment to read and comment on personal letters of their classmates, or to write a reply, etc.

Despite the special emphasis on development of writing and reading skills, blogs can also be used in the grammatical and lexical skills development. For example, each author of the blog can be given a task: using the number of lexical units studied to

Table 2. The range of skills developed via blogs.

Type of blog	Information for publishing in the blogs	Skills developed
Teacher's blog	course program; a home task; information about the material studied in a particular lesson; recommended sources for further study; links to information and reference to Internet resources; links to educational Internet resources and network tests on the topics studied	*Reading skills:* highlighting the necessary facts/information; extracting the necessary/interesting information; evaluating the importance of information
Personal student's blog	date and place of birth; family, relatives; hobbies and interests; friends; achievements in studies and/or sports; links to favorite Internet resources; photos and videos; comments of students and teachers; written personal letters; various types of essays; reviews of films and books	*Writing skills:* presenting personal information about yourself in writing (the author of the blog); representing the native country (town, school) and culture in a foreign language environment (the author of the blog); expressing opinions, agreement/disagreement in a non-categorical and non-aggressive form (commenting on the blog of a classmate or the opinions of other students); arguing the point of view using language means (commenting on a classmate's blog) *Reading skills:* highlighting the necessary facts/information; extracting the necessary information of interest; evaluating the importance of information
Students' group blog	topics for discussion; discussion (comments of students)	*Writing skills:* presenting the content of the read/listened text in abstracts or short messages; expressing an opinion, agreement/disagreement in a non-categorical and non-aggressive form; drawing an analogy, comparison, matching the available linguistic means; arguing the point of view using language means *Reading skills*: highlighting the necessary facts/information; separating the primary information from the secondary; determining the temporal and cause-effect relations between events and phenomena; summarizing the described facts/phenomena; evaluating the importance/novelty/reliability of information

make a story about the holidays spent or to discuss the movie or book. Each student can be given a task to write to the blogger (via the comments function) two special questions in the past tense (grammar), etc. In this sense, the methodological potential of blogs has no limits.

To sum it up, application of blogs in a foreign language teaching can contribute to the following aspects: motivating the use of a foreign language for communication in extracurricular time; the development of reading and writing skills (including those indicated in the modern educational standard); the development of skills of using a foreign language and the Internet to meet the cognitive interests of students; the development of skills of using a foreign language as a means of education and self-education; expression of personal opinion and its argumentation in the judgment of social issues which is not always feasible in the classroom; more effective discussion of the studied problem for further essay writing or a dialogue or monologue statement.

4   Stages of Teaching Writing Skills to Students via Blogs

Scholars describe several methods for students' writing skills development via blog technology. These methods vary on the learning objectives and the students' peculiarities [11, 16, 23]. In this paper we propose one of possible algorithms for organization of students project activities based on the blog technology and aimed at writing skills development. The algorithm of student writing skills development via blogs is presented in nine stages.

At the first stage there is an acquaintance of students with the project objectives in the classroom. The teacher explains the essence and the main stages of educational blog-based project activities to students, determines the theme of the upcoming project. After that teacher divides students into groups of 3–4 people, and explains to each group what they will do; explains what final result is expected; introduces students to the evaluation criteria; instructs students what algorithm of actions they should follow; introduces students to the list of topics to be developed. Students ask organizational questions; make suggestions about additional topics in the curriculum they would like to discuss.

The second stage is devoted to registration on the blog service and familiarizing with the rules of posting materials in the classroom. The teacher gives students the address of the blog server which will host the writing work; shows learners how to register on the selected blog server; explains students how to post materials in the blog. The students register on the blog server and try to post the text in the blog.

The third stage is the most important cause it deals with students' personal security. The teacher explains the rules of information security on the Internet to students.

At the fourth stage there is a selection of topics and material for publishing in personal blogs in class or distantly. Divided into small groups, students together with the teacher choose the genre and theme for a future project, participate in brainstorming, they also search and select the material. The teacher monitors the work of students, helps them in case of difficulties with choosing a topic for writing assignment. If necessary, students are looking for suitable material in various search engines, such as Rambler, Google, and Yahoo. Students are engaged in processing, systematization, analysis, synthesis of the material.

The fifth stage is writing and publishing a writing work in a personal blog distantly. Each of the project participants prepares their writing assignment in the Microsoft

Word editor and posts it in the personal blog. The teacher monitors the independent work of students, if necessary, provides assistance and advice online and monitors the publishing of student materials in personal blogs.

The sixth stage is discussion of writing assignments with classmates in personal blogs (remotely). Students alternately go to the personal blogs of classmates, study the published material and comment on it (content, structure, language correctness, etc.). The teacher monitors the independent work of students, monitors their participation in discussion of works in personal blogs.

The seventh stage is reaction of students to comments of classmates in personal blogs (remotely). The students respond to comments of classmates or make changes to their writing work, repost its modified version in a personal blog. The teacher monitors the independent work of students, monitors their participation in the discussion of written works in personal blogs.

The eighth stage is self-assessment. The students assess how they managed to reveal the essence of the discussed problem; try to understand what difficulties they faced and what they experienced during the project implementation, and tell what they will need to do to improve their work next time.

The ninth stage is evaluation by the teacher. The teacher evaluates the student work, according to the pre-defined criteria.

5 Conclusion

The methodological potential of modern ICTs in foreign language teaching is unlimited. Development of entirely new methods for teaching a foreign language with the modern ICTs allows us to review such processes as getting information, its processing and further use, taking into account the individual characteristics of each student, knowledge and student's skills. Application of ICTs as a pedagogical tool allows getting a qualitative education at a lower cost of time and teacher and student efforts. In this paper we considered the methodology for organization of students' writing skills development via blogs.

References

1. Arutyunova, N.D.: Diskurs. Lingvisticheskiy entsiklopedicheskiy slovar' [Discourse. Linguistic encyclopaedic dictionary]. Sov. Entsiklopediya, Moscow (1990). (in Russian)
2. Bloch, J.: Abdullah's Blogging: A Generation 1.5 student enters the blogosphere. Lang. Learn. Technol. **2**, 25–37 (2007)
3. Bloch, J., Crosby, C.: Blogging in Academic Writing Development. Handbook of Research on Computer-Enhanced Language Acquisition and Learning. Information Science Reference, New York (2007)
4. Common European Framework of Reference for languages: Learning, Teaching, Assessment. Cambridge University Press, Cambridge (2001)
5. Evstigneev, M.N.: Struktura IKT kompetentnosti uchitelya inostrannogo yazyka [The foreign language teacher ICT competence structure]. Yazyk i kul'tura – Language and culture **1**(13), 119–125 (2011). (in Russian)

6. Evstigneev, M.N.: Genezis i variativnost' ponyatijnogo soderzhaniya terminov v oblasti informatizatsii obrazovaniya [Genesis and variability of conceptual content of education informatization terms]. Yazyk i kul'tura – Language and culture **1**(21), 63–73 (2013). (in Russian)
7. Evstigneeva, I.A.: Razvitie diskursivnyh umenij obuchajushhihsja sredstvami sovremennyh informatsionnyh I kommunikatsionnyh tekhnologij [The development of students' discursive skills with the aid of Modern Information and Communication Technologies]. Inostrannye Yazyki v Shkole **2**, 17–21 (2014). (in Russian)
8. Federal'nyj gosudarstvennyj obrazovatel'nyj standart osnovnogo obshchego obrazovaniya [Federal State Educational Standard of Basic General Education]. Ministry of Education and Science, Moscow (2010). (in Russian)
9. Kennedy, K.: Writing with web logs. Technol. Learn. **2**, 11–14 (2003)
10. Lowe, C., Williams, T.: Into the Blogosphere: moving to the public: weblogs in the writing classroom. In: Gurak, L. et al. (eds.) Into the Blogosphere (2006). http://blog.lib.umn.edu/blogosphere. Accessed 20 Aug 2018
11. Markova, Yu.Yu.: Metodika razvitiya umenij pis'mennoj rechi studentov na osnove viki-tekhnologii (anglijskij yazyk, yazykovoj vuz) [Methodology of students' writing skills development on the basis of wiki technology]. Extended abstract of a Ph.D. theses. MGGU imeni M.A. Sholokhova (2011). (in Russian)
12. Primernye programmy po inostrannym yazykam: Anglijskij yazyk [Foreign language secondary school curriculum]. Ministry of Education and Science, Moscow (2010). (in Russian)
13. Safonova, V.V.: Kommunikativnaya kompetenciya: sovremennye podhody k mno-gourovnevomu opisaniyu v metodicheskih celyah [Communicative competence: modern approaches to multilevel description for methodical purposes]. Euroschool Press, Moscow (2004). (in Russian)
14. Savignon, S.J.: Communicative competence: theory and classroom practice. McGraw-Hill, New York (1997)
15. Sysoyev, P.V.: Informatizatsiya yazykovogo obrazovaniya: osnovnye napravleniya i perspektivy [Main directions and prospects of informatization of a foreign language education]. Inostrannye Yazyki v Shkole **2**, 2–19 (2012). (in Russian)
16. Sysoyev, P.V.: Blog-tekhnologiya v obuchenii inostrannomu yazyku [Blogs in teaching a foreign language]. Yazyk i kul'tura – Language and culture **4**(20), 115–127 (2012). (in Russian)
17. Sysoyev, P.V., Evstigneeva, I.A., Evstigneev, M.N.: Modern information and communication technologies in the development of learners' discourse skills. Proced. Soc. Behav. Sci. **154**, 214–219 (2014)
18. Sysoyev, P.V., Evstigneeva, I.A., Evstigneev, M.N.: The development of students' discourse skills via modern information and communication technologies. Proced. Soc. Behav. Sci. **200**, 114–121 (2015)
19. Sysoev, P.V., Evstigneev, M.N.: Tehnologii Veb 2.0: social'nyj servis blogov v obuchenii inostrannomu yazyku [Web 2.0 technologies in creating a virtual educational environment in teaching foreign languages]. Inostrannye Yazyki v Shkole **4**, 12–18 (2009). (in Russian)
20. Sysoyev, P.V., Evstigneev, M.N.: Foreign language teacher's competence in using information and communication technologies. Lang. Cult. **1**, 142–147 (2014)
21. Sysoyev, P.V., Evstigneev, M.N.: Foreign language teachers' competency and competence in using information and communication technologies. Proced. Soc. Behav. Sci. **154**, 82–86 (2014)

22.
    Sysoyev, P.V., Evstigneev, M.N.: Foreign language teachers' competency in using information and communication technologies. Proced. Soc. Behav. Sci. **200**, 157–161 (2015)
23. Sysoyev, P.V., Merzlyakov, K.A.: Ispol'zovanie metoda recenzirovaniya v obuchenii pis'mennoj rechi obuchayushchihsya na osnove blog-tekhnologii [Methods of Teaching Writing Skills to Students of International Relations Using Peer Review]. Moscow State University Bulletin. Series 19: Linguistics and intercultural communication **1**, 36–47 (2017). (in Russian)

E-Learning Course in a Foreign Language as a Means of Improving Well-Being Environment for Active Agers

Elena M. Pokrovskaya[1] , Lyudmila E. Lychkovskaya[1] ,
and Varvara A. Molodtsova[2(✉)]

[1] Tomsk State University of Control Systems and Radioelectronics,
Tomsk 634050, Russian Federation
{pemod,lef2001}@yandex.ru
[2] Moscow Polytechnic University, Moscow 107023, Russian Federation
v.a.molodtsova@gmail.com

Abstract. The article deals with an issue of ageing population. The authors define the notion of active third agers. The article describes the ageing policy in Russia. The problem is a contradiction between a significant increase in life expectancy, in other words, civilization entered the third age, and social prejudices that prevent older people from achieving a higher quality of life. An e-learning course in a foreign language is viewed as a means of improving well-being environment for active agers. The authors specify the role of a university in the ageing policy and comfortable barrier-free environment development taking the example of Tomsk State University of Control Systems and Radio-electronics that provides a high-quality platform for sociocultural changes. The project "English for Active Third Agers" is described in terms of its aim, objectives, and methodology. It facilitates the promotion of education importance throughout entire life (life-long learning), popularizes foreign language learning, and improves the quality of life for third agers.

Keywords: Active ager · Third age · Foreign language learning ·
E-learning course · Well-being

1 Introduction

Deep understanding of various socio - cultural trends and processes is one of the urgent tasks of modern science. Today, the demographic situation clearly illustrates that aging citizens are a dynamically growing part of the population around the world [2] and, at the same time, the ageing of society is a serious economic problem. According to the UN forecasts, by 2050, 22% of the land will have been retired [3], which requires an integrated approach - social, economic, and technological. The development of medicine allows us to hope that the age of "active old age", that is, a state when an elderly person can lead a more or less full life, will steadily increase. Automation of production enables aging people with a deteriorating physical condition to work. Researchers recognize in the current conditions several areas of work, namely, the development of issues related to biological age and health [6]; consideration of the specific features of

© Springer Nature Switzerland AG 2019
Z. Anikina (Ed.): GGSSH 2019, AISC 907, pp. 219–223, 2019.
https://doi.org/10.1007/978-3-030-11473-2_24

"successful" ageing [7], [11]; study of the emotional background, history of interactions, barriers and obstacles to the communication of the "third age" citizens [10]; development of methods and forms of work within the concept of lifelong learning [8].

It should be noted that the concept of "average age (50–65)" [1] and "third age (65 +)" is the conventional term for the state after retirement. With regard to Russian realities, the understanding of the third age is now being transformed and is more in line with the world practice. The activity of third-age citizens is the basis of the policy in the field of ageing (ageing policy) at international level [4]; it is implemented in diverse forms of manifestations characterized by a wide range and variability (leisure, travel, training, etc.).

2   Problem and Methods

Meanwhile, there is a contradiction between a significant increase in life expectancy (civilization entered the third age) and social prejudices that prevent older people from achieving a higher quality of life. Today, in the era of globalization and cosmopolitanism, the motivational dominant and self-presentation play a greater role in the ageing process, aimed at maintaining success and social recognition. In Russia, people of the third age have virtually no skills in organizing their leisure, or they have no other socially significant identification resources, except for garden plots. Therefore, it is necessary to maintain the real motives of well-being. One of the tools can be foreign language training with the use of an e-learning course. This kind of social innovation helps to overcome gerontophobia and ageism on the part of young and middle-aged people and build the environment of well-being.

In the context of the stated problem, the most relevant are the method of system analysis and project approach. The system analysis method allows solving the problem with a full consideration of the factors that influence it. The project approach, in turn, is effective due to maximum implementation of the third age people features, namely, the habit of the in-depth analysis of issues that are universal and go beyond their own lives; the realization that the real stock of knowledge is not sufficient for effective interaction with current society; the desire to satisfy cultural, creative, or intellectual thirst, etc.

Since comfortable barrier-free environment does not exist in itself, but is the result of systematic work on its development, we emphasize the leading role of educational discourse as a mediator of social and cultural transformations. Cultural and educational barrier-free environment, incorporated into the infrastructure cluster of the University, is a socio-economic, educational phenomenon: on its sites, both internal and remote, resources are concentrated in a number of key areas, related primarily to the problems of living of an individual in discrete world, humanization of society, humanitarization of cultural-value, socio-psychological environment of the University [5].

This practice is successfully implemented at Tomsk State University of Control Systems and Radioelectronics (TUSUR), one of the leading technical universities in Russia, where students are trained in the most relevant areas in the field of radio engineering, electronic and computer technology, programming, automation and control systems, information technology, information security, innovation, economics, and social work.

At present, TUSUR is:

- the first students' business incubator in Russia,
- the largest center of distant education beyond the Urals,
- a leader in the implementation of innovative development programs aimed at creating a continuous system of generation of new ideas, technologies, and business projects [12].

3 Project Description

We present the *English for Active Third Agers* project designed for active ageing. It promotes the importance of life-long learning, popularizes foreign language learning, and improves the quality of life for third agers. It is obvious that this project can be 'replicated' to the cities where the project participants work (Moscow, Moscow Polytechnic University; Kaliningrad, Baltic Federal University named after Immanuel Kant).

The purpose of the project is to develop a system of foreign language popularization as a means of communication for third agers using an e-learning course in the Foreign Language Department of the Tomsk State University of Control Systems and Radioelectronics (TUSUR) - http://lang.rk.tusur.ru/moodle) [9].

The tasks are as follows:

- to identify the target audience, to specify groups and levels of language competency;
- to develop the content of educational programmes;
- to define the main points of growth in individual trajectories of students;
- to review the approbation results;
- to come up with recommendations on the application of the e-learning course.

Based on the above-mentioned tasks, we note that the project is designed for people with different levels of language competency. An e-learning course allows relatively easy individualization of training trajectories since it incorporates various grammar, listening, reading, and writing tests, as well as different interactive tasks. When developing our course, we employed the most important principles of using emerging technologies in education, in particular, for the development of online courses, including: the principle of scientific character (optimization of the process of teaching materials selection), the principle of activity (opportunity to demonstrate intellectual activity), the principle of consciousness (selection of teaching materials), the principle of visualization (use of auditive and audiovisual components to develop skills of foreign-language speech comprehension); the principle of availability (organization of the educational process depending on the level of communicative competence development), the principle of systematicity and sequencing (structurally functional coherence of presentation of content).

The long-term effects of using the e-learning course are:

- medical and improving,
- cultural and educational,
- social and economic.

4 Conclusion

In conclusion, we note that the problem of the quality of life of the "third age" citizens is undoubtedly relevant for Russia today. Therefore, the issue should not remain unaddressed by domestic scholars. The project is considered, firstly, as a supporting tool aimed at increasing motivation and improving the life quality of the "third age" people, at developing the environment of well-being in society. Secondly, it is effective and methodically reasonable. Thirdly, the project has a pronounced applied character where the e-learning course *English for Active Third Agers* acts as its product.

Acknowledgements. *The English for Active Third Agers* project (application №СЗИ180000285) is supported by Vladimir Potanin Foundation and is the winner of the competition of socially important initiatives in 2018 in the *Professional Realization* nomination [13].

References

1. Koskinen, V., Ylilahti, M., Wilska, T.A.: "Healthy to heaven" - Middle-agers looking ahead in the context of wellness consumption. J. Aging Stud. **40**, 36–43 (2017)
2. Sabo, R.M.: Lifelong learning and library programming for third agers. Librar. Rev. **66**, 39–48 (2017)
3. Birger, P.: Tekhnologii pozhilogo obschestva [Technologies of ageing society]. http://polit.ru/article/2012/07/11/oldest/. Accessed 23 Aug 2018. (in Russian)
4. Power, E.R.: Housing, home ownership and the governance of ageing. Geogr. J. **183**, 233–246 (2017)
5. Pokrovskaya, E., Raitina, M.: University infrastructure as vector of region sustainable development. In: Proceedings on International Conference on Trends of Technologies and Innovations in Economic and Social Studies (TTIESS), pp. 546–550. Atlantis Press, Tomsk (2017)
6. Kafkova, M.P.: The "Real" old age and the transition between the third and fourth age. Sociologia **48**, 622–640 (2016)
7. Marshall, B.L., Rahman, M.: Celebrity, ageing and the construction of "third age' identities. Int. J. Cult. Stud. **18**, 577–593 (2015)
8. Talmage, C.A., Lacher, R.G., Pstross, M.: Captivating lifelong learners in the third age: lessons learned from a University-based Institute. Educ. Educ. Res. **65**, 232–249 (2015)
9. Pokrovskaya, E.M., Lychkovskaya, L.Y., Smirnova, O.A.: The activation model of university - employer interaction in the field of master's students' foreign language proficiency. In: Proceedings of the XVIIth International Conference on Linguistic and Cultural Studies (LKTI 2017), pp. 187–194. Springer International Publishing, Cham (2018)

10. Potekhina, I.P., Chizhov, D.V.: Potential of senior citizens as a component of national social capital (Russian central federal district case study). Monitoring Obshchestvennogo Mneniya: Ekonomichekie i Sotsial'Nye Peremeny **2**(132), 3–23 (2016)
11. Ricci Bitti, P.E., Zambianchi, M., Bitner, J.: Time perspective and positive aging. Time perspective theory; review, research and application: Essays in honor of Philip G. Zimbardo, 437–450 (2015)
12. Shelupanov, A.: TUSUR – centr innovatsii, gde sozdautsya tekhnologii, menyayuschie zhizn k luchshemu [Alexandr Shelupanov: TUSUR is a centre of innovations where technologies that change the life for the better are created]. http://www.sib-science.info/ru/heis/shelupanov-tusur-06022017. Accessed 23 June 2018. (in Russian)
13. Vladimir Potanin Foundation. http://www.fondpotanin.ru/novosti/2018-07-07/41915816. Accessed 2 Aug 2018. (in Russian)

Academic Conferences as a Tool for Language Skills Development in Non-language Majoring University Students

Elena M. Shulgina[✉] [ID], Irina S. Savitskaya [ID],
and Petr J. Mitchell [ID]

Tomsk State University, Lenin Avenue 36, 634050 Tomsk, Russian Federation
modestovna2@gmail.com,
{sais.08,peter_mitchell}@mail.ru

Abstract. In our globalized society, one of the key requirements in professional education is development of the communicative competence, including multicultural communication skills, the ability to advance products of intellectual property, and the results of scientific research and innovative developments in the international market. The authors work on a system of continuous education in foreign languages for various categories of students in higher education institutions. One of their research areas lies in finding out how quality may be ensured in educational services for students of a sufficient level of proficiency in foreign-language competence in order to communicate with representatives of other cultures, which is a priority task for any competitive higher education institution. The paper examines organization of students' conferences, with English as the working language, as an effective means for successful formation of foreign-language communicative competence in natural science students. The authors consider the sequence of forms of student work (group and individual) as the most effective strategy for forming their foreign-language competence. The paper provides the description of different activities involving preparation of reports in English, the results obtained in the process of preparing bachelor's degree students for participation in conferences, and analysis of their practical experience. The mentioned approach can be applied not only when preparing students for participation in students' academic conferences, but also in other kinds of activity connected with professional foreign language education (seminars, round tables, webinars, and others) both in higher educational institutions and specialized schools.

Keywords: Academic conference · Language skills · University students · Non-language major · Foreign language competence

1 Introduction

According to a new generation of the Federal State Educational Standards for higher education in the Russian Federation, the main objective of foreign language teaching at the tertiary level is to educate bachelor and master degree students in order to develop such level of foreign-language communicative competence which allows them to

© Springer Nature Switzerland AG 2019
Z. Anikina (Ed.): GGSSH 2019, AISC 907, pp. 224–236, 2019.
https://doi.org/10.1007/978-3-030-11473-2_25

effectively solve professional problems in the respective sphere and use language as communication medium with foreign experts. Under these conditions, the professional community of foreign language teachers carries out an ongoing search for innovative teaching methods which can create conditions for modernizing content and organizing the educational process by means of active involvement of students. Many researchers see a solution in the use of new effective educational technologies, including both computer tutorials [1–4] and pedagogical technologies [5–17].

All the authors note that modern educational environment is undergoing basic changes which affect, first of all, foreign language teaching methods. However, the use of academic conferences on English language teaching to non-language majoring bachelor and master degree students remains little studied.

Our hypothesis is that academic conferences can serve as a way of language teaching which allows us to create favorable conditions for students in order to realize the skills and abilities of practical language proficiency in situations of foreign-language professional and academic communication; namely, to make presentations in a foreign language and discuss topics connected with the student academic research.

This paper aims to identify the teaching potential of academic conferences held in English for the English language teaching to non-language majoring students (ESP). Emphasizing an important condition of a successful foreign-language educational activity for development of a foreign-language communicative competence (in our identified format) with the following sequence of student's work: (1) group work, (2) individual work, and (3) group work, it is necessary to highlight individual work as a link which needs special methodological development because special preparation both on the part of students and teacher is required at this stage of student activity.

1.1 Individual Work of Students as an Important Condition of the Effectiveness of a Foreign-Language Educational Activity

Let us consider the essence of individual work under new educational conditions. Considerable experience of researching student's individual work has been accumulated in the domestic literature. Bulanova-Toporkova [18], Zimnyaya [19], Kazakova [20] and Pidkasisty [21] determine the role and place of independent work in integrated educational process, approaches to content, and classification of types. Evdokimova [22], Zmievsky [23], Sysoyev [24] and others analyze the organizational and methodological aspects of student's individual work as a reserve for improving the educational process effectiveness.

Evdokimova highlights a new important condition for a successful independent educational activity – student autonomy. She remarks that the concept of student autonomy has something in common with the concept of students' individual work. It should be noted that the main difference is that a greater significance in development of student autonomy is placed on the student responsibility for the result of their educational activity: "If work supervision and regulation is gradually and consistently transferred to the student, this creates prerequisites for transition from educational activity supervised by the teacher to that of study, which is only carried out independently, without a teacher's supervision" [22, p. 29].

Individual work is defined by the Federal State Educational Standards for Higher Education as educational, research and socially significant activity of students, which is carried out without a direct participation of teachers, although managed by them, aimed to develop general and professional competences. Most researchers agree that the main characteristic of student's individual work is lack of the teacher's direct participation [21].

However, in our case, when preparing students for participation in an academic conference, lack of the teacher's control is impossible, since the teacher has to check the literacy and adequacy of the material in reports made independently by students. At the same time, the teacher's role comes down not to "supervisory authority", but preparation – they show students the path, whereas the students carry out searching and creative tasks. The students should choose a topic for presentation independently, evaluate information, and transform it for writing the report. The teacher's task is to prepare their students in advance for all kinds of activity in the situation when they perform their task independently: how to read (searching, viewing, fact-finding, studying and others) different types of specialized literature and scientific and technical information, summarize and structure the text, be able to work with special search engines, and others. In this case, the main marker of individual work is that the student's purpose consists in their awareness of the need for activity, and their ability:

- To understand a foreign-language literature for respective specialty;
- To separate the major from the minor;
- To see difficult places and clear them up;
- To summarize what has been read/seen;
- To predict the final result;
- To reveal professionally focused lexical units and grammatical phenomena characteristic for a certain type of discourse.

1.2   Foreign-Language Communicative Competence as a Compound Component of Professional Competence of Non-language Majoring Bachelor and Master Degree Students

Development of the foreign-language communicative competence presupposes versatile independent activity of students, including their work on themselves as a communicative and cognitive personality, because any process of communication implies an individual approach, the subject of discussion, and others [25].

According to the modern educational standards, a foreign language is included into the basic unit of the main educational program for Higher Education for Bachelor's degree. According to the requirements for teaching professionals of a foreign language in the Russian Federation, the higher education institutions face a task of creating the following common cultural competences (CC):

1) Ability to communicate (in speaking and writing form) in the Russian and foreign languages for the purpose of solving the tasks in standard situations of interpersonal and cross-cultural interaction (CC-5);
2) Ability to work in a team, tolerance towards social, ethnic, confessional, and cultural distinctions (CC-6). We cannot but underline that modern educational

programs tend to reduce classroom hours for foreign language teaching and increase the time for student's individual work. Therefore, the main educational program for non-language departments defines the student's individual work as a key component of educational process, which defines the formation of skills, abilities and knowledge, methods for cognitive activity, and guarantees interest in a creative work. According to the main educational program, the maximum academic load for a full-time student, including all types of classroom and out-of-class study, should not exceed 54 h a week; whereas the hours for student's individual work make up half of all classroom hours, but it can be increased at the teacher's behest.

Therefore, when teaching a foreign language to non-language majoring students, it is necessary to use such approaches and methods of teaching which would be aimed at mastering the professionally focused vocabulary, development of IT competence, formation of abilities and skills for professional texts translation [26]. They enhance creative abilities and stimulate motivation for learning [21, 25]. During a conference, the students form skills of purposeful observation and experimental set up; they go through the whole cycle of research activity – from definition of a problem to defense of the obtained results.

2 Research Methodology

2.1 Method

In this research, we rely on the experience of holding academic and practical conferences with natural science students as part of the project "Development of Foreign-Language Competence in Tomsk State University Students (TSU)". The experiment, which started in 2015, comprises 24 months and 6 semesters. Implementation of this project involves an intensive study of English, including not only an intensive course of practical learning, but also various out-of-class activities. It began in September 2015 and involved two non-language faculties – the Faculty of Physics and the Faculty of Chemistry. One hundred and sixty of the first-year bachelor students and twenty first-year master students participated in the research which lasted 2 semesters (8 months). The authors have carried out a qualitative analysis of the educational activity, methodological conditions and educational results of teaching using English-language academic and practical conferences as a means for activating individual work of non-language major students at TSU. As a part of this research, we attempted to answer the following research questions:

1. Which skills, abilities, and competences are developed in students when they participate in academic and research conferences?
2. What are the recommendations for organizing individual work when using the conference method?

2.2   Participants

The research was carried out as a part of the university project "Development of Foreign-language Competence in TSU Students". Eighty-five of the first-year bachelor students at the Faculty of Chemistry majoring at Chemistry and Fundamental and Applied Chemistry, eight first-year master program students and fourth-year bachelor students majoring at Chemistry, and also seventy-five bachelor students at the Faculty of Physics majoring at Physics and Information Systems and Technologies, twelve first-year master students and fourth-year bachelor students majoring at Physics participated in this research.

2.3   Academic and Practical Foreign-Language Conference as a Form of Improving the Quality of a Foreign Language Teaching

Academic and practical foreign-language conferences at higher educational institutions are focused on development of student motivation for independent cognitive activity and formation of universal research activities. Organization and holding of the conference is supervised by a specially created organizing committee which includes students from the experiment participants. The work is carried out in 3 steps: collection of applications for participation, work on the research projects, and the conference holding.

The conference was held on a special assigned study day and included reports and poster presentations. Students had been informed in advance on the participation requirements which consisted in the following: 5–7 min long presentations; the speaker has to designate the problem relevance, formulate the main idea, express their own opinion on the matter and be ready to provide reasoned answers to the questions from audience. The process of student preparation for the conference consisted of three main stages:

1 stage – Organizational (group work of students)
2 stage – Preparatory (individual work of students)
3 stage – Implementation (group work of students)

At the organizational stage, selection and formulation of the topic takes place. At the same time, the topic cannot be highly specialized, since first-year students are not familiar with the terminology and specifics of a future profession, so the subject has to correspond to the format of "popular" or entertaining science. In the course of work, the teacher needs to support the maximum level of activity in all students of the group because a foreign-language communicative competence is formed through personal activity experience. For example, students with a higher level of English help those with a lower level to understand the new material. At the same time, besides mutual training, there is a self-learning process since stronger students improve their knowledge by explaining. The most appropriate variant to guarantee involvement of the whole group is to use interactive teaching methods, which are understood as all kinds of activity, demanding a creative approach to selection of material and provide conditions for each student to show themselves [27].

2.4 Results

The first experimental English-language student conferences took place at Tomsk State University in April 2016 as part of the project "Development of Foreign-Language Competence by TSU Students". The conference "Take a Gander at Chemistry: Popularize, Experiment and Theorize" at the Faculty of Chemistry, and the 1st Conference of TSU Physics Students at the Faculty of Physics.

The main purpose of academic conferences was to allow future graduates to implement the previously mentioned competences, speech and practical language skills, and the ability to work independently in situations of foreign professional and scientific communication. The leading types of speech activity by participation in academic and research conferences were preparation and presentation of the poster or plenary reports in English and discussion of the reports and subjects connected with student research work. The conferences involved student presentations on different subjects in various fields of Physics and Chemistry. Initially, it was very difficult for both students and teachers. Preparation for presentations required the students to work independently.

Individual work has to have a definite purpose and students have to know how to do it, therefore it has to be followed by instructing, methodological, and organizational explanations from the teacher. Individual work has to correspond to student capabilities, i.e. if a student's vocabulary is insufficient, and they have insufficient knowledge of grammar, such students usually prefer the poster reports. Since various types of individual work have to be combined, the teacher's task is to manage the learning process and prevent difficulties which can arise when doing individual work. As individual work has to be of minimal cliché, it carries out the mission of developing cognitive abilities, student initiative, and creative abilities when preparing a conference presentation.

The experiment revealed that participation in conferences was especially difficult for the first-year bachelor degree students. This can be explained, on the one hand, by lack of skills for carrying out a scientific research, writing a research paper, and insufficient knowledge of specialized vocabulary. On the other hand, they have insufficient skills and abilities of public speaking in a foreign language and discussing professionally focused information. Delivering presentation in front of the audience is a stressful situation for them.

3 Discussion of Results

In April 2017, the conference "Take a Gander at Chemistry: Popularize, Experiment, Theorize" took place at the Faculty of Chemistry, and the conference "2nd Conference of TSU Physics Students" was held at the Faculty of Physics.

There were more reports at these two conferences which were done as a part of the First English-speaking forum for TSU students, which made 57% of total number of the forum speakers. More experienced second-year bachelor degree students made plenary reports, while the first-year students prepared more poster presentations, making 61% of the total number of presented poster presentations. In comparison with the previous conferences, those of 2017 comprised more independent practical research, and the experiments were made both during the performances (for example, chemical means for making ice cream) and broadcasted during performances from the laboratories (for example, spectral analysis).

In April 2018, as part of the Second English-speaking forum of students at TSU, "3rd Conference of TSU Physics Students" took place. In May 2018, the conference "Take a Gander at Chemistry: Popularize, Experiment, Theorize" was organized as a part of the XV International conference for students, university graduates, and young scientists "The Prospects of Fundamental Sciences Development". The number of speakers both with plenary reports and poster presentations at these conferences increased both in 2017 and 2018, in comparison with 2016. The results of increase in a number of students who delivered reports at the academic conferences are shown in Table 1.

Table 1. The results of increased number of students who delivered reports at the academic conferences.

Years	2016	2017	2018
Student Academic Conferences at the Faculty of Physics			
Number of speakers who delivered plenary reports	12	26	29
Number of speakers who delivered poster presentations	34	42	56
Student Academic Conferences at the Faculty of Chemistry			
Number of speakers who delivered plenary reports	9	22	38
Number of speakers who delivered poster presentations	2	4	13

Practical experience gained as a result of planning and holding the above conferences from the methodological point of view was analyzed. It can be noted that preparation for an academic conference puts a special emphasis on support of students individual work. Appropriate planning of individual work is an important and necessary task, which, if applied well, will allow the student to successfully participate in the conference. It is necessary to note that, in our case, not only a quantitative increase in students interested in participating in the conferences, but also a qualitative component of their reports was observed.

In our opinion, it is expedient to immerse students in the issue using videos from the Internet, introduce ideas by means of such interactive receptions as "heuristic conversation", "brainstorming" and others. It is advisable that the teacher learns which questions are more interesting for students within the current subject; creates active motivation; discovers the student initial impressions regarding this problem. It is important to reveal which difficulties are possible in the course of work (lexical, grammatical, and others). Finally, it is desirable that the student chooses a report subject independently. Further, the teacher and the student work together on the title of a report. To formulate the title correctly, it is necessary not only to see a problematic situation, but also specify possible ways and means for its solution. Therefore, symptomatology of problematic situations is emergence of a set of counter examples which involve a set of questions and give rise to feelings of doubt, uncertainty, and dissatisfaction with the available knowledge. The appearance of new rationally meaningful forms of theoretical knowledge organization is the result of escape from problematic situations [28]. Reliance on a problem in education allows us to develop analytical, critical thinking in students, as well as their ability to work independently with information flow, extract new knowledge, apply it in practice, and interact efficiently.

The idea of using rhetorical questions in the report titles seemed very successful. For example, such titles as: *How can we see? Forensic chemistry: Justice for all? Can we re-imagine chemistry education? What wind speed will lift you off the ground? Why does the rainbow have only seven colors?* confirm, in our opinion, the assumption that the ability to make contact with the audience is a very important factor for a successful participation. Success and the student self-assessment depend on the audience's reaction.

Thus, the criteria of successful implementation of this pedagogical task are the following:

- To understand the context and algorithm of a task performance;
- To understand the problem;
- To approach the task accomplishment;
- To be able to predict the result;
- To organize individual work of students according to the development of project activity for conference preparation;
- To be able to present the result of the performed task.

Collection and analysis of the material takes place at the Preparatory stage. At this stage students work hard individually. Therefore, in our opinion, students should not be required to work only with authentic sources, since this is a very labourious process, and preparation for a conference is limited to a number of hours permitted for contact work. Therefore, any acceptable options to promote the achievement of goals are possible. It is important to explain to the student that their proficiency in foreign and native languages are different, and that they should use adequate language means, appropriate grammatical constructions and vocabulary. In the main educational program for a bachelor degree, it is noted that the bachelor degree student has to possess the skills of reflection, self-assessment, and self-control. Consequently, we claim that a rigorous independent searching activity of students in the course of preparing their reports for a conference helps to improve their meta-informative abilities, thereby increasing self-control and understanding of the material while working with it. Thus, as a part of independent activity during the work on the student report, the teacher needs to accurately define the plan and the result of the student independent activity and ways of supervising this activity during work with authentic foreign-language information, namely:

- Ask a series of questions which require answers to be found in the course of work with the Internet resources;
- Accurately outline the problem which needs to be solved;
- Specify the sequence in the study of material;
- Report the deadline for a task performance;
- Specify a report form, after processing of the collected information;
- Define the evaluation criteria.

At this stage, it is expedient to involve, whenever possible, subject teachers from the main departments. There are unconditional advantages in such cooperation. First, it increases the academic and research value of the work. Second, it gives the first-year students a chance to get acquainted with different departments and their employees. Third, it increases motivation to learn foreign languages for professional purposes as the student sees how modern researchers use foreign literature and what opportunities are available to the scientist who knows a foreign language.

At the Implementation stage, the students deliver their reports to the audience followed by a group discussion which will considerably facilitate their further performance at a conference because it results in reduction of psychological tension. In addition, there is a new form of interaction between the students during organization of the student group work at the final stage. This form influences the development of their speech, communicativeness, cognition, and intelligence, and it leads to mutual enrichment. It is necessary to emphasize that group work is the most effective at the Implementation stage if the group discussion is preceded by an individual search.

After the conducted analysis of the formation of speech abilities, which are a part of foreign-language communicative competence, we defined the nomenclature of speech abilities of the non-language majoring students developed by means of the conference method. The nomenclature is presented in Table 2.

Table 2. The nomenclature of speech abilities of non-language majoring students developed by means of the conference method.

Formation of foreign-language communicative competence by means of the conference method	
Type of speech activity	Speech abilities
Listening	- **Perceiving** the main contents of authentic scientific and popular scientific texts with use of special lexicon in this scientific area; - **Selecting** significant/required information according to the objective in authentic scientific and popular scientific texts using special lexicon in this scientific area;
Reading	- **Selecting** significant/required information in authentic scientific and popular scientific texts; - **Carrying** out the search of information according to the objective; - **Understanding in detail** authentic scientific and popular scientific texts using special lexicon in this scientific area;

(*continued*)

Table 2. (*continued*)

Type of speech activity	Speech abilities
	Formation of foreign-language communicative competence by means of the conference method
Speaking	- **Specifying** information (clarification, paraphrasing, and others); - **Interrogating** the interlocutor, asking questions and answering them; - **Starting**, conducting/supporting and finishing a dialogue about what has been read; - **Stating** one's own point of view on what was seen or read; - **Builingd** a monologue description, monologue narration and monologue reasoning on what has been read; - **Beginning**, conducting/supporting and finishing dialogue (exchange of opinions), using special lexicon in this scientific area;
Writing	- **Writing** messages (briefly/in detail) on what was seen or read; - **Keeping** a record of the main ideas and facts (from video and text) on a scientific issue; - **Performing** written project tasks (written description of presentations, information booklets, posters, and others).

4 Conclusion

After the analysis of the experience gained in holding student conferences in a foreign language, we concluded that the student conference held in English is the most effective and rational form of organization (at all its stages) of student's individual work in a foreign language.

Thanks to the above detailed work algorithm of "group work – individual work – group work", students form the following abilities:

- Classifying the obtained information;
- Organizing the obtained information in coherent structures;
- Analyzing various situations;
- Checking, planning, and correlating the obtained information during cognitive activity;
- Predicting, anticipating, and considering consequences of decisions;
- Choosing and defining a strategy for task performance.

Moreover, preparation of a report for an academic conference solves one of the key problems in modern education – increasing the volume of student out-of-class individual work due to which the process of the personality self-regulation is formed, for which reason independent work becomes controllable by the learner. Thus, it is necessary for teachers of higher educational institutions, on the one hand, to promote the creative and cultural development of students at foreign language lessons, and, on the other hand, help future researchers to feel confident in situations of foreign-language professional communication. The experience of the teachers at TSU Faculty of Foreign

Languages showed that student's individual work in the course of preparation for student academic conferences activates various forms of perception and learning of study material. The combination of all its types helps to successfully solve the tasks which higher education institutions face.

In summary, we consider academic and practical conferences the most natural and harmonious form of student work in modern higher education institutions. From our point of view, modeling an academic conference and involvement of future bachelor and master students in its preparation wholly correspond to the aim of forming a foreign-language communicative competence and also contributes to the development of a versatile, modern, and competitive individual.

References

1. Gural', S.K., Lazareva, A.S.: Obespechenie kachestva obuchenija ustnoj inojazychnoj rechi sredstvami informatsionno-kommunikatsionnyh tehnologij [Providing quality of teaching foreign oral speech by means of informational and communicational technologies: manual]. Izd-vo Tom. un-ta, Tomsk (2007). (in Russian)
2. Sysoyev, P.V.: Sovremennye informatsionnye i kommunikatsionnye tehnologii: didaktich-eskie svojstva i funktsii [Modern informational and communicational technologies: didactic properties and functions]. Jazyk i kul'tura 1(17), 120–133 (2012). (in Russian)
3. Shul'gina, E.M.: Algoritm raboty` s texnologiej veb-kvest pri formirovanii inoyazy`chnoj kommunikativnoj kompetencii studentov [The algorithm of the webquest technology using for foreign language communicative competence formation of students]. Vestnik Tambovskogo gosudarstvennogo universiteta. Seriya: Gumanitarny`e nauki 9(125), 125–130 (2013). (in Russian)
4. Shul`gina, E.M.: Ispol`zovanie IKT v formirovanii umenij studentov rabotat` v sotrudnich-estve (na primere texnologii veb-kvest) [ICT using in student skills formation to work in teams (by the example of the webquest technology)]. In: Materialy XVI Mezhdunarodnoj nauchno-prakticheskoj konferencii. Vozmozhnosti razvitiya kraevedeniya i turizma Sibirskogo regiona i sopredel`ny`x territorij, posvyashhennoj pamyati Pochetnogo predsedatelya TOO RGO, professora Petra Andreevicha Okisheva, pp. 85–89. Izd-vo Tom. un-ta, Tomsk (2016). (in Russian)
5. Arijan, M.A.: Tehnologizatsija jazykovogo obrazovanija I professional'noe sovershenstvo-vanie uchitelja inostrannogo jazyka [Technologization of linguistic education and professional enhancement of foreign language teacher]. Res. J. Int. Stud. 3, 7–12 (2014). (in Russian)
6. Aylazyan, Y., Obdalova, O.: ESP Adult Course Implications for Professional Competence Development. Procedia – Soc. Behav. Sci. 154, 381–385 (2014)
7. Bezukladnikov, K.E., Novoselov, M.N.: Tehnologicheskoe obespechenie kompetentnos-tnogo podhoda k professional'noj podgotovke buduschego uchitelja inostrannogo jazyka [Technological support of competence approach to professional preparation of future foreign language teacher]. Pedagogicheskoe obrazovanie I nauka [Pedagogical Education and Science] 5, 104–107 (2014)
8. Gural', S.K., Nagel', O.V., Temnikova, I.G., Najman, E.A.: Obuchenie inojazychnomu diskursu na osnove kognitivno-orientirovannyh obrazovatel'nyh tehnologij [Foreign discourse teaching based on cognition-focused technologies]. Jazyk i kul'tura 4(20), 62–71 (2012). (in Russian)

9. Gural, S., Shulgina, E.: Socio-cognitive aspects in teaching foreign language discourse to university students. Procedia – Soc. Behav. Sci. **200**, 3–10 (2015)

10. Obdalova, O.A., Gural, S.K.: Konceptual'nye osnovy razrabotki obrazovatel'noj sredy dlya obucheniya mezhkul'turnoj kommunikacii [Conceptual Foundations for Educational Environment Development when Teaching Intercultural Communication]. Lang. Cult. **4**(20), 83–96 (2012). (in Russian)

11. Obdalova, O.: Modelling conditions for students' communication skills development by means of modern educational environment. In: Al-Mahrooqi, R., Denman, C. (eds.) English Language Instruction in EFL Contexts, pp. 73–91. Peter Lang International Academic Publishers, Bern (2015)

12. Obdalova, O.A.: Inoyazychnoe obrazovanie v 21 veke v kontekste sociokul'turnyh I pedagogicheskih innovacij [Foreign language education in the 21st century in the context of socio-cultural and pedagogical innovations]. Izd-vo Tom. un-ta, Tomsk (2014). (in Russian)

13. Savitskaya, I.S.: Razvitie professional'ny`x yazy`kovy`x kompetencij studentov v vedush-hem issledovatel`skom universitete (na materiale proekta 'Povy`shenie yazy`kovy`x kompetencij obuchayushhixsya v NI TGU (klaster estestvennonauchny`x fakul`tetov)') [Professional language competences development of students in leading research university (by the material of the project "Increasing of language competences of students in NR TSU (cluster of natural-scientific faculties)")]. In: Materialy XXVIII Mezhdunarodnoi nauchnoi konferencii Yazy`k i kul`tura, pp. 406–412. TGU, Tomsk (2018). (in Russian)

14. Shul`gina, E.M.: Autentichnost` kak odno iz metodicheskix uslovij pri formirovanii IKK studentov posredstvom texnologii veb-kvest [Authenticity as one of the didactic conditions for formation of foreign language competence in students via the webquest technology]. In: Materialy XXIV Mezhdunarodnoi nauchnoi konferencii Yazy`k i kul`tura, pp. 59–63. TGU, Tomsk (2013). (in Russian)

15. Sumtsova, O.V., Aikina, T.Y., Bolsunovskaya, L.M., Phillips, C., Zubkova, O.M., Mitchell, P.J.: Collaborative learning at engineering universities: benefits and challenges. Int. J. Emerg. Technol. Learn. **13**(1), 160–177 (2018)

16. Krivova, L., Imas, O., Moldovanova, E., Sulaymanova, V., Mitchell, P.J., Zolnikov, K.: Towards smart education and lifelong learning in Russia. Smart Innov. Syst. Technol. **70**, 357–383 (2018)

17. Zhigalev, B.A.: Koncepciya ocenki kachestva professional'nogo obrazovaniya v VUZe [The concept of assessing the quality of vocational education at the university]. Vestnik Nizhegorodskogo gosudarstvennogo lingvisticheskogo universiteta im. N.A. Dobrolyubova **10**, 176–184 (2010). (in Russian)

18. Bulanova-Toporkova, M.V.: Pedagogika i psikhologiya vy`sshej shkoly` [Pedagogy and Psychology of a higher educational institution]. Feniks, Rostov n/D (2002). (in Russian)

19. Zimnyaya, I.A.: Klyuchevy`e kompetencii - novaya paradigma rezul`tata obrazovaniya [Key competences – a new paradigm of the education result]. Vy`sshee obrazovanie segodnya **5**, 34–42 (2003). (in Russian)

20. Kazakova, A.G.: Osnovy` pedagogiki vy`sshej shkoly` [Fundamentals of pedagogy of higher school]. IPO Profizdat, Moscow (2000). (in Russian)

21. Pidkasisty`j, P.I.: Pedagogika [Pedagogy]. Ped. obshhestvo Rossii, Moscow (1998). (in Russian)

22. Evdokimova, M.G.: Sposoby` formirovaniya i razvitiya avtonomnosti uchashhixsya v processe ovladeniya IYa [Ways of formation and development of students' autonomy in the process of foreign language mastering]. MIET, Moscow (2010). (in Russian)

23.
Zmievskaya, E.V.: Uchebnaya delovaya igra v organizacii samostoyatel`noj raboty` studentov pedagogicheskix vuzov [Learning business games in organization of students' independent work of pedagogical institutions]. Ph.D. theses. MPSU, Moscow (2003). (in Russian)

24. Sysoyev, P.V.: Kul`turnoe samoopredelenie obuchayushhixsya v usloviyax yazy`kovogo polikul`turnogo obrazovaniya [Cultural self-identification of students in the conditions of language polycultural education]. Inostranny`e yazy`ki v shkole **4**, 14–20 (2004). (in Russian)

25. Shul`gina, E.M., Obdalova, O.A.: Organizaciya upravlyaemoj samostoyatel`noj dey-atel`nosti studentov posredstvom texnologii veb-kvest kak uslovie uspeshnosti formirova-niya inoyazy`chnoj kommunikativnoj kompetencii [Organization of students' guided independent activity via the webquest technology as a condition of successful formation of foreign language competence]. Vestnik Tomskogo gosudarstvennogo universiteta **376**, 162–167 (2013). (in Russian)

26. Abaeva, F.B.: Formirovanie i razvitie IK studentov estestvennonauchny`x fakul`tetov [Formation and development of foreign language competence in students of natural faculties]. Zhurnal Sovremenny`e problemy` nauki i obrazovaniya **6**, 13–16 (2014). (in Russian)

27. Verbiczkij, A.A.: Aktivnoe obuchenie v vy`sshej shkole: kontekstny`j podxod [Active learning in higher school: context approach]. Vy`sshaya shkola, Moscow (1991). (in Russian)

28. Zotova, A.F., Mironova, V.V., Razina, A.B. (eds.): Filosofiya [Philosophy], 2nd edn. Akademicheskij Proekt/Triksta, Moscow (2004). (in Russian)

Teaching English as a Foreign Language to Law Students Based on Content and Language Integrated Learning Approach

Pavel V. Sysoyev[1,2]([✉]) [iD] and Vladimir V. Zavyalov[1] [iD]

[1] Derzhavin Tambov State University, Tambov 392000, Russian Federation
psysoyev@yandex.ru, zavtmb@mail.ru
[2] Moscow Pedagogical State University, Moscow 119991, Russian Federation

Abstract. Students of all majors at Russian universities have to take a foreign language class. According to the Federal state educational standards for higher education, teaching a foreign language for professional communication to students of non-linguistic majors should be directed to their further specialization. However, the complexity of content selection for language teaching for professional purposes is that within many majors there are separate specializations. These specializations regulate and limit the future professional activities of university graduates. Analysis of the course curricula for "Foreign Language for Professional Communication" shows that in most schools the content of this course is the same for students of the same major, despite the differences in specialization. Obviously, this does not fully meet students' professional interests and needs. Content and Language Integrated Learning Approach aims to intertwine the study of the target language and the professional content. In this paper, the authors consider the features of the content of teaching writing skills to law students, who chose three different areas of specialization: state law, criminal law and civil law. The authors develop the list of relevant written documents for students, studying three areas of specialization, in accordance with the specifics of their future professional activities.

Keywords: Content and language integrated learning · Language for legal purposes · Language content

1 Introduction

The goal of teaching a foreign language to students of non-linguistic areas of study is the further development of their foreign language communicative competence in social and socio-cultural spheres of communication for interpersonal and intercultural interaction, as well as in the *professional* sphere of communication [3]. However, if scholars have come to a common opinion on the content of foreign language teaching for social and socio-cultural spheres of students' communication which is reflected in domestic and foreign regulatory legal documents and research [1, 4–8, 12, 14], the content selection of foreign language teaching for *professional communication* is still the subject of discussion and independent study in the field of foreign language teaching methods [13–17].

© Springer Nature Switzerland AG 2019
Z. Anikina (Ed.): GGSSH 2019, AISC 907, pp. 237–244, 2019.
https://doi.org/10.1007/978-3-030-11473-2_26

The curriculum for each non-linguistic area of specialization includes the course "Foreign language for professional communication". Depending on the university, this course is usually taught during 2–4 semesters. However, the complexity of determining the content for this course is due to the fact that within some majors there are separate areas of specialization that regulate the future professional activities of the educational program graduates. In this regard, the content of teaching a foreign language for professional communication to students in each area of specialization should differ from each other and be directed specifically to further specialization. The professional orientation of foreign language teaching will be observed in the selection of the instruction content for all types of speech activities, including writing. In this paper we are going to discuss characteristic features of the content selection for teaching writing skills to law students who chose three different areas of specialization: state law, criminal law and civil law.

2 Literature Review

Teaching a foreign language for professional communication using the Content and Language Integrated Learning (CLIL) Approach is a new direction in teaching a foreign language to non-linguistic major university students. Many scholars in their research argued for the benefits of this approach, highlighting its main advantage and possibility to prepare students for real life written and spoken communication on professional matters [13–17]. In their research Sysoyev and Zavyalov show that, according to this approach, a foreign language is viewed both as a *goal of teaching* and the *medium of learning* [15–17]. On the one hand, students still learn a foreign language and master language and communication skills which enable them to further develop their level of a foreign language communicative competence (levels B1 to B2). On the other hand, students learn professional content in the area of their specialization. Moreover, as research shows, the professional content of teaching should not necessarily replicate the content students learn during regular professional development courses in their first language. Ideally, via a foreign language for professional development students should acquire additional knowledge and experience in the area of their specialization.

To some extent, this approach of integrated study of a foreign language and content is not particularly new. In her instructional model Savignon [9] showed that all components of a foreign language communicative competence are intertwined. Learning and mastering language and communication skills, students learn knowledge about a foreign language and the native language countries and cultures. This model, being applied to teaching a foreign language for *social* and *socio-cultural* spheres of communication for interpersonal and intercultural interaction, can be extended to teaching a language for professional communication, and it has just recently started to be applied at the university level. Despite many advantages, one of the main obstacles for university teachers is to determine the course content for each area of specialization within one major.

3   Teaching Writing Skills to Students at Different Stages of Training

"Writing" is one of the main terms in their paper. In the research one can find many different definitions of "writing". It can be defined as a productive type of speech activity that provides expression and fixation of thought with the help of a system of graphic signs adopted in a particular language. Thus, this term reflects the technical side of speech fixation. It includes the formation of graphic, calligraphic, spelling and phonetic skills. Writing can also be defined as a process and result of communication in the form of text [10, 11, 14].

The content of teaching writing skills to secondary school students and university students is determined by the level of their foreign language proficiency (from A1 to C2 at a European school) [1], as well as the level of education (secondary school or linguistic/non-linguistic university). The content of foreign language teaching in general and writing skills in particular at all levels (from A1 to C2) in the social and socio-cultural spheres of communication has already been described in detail, reflected in modern sample programs for secondary school and university programs of foreign language courses and does not cause disputes in the methodological environment. By the end of the 9th grade secondary school students reach the Waystage level (A2) and must develop the ability to make extracts from texts, write short congratulations, fill out forms, and write personal letters. The content of teaching writing skills to the secondary school graduates (grade 11, the basic level - level B1) includes the development of additional skills to ask about the news in personal letters and give them; to tell about individual facts or events of life, expressing their judgments and feelings, to describe their plans for the future [5, 7]. The content of teaching writing skills to university students of non-linguistic majors (level B2) will include the development of skills to express in writing the information about yourself in the format adopted in the country of the language of study (CV, questionnaire, autobiography), as well as to summarize the content of the read/heard foreign text in theses, abstracts, reviews; to make notes of the lecture's main content; to use writing in a foreign language in research activities, fix and summarize essential information from different sources; to compile abstracts and a detailed plan of speech; to describe events/facts/phenomena; to inform/request information, expressing one's own opinion/judgment [1, 2, 7].

The analysis of the content of writing skills teaching at different stages and levels of training indicates the existence of continuity. So, if students learn to write personal letters at level A2, then at the level of B1 and B2 the quantitative and qualitative requirements for writing are increasing. In addition, with the transition from level to level, new writing skills are added, which must be mastered by students. With the transition from level to level, considering the cognitive abilities of students, the proportion of reproductive tasks gradually decreases and the proportion of problematic and creative tasks increases [8].

In addition to the social and socio-cultural spheres of communication, the content of teaching writing skills to students of linguistic majors (levels B2–C2) includes teaching the genres of academic and scientific discourse (various types of essays, statements, personal letters, business letters, applications for participation in conferences, grant competitions, reviews of scientific works, abstracts, scientific reports, course and degree works, etc.) [13].

The content of teaching writing skills to students of non-linguistic majors (levels B1–B2) in addition to the above-mentioned skills of social and socio-cultural spheres of communication should reflect the specific character of the future professional activities of graduates. Let's consider the content of teaching writing skills to students, majoring in law.

4   English Language Course Content for Law Students for Development of Writing Skills

Students, who major in law, may choose one of three basic possible areas of further specialization: state law, criminal law and civil law. Each area of specialization is aimed at a professional development of students and their preparation for professional activities in a particular legal field. In particular, state law specialization is aimed at development of competencies necessary for work in government institutions; criminal law - on training specialists in the field of criminal law regulation of public relations; civil law - on the training of lawyers involved in property relations, as well as related personal non-property.

The curriculum for "Law" major includes two types of courses: *general professional courses*, studied by all law students, despite their area of further specialization, and *specific advanced professional courses* which differ from one area of specialization to another, and which reflect the features of future professional activity of graduates. In Table 1 a list of the most common courses, studied by law students, is given.

Table 1. List of courses taught to law students.

Type of course/Area of specialization	State law	Criminal law	Civil law
Advanced professional courses	State and Municipal Law Problems of administrative responsibility Suffrage Customs Law Professional culture of civil servants Problems of state and municipal service	Criminal legal regulation of national security Problems of qualification of crimes Problems of criminal law Prosecutorial supervision	Administrative Proceedings Extrajudicial forms of protection Problems of Civil Law Intellectual Property Law
General professional disciplines taught to students of all areas of specialization	Constitutional Law, Municipal Law, Administrative Law, Environmental Law, International Law, Criminal Law, Criminal Procedure, Law Enforcement, Civil Law, Civil Procedure, Arbitration, Family Law, Labor Law, Tax Law		

Taking into the account that the course "Foreign language for professional communication" is aimed at students' further professional specialization, the content of the course in general and of teaching writing skills in particular should reflect the specifics of the future professional activity of graduates for each area of specialization. Based on their education, students who studied State law are in demand in the Federal institutions of state power of the Russian Federation, state authorities of the subjects of the Russian Federation, local self-government institutions, etc.; Criminal law - the Ministry of Internal Affairs, the Prosecutor's Office of the Russian Federation, the Federal Service of Execution and Punishment, the Federal Bailiff Service; Civil law - advocacy, notary, courts, etc. In this regard, the specific features of the future professional activities of the graduates will determine the content of teaching of the writing skills to students of each

Table 2. The content of teaching writing skills to law students, studying three areas of specialization.

Specialization	Writing skills
State law	*Ability to write/compose:* - Decrees, orders, resolutions; - Agreements between the subjects of the Russian Federation; - Extract of EGRN (Uniform State Register of Taxpayers), EGRUL (Uniform State Register of Legal Entities); - Resolutions, orders of local self-government bodies; - Resolutions, protocols of state bodies of the subjects of the Russian Federation; - Job descriptions for employees; - Preparation of reports, plans, methodological recommendations, concepts, declarations of federal authorities of the Russian Federation; - Draft normative legal acts; - Decision on refusal to provide requested information from the Unified State Register of Real Estate; - Extract from EGRN on the cadastral value of the property; - Extract from EGRN about the main characteristics and registered rights to the real estate object (land plot); - Orders of the federal bodies of state power; - Requests for the availability of property; -Correspondence with arbitration managers; - Correspondence with legal entities concerning bankruptcy; - Documents (correspondence, requests, extracts from the register) on the provision of information from the register of licenses for carrying out activities for the manufacture of printed products protected from forgery (security paper blank sheets), as well as trade in these products
Criminal law	*Ability to write/compose:* -Protocols, resolutions, orders, reports, inquiries, notices, summons, inspection materials, and pack inventories; -Photographs to the inspection protocols of objects (documents); -Cards of state statistical accounting forms 1, 1.1, 1.2, 2, 3, 4, 5, about results of the crime investigation;

(*continued*)

Table 2. (*continued*)

Specialization	Writing skills
	-Protocols: making a verbal statement of a crime, inspection of the scene of the incident, detaining the suspect, conducting a personal search, interrogation of the suspect, confrontation, interrogating the witness, interrogation of the victim, investigative experiment, verification of testimony on the spot, interrogation of the minor victim [witness], acquaintance of the accused and his defender with the materials of the criminal case, appearance with confession, inspection of objects (documents); -Reports, requests, notifications, and inventories; -Notifications: termination of the criminal case due to the expiration of the statute of limitations, termination of the investigative actions, and carrying out a personal search; -Decisions: on the initiation of a criminal case and its acceptance for proceedings, on the adoption of a criminal case to proceed, on initiating a criminal case and transferring it to the head of the investigative body to determine the investigative jurisdiction, on refusal to initiate a criminal case, on seizure and transfer of a criminal case, on non-disclosure of the fact of a suspect detention, on combining of criminal cases, on allocation of a criminal case, on allocation of criminal case materials on a separate production, on the recognition and inclusion of material evidence in the criminal case, on recognition as the victim, on attraction as the accused, on election of a measure of restraint, on the appointment of forensic examination; -Subscription of non-disclosure of the data of preliminary investigation, undertaking of 'not to leave place of residence' and proper behavior; - The summons for interrogation; - Obligation of appearing; - Indictment
Civil law	*Ability to write/compose:* - Contracts: purchase and sale of movable/immovable property, purchase and sale of apartments using credit funds, pledge of immovable property, lease, cession, supply of goods, donation, annuity, rent, loan agreement, orders, commissions, organized tenders, rental of premises, license agreements, the use of neighboring rights, the alienation of exclusive rights, the creation of results of intellectual activity, the performance of research, development and technological works; - Act of property receipt and transfer; -Development of an employment contract (both standard and considering the peculiarities of work in certain categories for employees); - Constituent documents of a legal entity; - Statement of claim in civil, arbitration, and administrative cases; - Appeals against the decision of the court on civil, arbitration, and administrative cases; - Statement of claim for the evidence recovery of the case; - The debtor's declaration of recognition as insolvent (bankrupt); - Protocols of the court session; - Court summons; - Writ of execution; - Marriage contract; - Documents on the issue of adoption; - Documents in the field of social security

of the areas of specialization (Table 2). Table 2 shows the content of teaching writing skills to students of three areas of specialization: state law, criminal law and civil law.

Materials in Table 2 indicate that in many ways there is a certain correlation between the development of writing skills to university students of all majors (levels B1-B2) and specific writing skills, taught to law students of three areas of specialization. For example, at B1-B2 levels the students should demonstrate the ability to fill out questionnaires, forms, etc. Law students, of B1-B2 levels should demonstrate the ability to fill out *special legal forms, applications, notifications,* etc.; or if B2 level students must learn how to write personal letters, asking for information and describing facts and information, then during the course of a foreign language for professional sphere, these skills will be refined and expanded, and students will learn to make official inquiries and conduct professional business correspondence by the nature of professional activity. In addition, the content of teaching writing skills to law students will also include development of the ability of writing and composing unique legal documents, such as contracts, protocols, orders, reports, resolutions, etc.

5   Conclusion

The content of the course "Foreign language for professional communication", which is taught to all non-linguistic major university students in Russia, is directed to further students' specialization in the area of their training. The complexity of selecting the content for this course is due to the fact that within one major there can be several different areas of specialization. In this paper we considered the characteristic features of the content of the course "Foreign language for professional communication" for students, majoring in Law, and the content for teaching writing skills to students of the state law, criminal law and civil law areas of specialization was suggested.

References

1. Common European Framework of Reference for languages: Learning, Teaching, Assessment. Cambridge University Press, Cambridge (2001)
2. van Ek, J.A., Trim, J.L.M.: Vantage. Cambridge University Press, Cambridge (2001)
3. Federal'nyj gosudarstvennyj obrazovatel'nyj standart vysshego obrazovaniya po napravleniyu podgotovki 40.03.01 « Yurisprudenciya » (uroven' bakalavriata) [Federal state educational standard of higher education in the field of preparation 40.03.01 "Law" (bachelor's level)]. Ministry of Science and Education, Moscow (2016). (in Russian)
4. Federal'nyj gosudarstvennyj obrazovatel'nyj standart vysshego obrazovaniya po napravleniyu podgotovki 45.03.02 « Lingvistika » (uroven' bakalavriata) [Federal state educational standard of higher education in the field of training 45.03.02 "Linguistics" (bachelor's level)]. Ministry of Science and Education, Moscow (2014). (in Russian)
5. Federal'nyj gosudarstvennyj obrazovatel'nyj standart osnovnogo obshchego obrazovaniya [Federal State Educational Standard of Basic General Education]. Ministry of Science and Education, Moscow (2010). (in Russian)
6. Obdalova, O., Gulbunskaya, E.: Cross-cultural component in non-linguistics EFL students teaching. Proc. Soc. Behav. Sci. **200**, 53–61 (2015)

7. Primernye programmy po inostrannym yazykam. Anglijskij yazyk [Foreign language secondary school curriculum]. Ministry of Science and Education, Moscow (2010). (in Russian)
8. Safonova, V.V.: Kommunikativnaya kompetenciya: sovremennye podhody k mnogourovnevomu opisaniyu v metodicheskih celyah [Communicative competence: modern approaches to multilevel description for methodical purposes]. Euroschool Press, Moscow (2004). (in Russian)
9. Savignon, S.J.: Communicative Competence: Theory and Classroom Practice. McGraw-Hill, New York (1997)
10. Shchukin, A.N., Frolova, G.M.: Metodika prepodavaniya inostrannyh yazykov [Methods in teaching a foreign language]. Academia, Moscow (2017). (in Russian)
11. Solovova, E.N.: Metodika obucheniya inostrannomu yazyku [Methods in teaching a foreign language]. Prosveshchenie Press, Moscow (2001). (in Russian)
12. Sumtsova, O.V., Aikina, T.Y., Bolsunovskaya, L.M., Phillips, C., Zubkova, O.M., Mitchell, P.J.: Collaborative learning at engineering universities. Int. J. Emerg. Technol. Learn. 1, 160–177 (2018)
13. Sysoyev, P.V., Amerkhanova, O.O.: Obuchenie pis'mennomu nauchnomu diskursu aspirantov na osnove tandem-metoda [Teaching graduate students research discourse via tandem-method]. YAZYK I KULTURA-LANGUAGE AND CULTURE 36, 149–169 (2016). (in Russian)
14. Sysoyev, P.V., Merzlyakov, K.A.: Ispol'zovanie metoda recenzirovaniya v obuchenii pis'mennoj rechi obuchayushchihsya na osnove blog-tekhnologii [Methods of Teaching Writing Skills to Students of International Relations Using Peer Review]. Moscow State University Bulletin. Series 19: Linguistics and intercultural communication 1, 36–47 (2017). (in Russian)
15. Sysoyev, P.V., Zavyalov, V.V.: Obuchenie inoyazychnomu pismennomu yuridicheskomu diskursu studentov napravleniya podgotovki yurisprudenciya [Teaching foreign language written legal discourse to law students]. YAZYK I KULTURA-LANGUAGE AND CULTURE 41, 308–326 (2018). (in Russian)
16. Sysoyev, P.V., Zavyalov, V.V.: Ehlektivnyj kurs "Introduction to law" v sisteme professionalno orientirovannogo obucheniya inostrannomu yazyku v starshih klassah [The elective language course "Introduction to law" in the system of professionally oriented foreign language teaching in senior high school]. Inoctrannye Yazyki v Shkole 7, 10–18 (2018). (In Russian)
17. Zavyalov, V.V.: Modeli obucheniya inostrannomu yazyku dlya professionalnyh celej studentov nelingvisticheskih napravlenij podgotovki [Models of teaching students of non-linguistic majors a foreign language for the professional purposes]. Derzhavin's Forum 6, 175–184 (2018). (in Russian)

Integrative Methodology of Teaching Translation and Interpreting

Elena V. Alikina⬤, Lyudmila V. Kushnina⬤,
Anton Yu. Naugolnykh⬤, and Kirill I. Falko⁽✉⁾⬤

Perm National Research Polytechnic University,
Perm 614990, Russian Federation
{elenaalikina,lkushnina,crispian}@yandex.ru,
falkokirill@gmail.com

Abstract. The article is devoted to the development of professional translators' education. The research considers two teaching models. One of them is based on the concept of translation space. The study demonstrates how teaching students to analyze various semantic fields of translation space promotes a gradual development of professional translator's competencies. The second model represents the potential of using interpreter's note-taking as a means of processing semantic information in the process of interpretation. The authors argue that teaching both translation and interpreting within a single curriculum is a common practice in the Russian universities. However, despite having common goals, these two professional activities each requires a certain set of professional competencies, as well as the application of different professional tools. According to the authors' viewpoint, reconstructing a professional reality in the classroom may promote the improvement of professional skills in the most natural educational environment. In conclusion, the authors describe methodological solutions to integrate translation and interpreting teaching models into the educational process by means of project-based learning.

Keywords: Translation · Interpreting · Teaching models · Translation space · Note-taking · Integration

1 Introduction

Translation practice is open for studies, covering the spheres of expertise from linguistics to didactics, from theories of verbal and mental activity to pedagogy, from translation theory to translation practice. Modern translation studies incorporate the achievements of anthropocentric, cognitive, discursive, synergetic and culture-centered approaches. If integrated, they will allow formulating the basic principles of a new translation paradigm. This goal can be achieved by the interdisciplinary study of translation activity. Furthermore, translation activity can be the basis for modeling the process of teaching translation and interpreting which is the subject of this paper.

© Springer Nature Switzerland AG 2019
Z. Anikina (Ed.): GGSSH 2019, AISC 907, pp. 245–251, 2019.
https://doi.org/10.1007/978-3-030-11473-2_27

2 Translation Teaching Model

Translation teaching model is based on the idea that the text for translation is polysemantic. Since only a small part of meanings in the text is explicit, the other remains implicit. This conclusion is based on the field principle, which is being actively developed in synergetic linguistics [1]. According to this, a text is synergetic by its nature as well as translation that also fits into synergetic processes. Thus, translation synergy can be defined as the ability of the text to increment new meanings depending on the host culture's norms and requirements, as well as the expectations and preferences of the recipients. It is the natural inclusion of the text into the host culture that represents the main result of that synergetic effect, ensuring a qualitative and harmonious translation.

We developed the translation teaching model based on the correlation of translation space levels of analysis with the levels of translator's competencies development. It should be noted that we distinguish six components of translation space, particularly the content of the text (which is the core of it), along with the author's field, translator's field, recipient's field, energy field and phatic field [2].

At the primary level, students are taught to translate the invariant content of the text, i.e. the core of the translation space. This is the initial level of the explicit foreign language text analysis, which helps to form students' pre-translational skills and abilities. To analyze the content of the text we draw students' attention to the thematic and thematic progression which is simple to decode for a beginner translator, however its misunderstanding leads to translation errors and consequently, to the factual meaning distortion.

Having recognized the need for building translator's abilities and skills of perceiving and generating thematic and thematic progression with a view of transposing the content of the text, we identify the core of the text. However, the text itself is not identical to its core, since it contains the information, which reveals the subject of translation activity.

The implementation of the second level of analysis is related to understanding the pretext and the formation of the communicative translator's competency. Due to the fact that the pretext is completely implicit, the translator's task is either to explicate it, which is not always appropriate, or leave it implicit, i.e. to preserve some manifestations of the pretext (allusions, presuppositions, etc.) in the text of translation. In other words, the translator aims at unraveling the author's intention preparing to generate a new text of his own.

At the next level of analysis students are taught to understand the subtext through the so-called meta-linguistic phenomena (i.e. texts about the text). Here the subtext is intentionally explicated by the author for further interpretation of new realities to other culture bearers, i.e. corresponds to the constructive communicative translator's competency.

The translator's competency formation, aimed at the pragmatic adaptation of the text, is among the general linguistic competencies of the fourth level, determined by the pragmatic potential of the text. The level is hypothetical since neither the author of the text nor the translator can fully assess the whole translation space in the mind of the recipient, as well as the formation of reflective meaning in the process of understanding the context. This is a matter of competency, aimed at pragmatic adaptation, and in some cases, a pragmatic super-task.

The fifth level involves the formation of the emotional translator's competency, which promotes the perception of the translation space energy field. The importance of this competency's development is in creating the emotional correlation between the original text and the text of translation. In other words, the latter should possess the same emotional tension, intonation and rhythmic pattern, and the same psyche, as the original one. This can be achieved by teaching poetic translation to students.

Finally, the sixth level promotes the development of the cognitive translator's competency, through the process of inter-text comprehension. Our observations show that this competency helps to analyze the phatic field of the text and to identify its cultural meaning to obtain a harmonious translation. It is at the level of inter-text analysis, where the translator's personality formation reaches its peak.

Having considered the author's translation teaching model, let us move towards modeling the process of teaching interpretation.

3   Interpretation Teaching Model

Interpretation includes all types of professional translation, based on verbal or non-verbal transmission of the original message, either orally or in writing, from the source language to the target one. As a conceptual framework we propose the idea that the choice of integrative basis for interpreting teaching model should consider the dominant difficulty of interpretation. Learning to overcome this difficulty allows students to master the necessary competencies for a specific type of interpretation. Thus, the dominant difficulty for sight interpretation lies in the simultaneous reading of the source text, with generating statements in the target language. Therefore, teaching sight interpretation should aim to develop quick text orientation skills in the translated text discourse. In simultaneous interpretation the dominant difficulty is the simultaneity of auditory perception in one language and speaking in the other. This process occurs in the conditions of a certain speech rate, which is why the simultaneous interpretation teaching methodology focuses on learning to distribute information, switch the attention and concentrate on the necessary data, as well as to develop speech compression skills, anticipation, etc.

Consecutive interpreting is the basic type of interpretation activity, which implies auditory verbal transmission of the original message, perceived phonetically in between semantically completed fragments of the text and is carried out by means of noting down the information. The dominant difficulty of consecutive interpreting is a significant volume of perceived oral information of a high semantic density, pronounced only once. To overcome this difficulty the interpreter applies for interpreter's note-taking.

According to our definition, interpreter's note-taking is a type of professional written semantic recording of the information in the process of perceiving, comprehending and understanding a long oral speech fragment, with the aim of generating an accurate and complete equivalent in the target language.

The issues of note-taking functionality are widely discussed in contemporary scientific publications. Iliescu Gheorghiu [3] attempts to estimate its efficiency in the process of overcoming cognitive difficulties of interpretation by matching certain criteria, such as abbreviations length, complexity/ease of symbols recording, ways of connectors identification, and scarcity/redundancy of recording. Bastin [4] analyzes the ways of marking and transferring cohesion in consecutive interpreting by studying the texts of notes. Abuin Gonzales [5] describes typical interpreter's difficulties, which include performing simultaneous actions and decoding the notes. Andres [6] highlights two important notation quality parameters, which are notation rate and readability. Szabo [7], Dam [8] determine several factors for notation efficiency, such as the language of recording, sphere of interpretation, work experience and interpreter's individual style. All the researchers agree that note-taking, besides being a professional tool, is also a sense-oriented phenomenon.

The multifunctional nature of note-taking allows us to consider it as an integrating bond in space and time for teaching interpreting. This can be applied at the pre-university, university and postgraduate levels of education, within any set of disciplines of motivational, basic and professional interpreter and pedagogical levels. As for the integration mechanisms, we propose to use the vertical continuity approach (interrelation of the stages, degrees and levels of education in time), horizontal inter-subjectivity approach (interrelation within one education space) and radial expansion approach (the total of integrated processes' dynamics).

Vertical continuity is reflected in the integration of note-taking into all of the stages and levels of translation training program, which is the basis for continuity.

Horizontal inter-subjectivity manifests itself in one of the stages or levels of education, where various elements of interpreter's note-taking (methods, techniques and tools) are mastered within certain disciplines.

Finally, radial expansion is a gradual, spiral-like transition from inter-semiotic to inter-lingual interpretation, through intra-lingual interpretation in the native and foreign languages. The dynamics of educational process in space and time is reflected through a gradual expansion of professional competencies and the improvement of professional qualities, as well as the development of individual style of writing.

Considering training of future translators as an integrated educational process aimed at forming a professional and harmonious personality, it is necessary to find the ways for teaching the two universal types of professional activity – translation and interpreting.

4   Integrative Model of Teaching Translation and Interpreting

Although teaching translation and interpreting within a single curriculum or a certificate program is not typical of most western institutes it is a standard practice in Russian universities. Being positioned by scholars at opposite poles, translation and interpreting, however, are alike from the perspective of functional approach, in the sense that both imply information exchange using different verbal codes as an ultimate objective. Both are focused on transmission of a definite message from a source to a target culture and both require specific competencies, though indeed translators usually operate in a more artificial environment – mastering technology – while interpreters are engaged into direct communication between the parties.

Acknowledging the fact that translators and interpreters do need unique competencies in their professional development, we here argue that teaching translation and interpreting can go hand in hand if we find a proper integration axis. As González Davies puts it, "It could almost be suggested that, nowadays, the "read and translate" directive to teach translation is probably as obsolete and unproductive as the Grammar-Translation Method is to teach a foreign language... New paths should be explored instead of keeping to one approach to translation or to its teaching" [9, p. 3].

It is a matter of record that the language industry has been driven by translation and localization projects undertaken by vendors and freelancers usually scattered across vast areas and communicating with LSPs solely via the Internet. Some professionals may specialize solely in translation, never undertaking interpreting assignments. Others are natural born interpreters. This is a common reality of outsourcing. However, if we now turn to professional activities of a full-time staff member charged with cross-language communication duties we will see quite a different picture. For an in-house mediator translation and interpreting perfectly fit in the process of handling business information.

To develop the point, let us outline the role of an in-house translator/interpreter who reports to, or shares his/her duties with, an import manager. The latter is charged with a plethora of tasks ensuring a smooth transfer of international materials, commodities and products for corporate needs. Even in its simplest forms importation would require repeated contacts with contractors/suppliers (in our example located abroad and speaking another language) and intense paperwork with translation being a significant part thereof.

A mediator may get involved at the initial stage when corporate websites of counterpart companies are surfed to retrieve such information as general business terms and conditions, company's history, product range, contact information, etc. Some parts of the websites may be translated or summarized to expand the database of potential suppliers. At the next stage, first contacts may be established via email or phone, followed by negotiations at the foreign office. International trade contacts may be another alternative where interpreting dominates. If a deal is made, the next stage would embrace signing a contract (possibly, bilingual) and processing such input documents as letter of intention, proforma invoice, insurance policy, packing list, bill of lading, etc. Customs clearance and producing sales literature in a target language usually requires mediation too. Translation of operation manuals and product brochures based on original product description are all indispensable. Naturally, the entire process is accompanied by business correspondence and/or phone calls from first inquiries to resolving quality claims, translation and interpreting combining and switching each other all the time as long as communication proceeds.

Thus, we see that although the stages, or cycles, of goods transfer are specific they are connected by the same context, serving a single goal. Bringing this context with all its steps, links, tasks, roles, actions into the classroom means reconstruction of professional reality – an inevitable effort targeted at training a highly qualified mediator.

The positive sides are:

1. Holistic view of the mediation process 'from inside', adding to project thinking.
2. Good variety of reading, writing, speaking, listening activities within mediation tasks.
3. Exposed intertextuality fostering text/discourse perception, understanding and processing.
4. Lexical abundance enabling students to enrich business and financial vocabulary along with professional terminology, depending on the products chosen as 'imported commodities'. Special emphasis is put on making bilingual/multilingual glossaries.
5. Adjustable ratio of translation and interpreting tasks of numerous sources linked by the single background context.

By the same manner other 'integration axes' (contexts) may be revealed and reconstructed. For instance, an interpreter of a lecture does some preliminary work reading the available publications of the lecturer and perhaps translating some key parts and searching approved terminology equivalents. An interpreter of an interview with a politician would need some knowledge of the program that may be partly translated beforehand. To interpret product presentations careful reading and translation of user guides or product brochures would be much helpful. The challenges are here but the real-life teaching experience has demonstrated that most students make faster progress in translation/interpreting when context-based education technology is introduced. Driven by sound motivation, the learners acquire comprehensive knowledge of a certain professional domain, simultaneously improving translation/interpreting skills in the most natural way, i.e. by doing what real professionals do.

5   Conclusion

The need for a common solution for both the linguistic and pedagogical problems within the study is determined by the unity of translation education goal-set. We define it as the training of a professional translator or interpreter who possesses flexible thinking and shaped worldview. The translation and interpreting teaching models (linguistic model of translation space analysis and pedagogical integrated model of interpreter's note-taking) illustrated above, are based on the author's concepts of translator's competencies development. Moreover, it can be applied within the project organization of educational process, which makes it possible to establish common conceptual links within it.

References

1. Haken, H.: Synergetik in der Psychologie. Hogrefe, Göttingen (2010)
2. Kushnina, L.: Les langues et les cultures dans l'espace traductif. Atelier de Traduction 21, 61–74 (2014)
3. Iliescu Gheorghiu, C.: Introduccion a la interpretacion. La modalidad consecutive. Universidad de Alicante, Alicante (2001)

4. Bastin, G.: Les marqueurs de cohérence en interprétation consécutive. Interpret. Newsl. **2**, 175–187 (2003)
5. Abuin Gonzales, M.: El proceso de interpretacion consecutiva: un estudio del binomio problema/estrategia. Interlingua, Editorial Comares Granada (2007)
6. Andres, D.: Konsekutivdolmetschen und Notation. http://www.fask.uni-mainz.de/ dolmetschwissenschaft/Dateien/Konsekutivdolmetschen_und_Notation_DoerteAndres.pdf. Last accessed 17 Sept 2018
7. Szabó, C.: Language choice in note-taking for consecutive interpreting. Interpreting **8**(2), 129–147 (2006)
8. Dam, H.V.: What makes interpreters' notes efficient? Features of (non)efficiency in interpreters' notes for consecutive. In: Gambier, Y., Shlesinger, M., Stolze, R. (eds.) Doubts and Directions in Translation Studies: EST Congress 2004, pp. 183–198. John Benjamins, Amsterdam/Philadelphia (2007)
9. González Davies, M.: Multiple Voices in the Translation Classroom: Activities, Tasks and Projects. John Benjamins Publishing Company, Amsterdam/Philadelphia (2004)

Lifelong Learning

The Use of MOOCs for Professional Development of Translators

Lidiia I. Agafonova$^{(\boxtimes)}$ (ID), Natalia A. Abrosimova (ID),
and Irina S. Vatskovskaya (ID)

Herzen University, Mojka river emb. 48,
191186 St. Petersburg, Russian Federation
liagafonova@herzen.spb.ru, naabrosimova@gmail.com,
irinavable@gmail.com

Abstract. A Massive Open Online Course has attracted a lot of attention in the last couple of years. This article deals with use of MOOCs as a resource for professional development of translator skills. MOOCs offer self-study and self-paced models that can be used as a resource for translators' autonomous learning. MOOCs can increase the translators' subject-matter expertise. The emergence and use of MOOCs for translator's professional development is still uncommon, but on the verge of gaining a foothold. The aim of this paper is to review current trends in MOOCs in Europe and the U.S. in comparison with the Russian Federation. The paper contemplates the benefits of using MOOCs for professional development of translators. We offer a reflection on creation and utilization of the online courses for translators designed for the first time on Udemy platform in the Russian Federation, at the Herzen State Pedagogical University of Russia (Herzen University). This paper further discusses the preliminary findings related to MOOCs effectiveness of learning outcomes and its impact on translators and calls for more practical studies and explorative research.

Keywords: Massive Open Online Courses (MOOCs) ·
Higher education institutions · Online learning · Professional development ·
Translator development · Continuing education

1 Introduction

Massive Open Online Courses (MOOCs) present ample opportunities for teaching people of different ages and having different educational experiences. Participation in MOOCs is becoming large scaled. According to some independent studies, European and the U.S. Higher Education Institutions (HEIs) are more involved in MOOCs compared to Russian Institutions [1, 2]. European and the U.S. HEIs are confident regarding MOOCs development and implementation, have a more positive attitude and experiences towards MOOCs.

There is no experience of mass application of MOOCs for professional training and development of translators in Russian HEIs. There is also no information on the effectiveness of such MOOCs integration into the educational process of Russian HEIs.

© Springer Nature Switzerland AG 2019
Z. Anikina (Ed.): GGSSH 2019, AISC 907, pp. 255–264, 2019.
https://doi.org/10.1007/978-3-030-11473-2_28

In general, the most of current MOOCs being used to teach all subjects, not only translators skills, do not comply with the Federal State Educational Standards, therefore their full integration into the curriculum of the Russian HEIs is not possible. MOOCs are integrated into the curriculum on a voluntary basis in the Russian HEIs, in most cases on the initiative of individual teachers.

The most promising direction in Russia is to utilize MOOCs in a blended format in traditional classroom settings, and as a professional retraining and improvement of professional skills. Professional translator development is an established and growing research field. All translators should be dedicated to continuous professional development. Information is constantly being produced and translators act as agents to international or digital communication. Translators have to be up-to-date on the changing aspects of translation industry. Translators are believed to be ideal candidates for MOOCs, since they are often already working as translators when they decide to be enrolled for a translation course. Moreover, as long-life education is important in the profession of translator, MOOCs can be an interesting alternative to the traditional higher educational education for those who cannot attend a classroom, that's why self-specializing can be a reality for a lot of highly motivated professionals.

2   Literature Review

Education in general, including language education, is experiencing a period of change [3], with traditional and new models of education being under discussion incl. MOOCs which are types of open educational resources (OER). They are open for anyone to access, are mainly but not always free of charge, with unlimited participation [4, 5] and such key features as the accessibility and flexibility [6].

The first reference of the MOOCs appeared in 2008 when the term was created by Dave Cormier and Bryan Alexander [7, 8] to describe the essence of an open online course entitled "Connectivism and Connective Knowledge" at the University of Manitoba in Canada to attract as many students as possible from all over the world who could not pay tuition fees [9]. In 2013, MOOCs were named among the 30 most promising trends in the development of education until 2028 [10] as they contribute to the democratization of educational process, creation of free qualitative open educational resources; provide great opportunities for non-traditional forms of teaching approaches and learner-centered pedagogy; eliminate territorial and temporary barriers, and allow students to study beyond one university.

MOOCs can be classified into xMOOCs and cMOOCs [11]. The xMOOCs adopt a behaviorist approach, while the cMOOCs are based on the connectionist theory [3]. In short, "cMOOCs focus on the creation and generation of knowledge while xMOOCs focus on duplication of knowledge" [3, p. 341]. However, some scholars [12] believe that nowadays there is a blend of cMOOC and xMOOC in the courses offered worldwide.

Some research has been done to present experiences and statistics in the production and supply of MOOCs. One of them was the report "4 years of Massive Open Online Courses" [13] published by Harvard and the Massachusetts Institute of Technology at the end of 2016. The review was conducted from fall 2012 to summer 2016 providing the statistics on 290 courses, 245,000 certificates, 4.5 million participants, 28 million hours of participation and 2.3 billion online events. The main findings covered in the above report are as follows: a) the pupils of these classes are people of different status with different motivation (men and women, from 0 to 60+ years, people with higher education and without, teachers and people who have nothing to do with pedagogy); b) a student usually spends 29 h to work with the online course, and every hundredth earns a certificate, spending less than 23 min on the Internet; c) more than half of the respondents report a desire to receive a certificate, but a typical course certifies 30% of these respondents; d) informatics courses are the largest (among science, history, health and other subjects), in addition, they send more participants to other disciplinary areas than they receive themselves; e) 32% of respondents report they are or were teachers, and 19% of them took part in courses on the subject they themselves teach [13].

According to another survey [14] conducted during the implementation of the HOME project (Higher Education Online: MOOCs the European way), partly funded by the European Commission's Lifelong Learning Programme in October–December 2014, in which 67 institutions out of 22 European countries responded representing about 2.8 millions of students, most European universities, unlike the American ones, do not agree with the opinion that official MOOCs certification might be an alternative to existing diplomas of higher education. However, 80% of European universities agree that MOOCs help colleges and universities to get acquainted with online pedagogy, and only 44% of American universities supported this idea; 50% of European universities perceive MOOCs as a new approach to implementation of courses and disciplines, while most American universities are neutral about this thesis; 66% of American universities note that it is still too early to say that MOOCs meet the goals and objectives of higher education, and in Europe 58% of universities believe that MOOCs correspond to today's higher education goals [14].

As for the Russian Federation, one of analogues of the MOOC platforms is the Information Technologies Internet University project (http://www.intuit.ru/), which provides free courses via the National Open University INTUIT. Another Russian electronic online education platform, developed in 2013 by the MOOC technologies is Universarium (http://universarium.org/), which offers free courses conducted according to the e-learning educational standards. Other successful MOOC platforms in Russia are Courson (https://www.courson.ru/) - the Russian analogue of Udemy, Lectorium (https://www.lektorium.tv/) - the educational outreach project, Lingualeo (https://lingualeo.com/ru) - leader in foreign languages training with 13 million users, Geek-Brains (https://geekbrains.ru/) - a flagship in programming training.

In 2018, a valuable study was carried out to provide the SWOT analysis, in which strengths, weaknesses, opportunities and threats of MOOCs for higher education were examined within all stakeholders' perspectives [15]. Among the MOOC strengths, accessibility, life-long learning, online learning communities, experimentation and brand extension were named. To the MOOC weaknesses, according to the scholars, belong dropout rates, expensive infrastructure, pedagogy and assessment. The MOOC opportunities include expand reach, collaborative learning, individualized learning and alternative education, while the MOOC threads consist of sustainability, quality education, business model, fuddy-duddy instructors, and identity, credential and degree [15].

3   Methodological Framework

3.1   Relevance of Research

The relevance of research is defined by the need to create curriculum documents appropriate to the challenges of modern times and the level of information and technology development of a society that contribute to the formation of an innovative personality type who strives for continuous improvement, finds and generates new knowledge independently, ready to continue the process of self-improvement throughout life, and the need for a comprehensive analysis of a new phenomenon in the world of education - Massive Open Online Courses - in the context of improving the quality of education of translators, both beginners and professionals.

3.2   Objective of the Paper

The objective of this paper is the theoretical justification and practical study of the feasibility of MOOCs using for translators. This research is aimed at helping key specialists to get a better understanding of MOOCs, trends towards openness of higher education, and the importance of MOOCs for translator training.

3.3   Methods of the Research

In the course of this study we used the method of data collecting and critical analysis of academic literature on ICT in education, quantitative approach to the study of existing MOOC platforms and MOOCs in the USA, Europe and Russia; the qualitative approach to the study of MOOCs for professional development of translators, and description of own experience in developing a MOOC for translators.

3.4   Factual Material of the Research

The factual material was the open and paid online courses intended for beginning and professional translators, placed on the MOOC platforms.

3.5   Practical Value of the Research

Practical value of the research consists in the possibility of using MOOCs as both basic and secondary means of individual study in the fundamentals of translation, specialized translation (legal, medical, etc.), and cognate disciplines, for the best study of the main educational direction.

4   Findings

4.1   Analysis of Using MOOCs for Teaching Translation

With the help of the specialized Class Central search engine (https://www.class-central.com) and Europe MOOCs and Free Online Courses search engine (https://www.mooc-list.com), designed to search for MOOCs using the given parameters, we carried out the analysis of the MOOCs for teaching translators (and foreign languages in general) on major international platforms. Besides, we did the same search on each individual international platform, as well as on the MOOC platforms in Russia, not included in the Class Central search engine. The statistics obtained in the middle of August 2018 is given in Table 1. There were no language courses found on Udacity, INTUIT, Universarium, Lectorium, Courson, that's why those MOOC platforms were not included in Table 1.

Table 1. MOOCs Overview.

MOOC provider	Search results for "translator", "translation", "translating"	Search results for "English for…"	Search results for the category "Language courses" (all languages/English)	Search results for courses developed for professional level students (all languages/English)
Udemy	14	808	2203/928	71/25
Coursera	2	137	192/141	2/230
FutureLearn	1	52	72/8	No such search criterion
edX	0	34	42/34	5/3
Total	17	1031	2519/1112	78/58

As the statistics in the table shows, Massive Open Online Courses teaching translating skills on Udemy, Coursera, FutureLearn and edX comprise only 0.67% of all 2519 language courses, massive online professional level English courses teaching translating skills amount 21.7% of all professional level courses, and 13 massive open online translation courses out of 17 ones are English/another language courses. As far as our search findings show, there are only 2 Russian/English MOOCs teaching translating skills, and those two courses are the courses developed at Herzen University, St. Petersburg, Russia. The titles of those 17 courses are listed below: a) Udemy: Translating Official Documents (Part I), Translating Official Documents (Part II), Website Localization for Translators, How to be a Successful Freelance Translator, The Business Side to Becoming a Successful Translator, Marketing for Translators, Ethical Codes for Translators and Interpreters, Financial Translation (Introduction to Financial Translation), Automated Translations with R and Google Translate API, Launching your Freelance Translating Business, Introduction to Translation Business Management, The Business Side to Becoming a Successful Translator, A Practical Introduction to Statistical Machine Translation, Introduction to Translation Project Management, b) Coursera: Translation in Practice, Principles and Practice of Computer-Aided Translation; c) FutureLearn: Working with Translation: Theory and Practice.

4.2   MOOCs for Professional Development

MOOCs can be utilized in the area of a translator professional development and provide resources for the 21st century professionals, they are free, delivered to the masses and accessible anywhere around the world [16]. Professional development can allow educators to teach an innovative curriculum that will promote engagement and progress among learners [17]. MOOCs are an emerging technology that has caught the attention of corporations, administrators, educators and learners alike and has compelled them to look at learning, teaching, and education through a different lens [18]. Furthermore, online learning environments such as MOOCs can give working professionals the opportunity to get professional development when it is convenient for them [19].

MOOCs can combine the most effective aspects of online learning and personal learning networks within learning communities [20], encourage cooperation among participants and also promote social constructivism and connectivism [21]. Similar to the experience students gain while taking a MOOC, educators who take a MOOC can control the content learners consume and work at their own pace [22]. Because feedback is not obtained instantaneously from the instructors teaching the MOOC, learners often look to get guidance and support from their peers [23]. Additionally, MOOC platforms like Udacity, Coursera, and edX provide online forums that allow students to work collaboratively, provide professional insight, share new ideas, and construct common knowledge [23]. One of the benefits from MOOCs is the ability for professionals to join community conversations about topics that interest them. MOOCs can provide professionals an abundance of resources on a topic of interest [24] and can create a large community of teaching professionals [22]. Moreover, MOOCs can allow professionals to learn from other participants and experts in the field and adopt new ideas from those teaching the MOOC [25]. Because different professionals from all over the world can sign on to take a MOOC, they can have discussions that are directly related to the online course or are generated from the students taking part in the course [20]. MOOCs promote professional development to be done in communities of practice while connecting professionals with other like-minded people like themselves [26]. Educators that receive professional development are motivated to help their students improve their outcomes [27].

4.3   Herzen University Experience in Developing a MOOC for Professional Translators

In 2017–2018, two MOOCs on translation of official documents "Translating Official Documents (Part I)" [28] and "Translating Official Documents (Part II)" [29] were offered by professors from Herzen university on Udemy platform (the language pair: Russian-English). As shown earlier in Table 1, these courses are the first courses designed at major international platform with the language pair Russian - English. The courses can be utilized as stand-alone courses or integrated into the curriculum of Russian HEIs within the subject "Official Documents Translation" in a blended format in the future.

The courses are targeted at beginning and professional translators, teaching the translation of official documents. Translators can develop the skills and techniques necessary to carry out an official translation. The authors of the courses tried to systematize knowledge in translating official documents based on authoritative sources, opinions of authoritative colleagues, lawyers and notaries. In the online course "Translating Official Documents (Part I)" students get acquainted with various features of personal documents (a passport, birth certificate, marriage certificate, divorce certificate, driving license, etc.). In the online course "Translation of official documents (Part II)" students get acquainted with various features of document legalization, apostille, translation of proper names and company abbreviations, peculiarities of translating charters, contracts, powers of attorney and other official documents. Authentic texts presented in courses are most often found in the practice of notarized translation of official documents.

Over 243 translators registered for the courses (the age range of 21 to 45). The courses are presented entirely online. Translators either beginning or professional have to enrol for the courses and follow all the online materials and complete the online course assignments. Course students utilized interactive materials like video lectures, presentations and quizzes provided in MOOCs. Students viewed the MOOCs videos a total of 1,113 times, with a median of 20 views per student. Majority of videos were viewed in entirety by a third of participants and more than half of length of the videos - by three-quarters of participants. The findings show that students were actively engaged in online discussion forums. Students expressed satisfaction as they received direct and immediate feedback from online instructors. The course authors also observed that 62% of students used the discussion forums for more than one session, while 25% of the students visit the forums just for one session, and the rest of students did not visit the forums at all. The course authors identified that 25% of students completed the entire MOOC, 65% of students completed half of MOOCs materials and necessary assignments. The remaining students met the basic requirement of 10% participation in the MOOC. Translators learning outcomes, and namely students' pass rates, scores on common tests, and grades were measured by the authors of the course. The overall course pass rate was 33.3%, and there were slight variations across the courses.

The course authors conducted a survey. The impact of MOOCs was almost equal or slightly better than in class settings. The conducted survey showed translators were highly motivated to improve their professional skills in this field. Translators improved their skills in translation of official documents. The motivation and perseverance of students played a significant role in completing the courses. The survey showed that the overall satisfaction of the online courses was positive. There is a significant prospect of using MOOCs and interactive online technologies in the continuous education of translators.

5 Discussion

MOOCs can serve as a unique professional development opportunity for translators, promote professional growth while providing new knowledge and skills. MOOCs can give translators ample opportunities to learn new concepts and enhance their professional skills in various subject areas. Translators increase their professional network by following courses that can include participants from all over the world. MOOCs can provide an extended network in which translators can create a peer community for learning with like-minded peers in similar situations, that can further increase the learning effects of a MOOC. The future of MOOCs for translator development is unsure and debatable. MOOCs with a strong collaborative focus help translators to develop professional skills. MOOCs offer individual work and self-paced models that can be used as a resource for translators' autonomous learning. Long-term effects and sustainability of MOOCs are debatable, but a consensus exists that such courses are here to stay, regardless if they are disruptive or merely transformative. The utilization of MOOCs for professional development of translators is relatively novel and unstudied. The combination of MOOCs and a translator's development seems to offer an obvious no-lose situation. Translators can receive high quality professional development for free. The combination of MOOCs and professional development of translators warrants more empirical and analytical research in the future in order to better study the potential success and hazards.

6 Conclusion

The analysis leads to the following conclusions: Russia is less confident in developing and implementing MOOCs in Higher Education Institutions in comparison with Europe and the U.S. Most of the open online translation courses presented on the international platforms are English/another language courses. The findings show there are only a few Russian/English MOOCs teaching translating skills on the international platforms. The existing problem of using MOOCs for professional translator development is that many courses may be more general and not be specifically directed towards beginning or professional translators. MOOC providers are creating courses specifically for professional development of translators. However, specific course offerings for translators will need to increase to ensure success. More research should be conducted on the development of Russian/English MOOCs for teaching translating skills, how MOOCs can be used for professional development of beginning and professional translators, on the effectiveness of MOOC's utilization, optimization of student engagement and better learning outcomes, potential of MOOCs usage in HEIs. The conclusion is that MOOCs can be a cost- and resource-effective means to deliver high quality education to further professional training and development of translators.

References

1. Darco, J., Schuwer, R.: Institutional MOOC strategies in Europe (2015). http://eadtu.eu/documents/Publications/OEenM/Institutional_MOOC_strategies_in_Europe.pdf. Accessed 12 Aug 2018
2. Porto Declaration on European MOOCs. http://home.eadtu.eu/images/News/Porto_Declaration_on_European_MOOCs_Final.pdf. Accessed 9 Aug 2018
3. Mihaescu, V., Vasiu, R., Andone, D.: Developing a MOOC: the romanian experience. In: Proceedings of the European Conference on e-Learning, ECEL, pp. 339–346. Academic Conferences and Publishing International Limited, Copenhagen (2014)

4. Pujar, S.M., Bansode, S.Y.: MOOCs and LIS education: a massive opportunity or challenge. Ann. Libr. Inform. Stud. **61**(1), 74–78 (2014)
5. Spector, J.M.: Remarks on MOOCS and Mini-MOOCS. Education Tech. Research Dev. **62** (3), 385–392 (2014)
6. Parkinson, D.: Implications of a new form of online education. Nurs. Times **110**(13), 15–17 (2014)
7. Iqbal, S., Zang, X., Zhu, Y., Chen, Y.Y., Zhao, J.: On the impact of MOOCs on engineering education. In: Proceedings of 2014 IEEE International Conference on MOOCs, Innovation and Technology in Education, IEEE MITE 2014, pp. 187. IEEE, Patiala (2014)
8. Stuchlíková, L., Kosa, A.: Massive Open Online Courses - challenges and solutions in engineering education. In: Proceedings of 11th IEEE International Conference on Emerging e-Learning Technologies and Applications, ICETA 2013, pp. 359–364. ICETA, Stara Lesna (2013)
9. Cormier, D., Siemens, G.: Through the open door: open courses as research, learning, and engagement. Educause **45**(4), 30–39 (2010)
10. Heick, T.: 30 incredible ways technology will change education by 2028. http://www.teachthought.com/trends/30-incredible-waystechnology-will-change-education-by-2028. Accessed 15 Aug 2018
11. Comeau, J.D., Cheng, T.L.: Digital "tsunami" in higher education: democratisation movement towards open and free education. Turk. Online J. Distance Educ. **14**(3), 198–224 (2013)
12. Martín-Monje, E., Barcena, E., Read, T.: La interaccion entre companeros y el feedback lingüístico en los COMA de lenguas extranjeras. Profesorado **18**(1), 167–183 (2014)
13. Chuang, I., Ho, A.: HarvardX and MITx: four years of open online courses. http://docs.wixstatic.com/ugd/6e7a17_bc48ba9438ef4759a22d3ddc30a66adf.pdf. Accessed 19 Aug 2018
14. Jansen, D., Schuwer, R.: Institutional MOOC strategies in Europe. Status report based on a mapping survey conducted in October–December 2014. http://www.eadtu.eu/documents/Publications/OEenM/Institutional_MOOC_strategies_in_Europe.pdf. Accessed 18 Aug 2018
15. Pilli, O., Admiraal, W., Salli, A.: MOOCs: innovation or stagnation? Turk. Online J. Distance Educ. **19**(3), 169–181 (2018)
16. Davis, M.: Summer professional development with MOOCs. https://www.edutopia.org/blog/summer-pd-moocs-matt-davis. Accessed 19 Aug 2018
17. Donaldson, C.: The Khan Academy: changing the face of education. https://www.education.com/magazine/article/khan-academy. Accessed 22 July 2018
18. Fischer, G.: Beyond hype and underestimation: identifying research challenges for the future of MOOCs. Distance Educ. **35**(2), 149–158 (2014)
19. Morris, L.: MOOCs, emerging technologies, and quality. Innov. High. Educ. **38**(5), 251–252 (2013)
20. Clarke, T.: The advance of the MOOCs (Massive Open Online Courses). Education + Training **55**(4/5), 403–413 (2013)
21. Clarà, M., Barberà, E.: Learning online: Massive Open Online Courses (MOOCs), connectivism, and cultural psychology. Distance Educ. **34**(1), 129–136 (2013)
22. Ferdig, R.: What Massive Open Online Courses have to offer K-12 teachers and students. http://media.mivu.org/institute/pdf/mooc_report.pdf. Accessed 1 July 2018
23. Li, N., Verma, H., Skevi, A., Zufferey, G., Blom, J., Dillenbourg, P.: Watching MOOCs together: investigating co-located MOOC study groups. Distance Educ. **35**(2), 217–233 (2014)

24. Aguaded-Gómez, J.I.: The MOOC revolution: a new form of education from the technological paradigm? Comunicar **41**(17), 7–8 (2013)
25. Nkuyubwatsi, B.: Evaluation of Massive Open Online Courses (MOOCs) from the learner's perspective. In: Proceeding of 12th European Conference on e-Learning ECEL-2013, pp. 166–181. Academic Conferences and Publishing International Sophie Antipolis, Paris (2013)
26. Wenger, E.: Communities of Practice: Learning, Meaning, and Identity. Cambridge University Press, Cambridge (1998)
27. Wolf, P.D.: Best practices in the training of faculty to teach online. J. Comput. High. Educ. **17**(2), 47–78 (2006)
28. Vatskovskaya, I.: Translating official documents (Part I). https://www.udemy.com/official_documents_translation/. Accessed 9 Sep 2018
29. Abrosimova, N.: Translating official documents (Part II). https://www.udemy.com/2-cdeiqr/. Accessed 9 Sep 2018

Learning Environment in Russian Universities for Developing Researchers' EAP Writing Skills

Elena A. Melekhina[✉] and Irina A. Kazachikhina

Novosibirsk State Technical University, Novosibirsk 630073,
Russian Federation
{melexina, kazachixina}@corp.nstu.ru

Abstract. Raising awareness of the university staff about the need to strengthen their academic writing skills in English has recently become a challenge to university administrations, EFL teachers and researchers across Russia. National research institutes, federal universities and flagship universities value researcher communication skills in English as an integral part of academic excellence, and initiate research for theory and practice of teaching EAP. The mixed methods study identified learning conditions stimulating the publishing activity of researchers through the educational context of Novosibirsk State Technical University (NSTU). It explored needs of 102 researchers before launching an in-service training course in academic writing in English and 21 course participants afterward. The results showed that to be able to write articles for leading international scientific journals the non-native English researchers require high proficiency in EAP writing, research findings of their foreign colleagues and highly-qualified language support in writing research articles in English. The research also revealed that participants in EAP in-service training courses are fully aware of the need to develop their general speaking skills as a step to EAP, and importance of relevant learning modes and activities for their success in writing articles in English. The research revealed learning conditions necessary to provide if universities are interested in stimulating the publishing activity and mainstreaming internationalization: electronic environment with full access to credible research-related resources, writing centers with a variety of EAP in-service training courses and services, highly-qualified EAP teachers experienced in publishing their research, and research collaboration of NNE and native English researchers.

Keywords: English for academic purposes (EAP) ·
Non-native English (NNE) · Internationalization · Publication activity ·
Learning environment

1 Introduction

1.1 Global Context

Internationalization as an increasing global trend in scholarly communication and higher education continues to challenge non-native English speaking (NNE) researchers to possess academic writing skills in English, conforming to those of a native speaker. That

Z. Anikina (Ed.): GGSSH 2019, AISC 907, pp. 265–275, 2019.
https://doi.org/10.1007/978-3-030-11473-2_29

can be literally a barrier that stands in the way of dealing with so-called 'borderless issues' which is only a dimension of internationalization but an equally important one. Being a medium of written instructions, the English language plays a key role in dealing with 'borderless issues', as Knight has put it, describing the dimension of internationalization this study relates to:

> Issues such as the degradation of the environment, population growth, security, global warming, immigration, terrorism, human rights, and health epidemics are without borders, they require international collaboration and cooperation to find policies and strategies that will mitigate negative effects and lead to positive solutions.
> Multilateral government agencies, international nongovernmental organizations, national governments, the private sector and also the higher education sector all have a role to play at national and international levels in addressing these trends. The role that higher education plays in researching, teaching about, and analyzing these areas needs to be given greater attention and prominence in public policy debates [11, p. 9].

In the early 1980s when internationalization began to transform national education systems and science, obviously, English-speaking countries were prepared to the challenge in terms of research and practice of teaching EAP. 'International education', 'international cooperation', 'foreign students', 'correspondence education', 'language studies', 'English for Academic Purposes', were not just terms but realities and educational and teaching traditions [11]. Only at present Russian educationists value writing centers for teaching academic writing to students and university staff, whereas '[i]n North America writing labs appeared more than one hundred years ago' [17, p. 66]. The British Council 'as part of its global commitment to the internationalization of higher education' has been disseminating the UK best practices in teaching methodology, material development and EFL training programs throughout the world [8, p. 9]. As internationalization processes increase, the number and quality of publications become an important factor defining the efficiency of the university faculty, which is in line with Lee who, describing the situation with publication activity in Hong Kong University, writes that:

> In many parts of the world, universities are scrambling for higher international rankings, and in order to prosper they need to boost their research profiles, which in turn depend on the research output of their academics [12, p. 250].

That is why universities create incentives to encourage staff to develop EAP writing skills to internationally share research achievements, and 'to participate in academic conversations that go beyond geopolitical borders' [2]. Such incentives include establishing writing centers, adopting writing across curriculum approach [20] starting 'writing-about-writing' courses [19], offering subsidized access to subscription databases and digital libraries such as Scopus and Web of Science [21], organizing conferences and seminars on EAP writing methodology [3].

1.2 Russian Context

Summarizing the findings in the field of English language teaching in Russian universities Frumina and West report about 'a long and depressing catalogue of factors standing in the way of internationalization' and among them low English language proficiency of university staff and students, shortcomings in teaching of EAP, absence of adequate teaching and learning materials, assessment systems, and poor English language proficiency of academics [8, p. 53–58].

Over the past three decades Russian universities have undergone a series of reforms and modernization carried out for a variety of reasons, such as globalization of higher education with rapidly changing economy and consequent growth of labor and academic mobility; digitization which has been changing traditional models of education into innovative ones; internationalization with strong competition for higher positions in the world university ranking which puts considerable emphasis on the research results and publication activity of the staff. The changes have been implementing due to a new policy of internationalization developed by the government of Russia.

To meet the growing needs of university researchers for participating in the global academic community Russian universities open Writing Centers offering access to resources, providing training courses and individual consultations on a regular basis. The leaders in the process are Higher School of Economics, National University of Science and Technology 'MISiS', Moscow School of Social and Economic Sciences and Samara State University. The centers have become places for research collaboration and academic self-development in research EAP skills.

1.3 NSTU Context

Being in the mainstream of reforms, NSTU[1] is striving to improve fairly strong positions holding in the world university rankings, such as QS or THE [14]. Last year it won the competition and became the flagship university of Novosibirsk region. The university worked out the program of strategic development which includes quantitative and qualitative criteria of prospective growth. According to the program, the number of the faculty's publications in the 'top' international research journals should increase twofold during the period of five years. This challenging task is being accomplished by the joint efforts of the Scientific Library, the Department of the Foreign Languages for the Humanities, and the faculty themselves.

This paper presents part of the findings of the study that was funded by NSTU within the frame of the program for university strategic development.

In this study, we consider conditions as an integrated part of the informational and educational environment of the university. According to the dictionary, one of the many definitions of the word 'condition' relevant to the study is 'a particular mode of being of a person or a thing; existing state; situation with respect to circumstances' [18, p. 141], while 'environment' is 'the aggregate of surrounding things, conditions, or influences' [18, p. 225]. An informational and educational environment is understood as a 'perspective educational environment, in which information and communication

---

[1] Originally founded as an electrical engineering institute it was reorganized in 1992 in the technical university. As one of the largest universities in Siberia, it offers about 120 bachelor, master, doctoral, and post-doctoral programs in technical, economic, and humanitarian fields. The number of students exceeds 25 thousand. Over 3000 faculty members and employees work at NSTU. [http://en.nstu.ru/about_nstu/info/index.php?sphrase_id=41802].

processes are deployed in both traditional and virtual (electronic) formats, causing qualitative changes in the scientific, educational, social and cultural problems' [10].

Thus, the study addressed the following research questions (RQ):

1. What needs for developing EAP writing skills do NSTU researchers have?
2. What conditions are necessary to stimulate the publication activity of NSTU researchers in leading international scientific journals?

2 Research Design

2.1 Methods

Studies for practical reasons require mixed methods to provide 'a deeper understanding of the change in the instructional practice' [6, p. 6]. Thus, we applied quantitative and qualitative methods to the research design [7].

The research was carried out in two stages. During the first stage, a questionnaire (Q1) was developed and circulated among NSTU researchers, teachers, and post-graduate students. Based on the results of the data analysis, an in-service training program 'Academic Writing in English: Fundamentals of Theory and Practice' was developed. During the second stage, in 3-month completion of the in-service training course the trainees' responses to the questionnaire (Q2) were collected.

2.2 Participants

During the first stage of the study a number of 102 NSTU teachers and students were involved in exploring their needs to increase their academic writing skills for publishing in English: 29 (28.4%) undergraduate students, 13 (12.7%) master program students, 15 (14.7%) postgraduate program students and 45 (44.1%) university researchers.

During the next stage, the study involved 21 NSTU researchers who attended an in-service training program for developing EFL academic writing skills for producing a manuscript to be published in English. The trainees met the following requirements: all of them were involved in research; they obtained results to be published (literature review also included; the levels of English language proficiency of the trainees ranged from B2 (B2 + , B2 ++) to C1. Research interests of the trainees differed: 18 (85.7%) were in different areas of engineering and 3 (14.3%) in social sciences. Participants ranged in age from mid-20s to late 60s. Clearly, they had different academic degrees, titles, and positions at the university. There were 4 professors and doctors of sciences[2] (19%), 10 associate professors and candidates of sciences[3] (47.6%), and 6 trainees (28.6%) were postgraduate program students and 1 (4,8%) was a master program student.

---

[2] A post-doctoral degree is given to reflect second advanced research qualifications or higher doctorates according to the International Standard Classification of Education (ISCED) 2011.

[3] A doctoral degree or equivalent according to ISCED 2011.

The group was heterogeneous in many aspects, though the participants were not selected by the researchers. They were free to apply for the in-service training program they were professionally interested in, and if all the program requirements were met, they became trainees on the program. During the first exam session, the trainees were puzzled by the group heterogeneity. It was evident that they differed in ages, status, science fields, research expertise, publishing experience, etc. Eventually, they realized that they intended to achieve the same learning goal which was to improve skills for writing articles in English. They also realized that their different features, such as research interests, learning styles, professional skills and many others, could serve to their mutual cooperation and success through the course.

2.3   Data Collection Instruments

Questionnaire 1 used during the first stage of the research was designed and administered in Russian to collect data quickly and without language difficulties for the participants. The survey instrument comprised 10 items based on closed (multiple choice (3), rating (1), yes/no (4)) and open (free-form (2)) types. The survey instrument provided a logical flow of questions focused on the aim of the survey. We examined participants' publishing activity in English and EAP learning experience. They were asked about difficulties they encountered producing manuscripts in English, and ways to avoid them. The data obtained helped to identify needs of NSTU students and teachers for developing EAP writing skills.

Questionnaire 2 used on completion of the in-service training course was designed in electronic format and administered in Russian to make the participants feel free in expressing their opinions. The survey guaranteed anonymous responses. The survey instrument comprised three sections of 10 items. Data obtained through the survey instrument were expected to be evaluative and reflexive in nature. We addressed to the participants' goals and expectations from the course, self-evaluation of their performance through the course, and feedback on it.

3   Results

3.1   Analysis of the Data (Q1)

As the results show almost all researchers and research-oriented students would like to publish their research in international journals (68.8%) but not practice-oriented students (20.3%), and there were some university teachers and students who did not know whether they need it or not (11%).

Unfortunately, students are likely not to be taught how to write research papers in English if they study at the technical university. The research shows that only 5 out of 100 students happen to be lucky, whereas every second student studies business correspondence and informal letter writing, and every fifth one studies essay writing in English.

Nevertheless, almost half of the university staff takes part in local conferences with English as a language of instructions held by the departments of foreign languages at universities. But only 8% of the researchers experienced preparing and submitting manuscripts to conference proceedings in Scopus and Web of Science. Moreover, no respondents published in high-rank international academic journals.

Those who are interested in presenting their research to the international academic community are aware of the difficulties they have to struggle. The difficulties can be categorized as follows:

- Poor academic writing skills in English,
- Inappropriate use of subject-specific terms,
- Differences between Russian and English academic writing traditions,
- Inappropriate content and structure required for article sections,
- Difficulties in applying different referencing systems,
- Lack of access to relevant credible full-text articles in English.

The NSTU researchers believe there can be provided some favorable conditions that can help overcome the above difficulties and, therefore, improve the current situation with low publication activity of the researchers in general. The data obtained relate to the conditions below:

- *'To develop an ESP course and make it a part of bachelor and master programs',*
- *'To launch an in-service training course in English for publications',*
- *'To create an open-source database covering issues in academic writing in English',*
- *'To train tutors who would direct the process of writing articles in English',*
- *'To provide opportunities to the university staff to attend courses in English Speaking',*
- *'To organize some meetings or activities to communicate informally with foreign colleagues'.*

3.2   Analysis of the Data (Q2)

At the end of the in-service training course, 63.4% of the university researchers achieved their goals related to the aim of the in-service program partially, whereas 34.4% achieved them completely. The goals described by the trainees can be categorized as short- and long-term. Most answers (63.3%) given to the open-ended question contain both types of goals (e.g. *'develop my written literacy skills in English, and correct the draft of my research manuscript'*), about one third of the goals (27.3%) are short-term (e.g. *'finish my manuscript and edit it according to the requirements of IEEE journal'*). Clearly, all of them relate to the aims of the course which are 'recognize cultural norms, rhetorical patterns, and common language templates used in the English language academic writing and begin applying them in writing' and 'develop a final draft of the research paper based on original research'[4], but focus on different aspects of academic writing. The requirements of journals of the trainees' particular

---

[4] E-course of the in-service training program 'Academic Writing in English: Fundamentals of Theory and Practice'. Novosibirsk State Technical University 1994–2018. Institute of Distance Learning. Homepage, https://dispace.edu.nstu.ru/didesk/course/show/7370, last accessed 2018/08/02.

interests and the language component of the articles already published there rank the highest (54.5%), language standards used in English language academic writing for research rank the second high (45.4%), and then comes a typical structure of research articles (27.3%). Interestingly, *'improving my English spoken skills'*, *'use English speaking with my colleagues'*) is one of the goals on the course in written English (18%).

On the course completion, there were no trainees who were not satisfied with what they achieved during the course, and those who were absolutely happy. Nevertheless, 36.4% of trainees rated their results as high ones, and an equal percentage of trainees as very good ones, though 27.3% feel far from satisfied yet.

When the trainees were asked to encourage and motivate themselves by praising themselves, only one person refused to do it, others (9%) were proud only for what they achieved through the course, when the majority (95%) for what they achieved and intended to do in the future. The quotes below give an impressionistic description of researchers' attitudes, beliefs and intentions:

'...I'll finish this article.'
'...I'll try to communicate in English with my students and my foreign colleagues.'
'I managed to meet the deadlines and complete the final assignment. Now I have a version of the article that can be submitted (with some improvements).'
'....I am going 'to polish' the final draft of my article and submit it to a journal.'
'I have restructured my article according to the requirements of the journal I am interested in...'
'Journals of Web of Science and Scopus level... now I feel I'm ready to try.'

According to the data, there is unanimous agreement among the trainees that they succeeded in producing final drafts of the research papers in English thanks to 5 external factors ranking as below:

1. An electronic database tailored for the course (e.g. *'an e-course'*, *'e-readings available in the e-course'*, *links to additional open-access resources'*, *'links to full-text articles of high-rank journals'*, *e-materials on academic writing'*, *'summaries of classroom discussions uploaded to the e-course afterwards'*, *'keeping in touch with trainers by e-mail'*).
2. EAP with focus on writing research articles (e.g. *'theory of academic writing'*, *'grammar peculiarities'*, *'examples of 'good' articles'*, *'linguistic features of writing articles'*, *'doing language exercises'*).
3. Trainers managing the process of writing an article (e.g. *'highly-qualified trainers*, *'trainers with experience in writing articles in English'*, *'encouraging'*, *'enthusiastic'*, *'ready to help'*, *'responsive'*, *'giving feedback'*).
4. A variety of learning modes and activities *(e.g. '...a detailed analysis of the language and structure of the articles guided by the mentors. Without mentors, I couldn't do it on my own.', '... experienced self- and peer-reviewing', 'individual work first then with colleagues together', 'comparison of articles published in the same journal', 'discussions of the language features of native and non-native English writers',).*

5. Scientific collaboration (e.g. *'communication with other course participants'*, *'team spirit'*, *'lively dialogue'*, *'partnership'*, *'spirit of collaboration'*, *'feeling that you are a part of the global family'*).

Table 1. Needs of the in-service training course valued by the trainees.

Features of the in-service training course valued by the trainees	Needs implied	Percentage of participants
'useful', *'relevant'*	Need for the course in EAP writing for publications	95.2%
'interesting', *'promoting'*, *'developing'*, *'making think'*, *making find ways out'*, *'active'*, *'improving'*, *'making it real'*	Need for stimulating writing in English	76.2%
'task-oriented', *'product-oriented'*, *'practice-oriented'*, *'rich in theory'*, *'comprehensive'*, *'well-organized'*	Need for relevant content and proper organization of the course	57.1%
'warm', *'harmonious'*, *'friendly*, *'attentive'*	Need for respect and support	47.6%
'time-consuming'	Need for allocating extra time	28.6%

The following Table presents data on features of the in-service training course which were valued by the trainees (Table 1). We categorized them in terms of needs.

The trainees (85.7%) named time factor as the most negative one in achieving their goals through the course (e.g. *'lack of time for redrafting'*, *'difficult to work and study simultaneously'*, *'time not convenient for attending the classes'*, *'all tasks need much more time to do them properly'*), and 23.8% of the trainees named difficulties with speaking and listening comprehension as the reason made them feel not comfortable (*'my poor English speaking skills'*, *'I'm afraid I speak too slow'*, *'too fluent speech to understand'*).

4   Conclusion

The situation with developing EAP writing skills is improving in Russia. The positive changes in higher education have started with a series of resolutions about strategic innovational policy passed by the Government of the Russian Federation [9]. The concrete measures have been taken at the government level, and now universities are seeking their own ways to implement them. 'Russia is a huge country with a huge university problem, and the problems of teaching English in these universities are also huge' [8, p. 58]. The necessity of implementing innovations has resulted in initiating research projects addressing the issues. As the educational context of universities across Russia differs greatly, demand for identifying needs of the researchers who are literally responsible for innovations is constantly increasing.

Our research explored and identified the researchers' needs for developing EAP writing skills. Achievement of the level to produce articles for leading international scientific journals NNES requires a high EAP proficiency from the researchers, research findings of foreign colleagues relevant to the fields of Russian researchers, and highly-qualified language support on the way of writing research articles in English. In addition, the research has revealed the importance of learning modes and activities that make researchers succeed through the process of writing articles. Finally, the researchers are fully aware of the need to improve their general speaking skills (speaking in particular) as a step to developing EAP writing.

Obtained within the frame of the program of university strategic development, the findings call the attention of administrators, educators, English language teachers and researchers to the existence of potential possibilities available in modern universities for creating conditions which can meet the researchers' needs. This research revealed that the below conditions would allow Russian universities to mainstream internationalization:

1. The universities are able to allocate or integrate some e-space for academic literacy in English required to their researchers. It aims at providing researchers with full access to credible research-related resources. This can be taken as a starting point for developing electronic environment afterward to meet the strategic needs and financial potential of universities.
2. Clearly, without organizing writing centers (departments of EAP, schools for writers and other) which provide a variety of EAP in-service training courses and other services learning universities will definitely impede their growth, taking into consideration the challenges NNE researchers have with publishing in English.
3. Teaching staff of the writing centers are expected to be of high EAP expertise, experienced in publishing internationally, proofreading, editing, and translation. Writing and Communication Center (WCC) in New Economic School [1] provides the best practice of a model for university students and academics writers.
4. A research collaboration of NNE and native English researchers is required to model 'natural' English speaking environment. The researchers are fully aware that common activities for achieving shared goals can stimulate for speaking and writing in English. Using language as a means of real communication outside the classroom contributes to language learner's success most. The international collaboration can result in sharing scientific knowledge and contributing economic prosperity of the countries which is of no less importance [15, 16].

These conditions are feasible in any university, but they can be provided in different ways. However, to implement such long-term and university-scale innovations is more difficult than identify and present. Fortunately, recent steps are taken by some universities. To support EAP professional and newly organized writing centers across Russia the National Writing Centers Consortium (NRCC) was founded in 2016 [4, 13]. WRCC has started to disseminate best practices in teaching EAP. All activities of WRCC are aimed at introducing academic writing in Russian education, and, therefore, 'engaging all stakeholders in the internationalization of Russian science and education' [5, p. 35].

The implications of the study can be important for the Russian and NNE educators, and the results can be used as a rationale for creating favorable conditions of the electronic and learning environment to stimulate publishing activity of the university faculty.

Acknowledgements. This study is funded by Novosibirsk State Technical University (Research #3.2.3.2, Block #3.2.3. Maintaining NSTU Science Reputation, Strategic Development Program of NSTU as a regional flagship university, 2017–2021).

References

1. Academic Writing Centre. National Research University Higher School of Economics. HSE 1993–2018. https://academics.hse.ru/en/awc/. Accessed 12 Aug 2018
2. Academic Writing for Publication. Workshop materials for trainers working with university (2016). https://id.usembassy.gov/education-culture/regional-english-language-office/. Accessed 12 Aug 2018
3. Academic Writing Talks and Events. http://eapconference.misis.ru/2017-2/academic-writing-talks-and-events-2/. Accessed 12 Aug 2018
4. Bazanova, E.M., Korotkina, I.B.: Rossiyskiy konsortsium tsentrov pisma [Russian Writing Centers Consortium]. Vysshee obrazovanie v Rossii [Higher Education in Russia] 4(211), 50–57 (2017). (in Russian)
5. Bazanova, E., Starostenkov, N.: Nacional'nyj konsorcium centrov pis'ma: panaceya ili placebo dlya poteryannogo pokoleniya rossijskih uchyonyh? [National Writing Centres Consortium: panacea or placebo for the lost generation of Russian researches?]. Universitetskaya kniga [University Book], vol. 3, pp. 27–35 (2018). (in Russian)
6. Berg, M.A., Huang, J.: Functional Linguist 2(5), 2–21 (2015)
7. Creswell, J.W.: Research Design: Qualitative, Quantitative, and Mixed Methods Approaches. Sage Publishing, Thousand Oaks, CA (2014)
8. Frumina, E., West, R.: Internalisation of Russian Higher Education. British Council, Moscow (2012)
9. Furin, A.G.: K voprosu o gosudarstvennoy podderzhke innovatsionnoy deyatel'nosti v sisteme obrazovaniya [To the issue of state support for innovations in education]. Internet-zhurnal Naukovedenie [Science Studies Internet-journal] 7(3/28), 1–14 (2015). (in Russian)
10. High-Tech Electronic Educational Environment. International Scientific-Practical Conference. Herzen University. Homepage. http://reading.herzen.spb.ru/hitech/. Last accessed 16 Aug 2018
11. Knight, J.: Higher Education in Turmoil: The Changing World of Internationalization. Sense Publishers, Amsterdam (2008)
12. Lee, I.: Publish or perish: the myth and reality of academic publishing. Lang. Teach. **47**, 250–261 (2014)
13. National Writing Centers Consortium Homepage. https://nwcc-consortium.ru/en/about-us/. Accessed 18 Aug 2018
14. NSTU in University Rankings. Novosibirsk State Technical University Homepage. http://en.nstu.ru/about_nstu/rankings/. Last accessed 12 Aug 2018
15. Salmi, J.: Sozdanie universitetov mirovogo klassa [Establishing World-Class Universities] Koroleva, T.M. (tr.). Ves mir, Moscow (2009). (in Russian)

16. Sonnenwald, D.H.: Scientific collaboration: a synthesis of challenges and strategies. In: Cronin, B. (ed.) Annual Review of Information Science and Technology, vol. 41, pp. 643–681. Information Today, Inc., Medford, NJ (2007)

17. Squires, L.A.: The NES writing and communication center: the case for student-oriented writing centers in Russia. Vysshee obrazovanie v Rossii [Higher Education in Russia] **8**(9), 66–73 (2016)

18. Meine, F.J. (ed.): The New Webster's Encyclopedic Dictionary of the English Language. Gramercy Books, New York (1997)

19. Wardle, E., Downs, D.: Reflecting Back and Looking Forward: Revisiting Teaching about Writing, Righting Misconceptions Five Years On. Compos. Forum **27**. Spring (2013). http://compositionforum.com/issue/27/reflecting-back.php. Accessed 18 May 2018

20. Werner, C.: Constructing Student Learning through Faculty Development: Writing Experts, Writing Centers, and Faculty Resources. The CEA Forum, Summer/Fall, 79–92 (2013)

21. Yessirkepov, M., Nurmashev, B., Anartayeva, M.: Scopus-based analysis of publication in kazakhstan. J. Korean Med. Sci. **30**(12), 1915–1919 (2015)

Preschool Teacher's Training in Professional Self-analysis: The Russian Arctic Region Experience

Maria Druzhinina[1]([✉]) [iD], Olga Morozova[2] [iD], Elena Donchenko[2] [iD],
and Olga Istomina[3] [iD]

[1] Northern (Arctic) Federal University named after M.V. Lomonosov, 163000 Arkhangelsk, Russian Federation
m.druzhinina@narfu.ru

[2] Northern (Arctic) Federal University named after M.V. Lomonosov, Arkhangelsk Pedagogical College, 163000 Arkhangelsk, Russian Federation
{olgamorozz2016,elena.donchenkoea}@yandex.ru

[3] Northern (Arctic) Federal University named after M.V. Lomonosov, Arkhangelsk Regional Institute of Open Education, 163000 Arkhangelsk, Russian Federation
olgaist03ll@yandex.ru

Abstract. The article is devoted to the issues of teacher training in professional self-analysis. Particular attention is paid to the investigations done by international research groups and Russian scholars in the areas of normative documents and educational practices.

The authors present a system for training specialists in the field of Arkhangelsk region preschool education. It was shown that teachers' professional self-analysis skills contribute to the development of their competence and the improvement of educational quality.

The article covers the interim results of a network project implemented by professional educational organizations of the Russian Arkhangelsk region. The primary study of professional educational organizations experience and the exchange of experience at different levels (higher, secondary, additional) has shown that modern education system demands effective methods of preschool teachers training in order to form the capacity for professional self-analysis. Training of educators for self-analysis is a complex process which includes theoretical knowledge and practical application. The authors have substantiated the role and necessity of pedagogical practice in the process of preschool teachers' education.

The paper proposes the use of recommendations for students. The recommendations are developed in the form of self-analysis cards applicable for self-analysis training. In the course of the pedagogical research an approbation of the methodical recommendations has been carried out. The approbation has resulted in a set of data obtained from the survey of the students who had used the proposed recommendations for self-analysis of their own professional activities. The data has proved that the recommendations are effective.

Keywords: Professional education · Preschool teacher's training ·
Professional pedagogical activity · Self-analysis of professional activity

© Springer Nature Switzerland AG 2019
Z. Anikina (Ed.): GGSSH 2019, AISC 907, pp. 276–291, 2019.
https://doi.org/10.1007/978-3-030-11473-2_30

1 Introduction

At present, the problem of providing kindergartens with professional teachers, who are capable of effectively solving educational problems, is urgent. The complexity of solving this problem is due to high requirements to the quality preschool education in Russia. In the conditions of reforming the state policy and updating the normative base in the field of preschool education, searching the ways of solving staffing and teacher's professional competence problems becomes even more urgent. Accordingly, the role of teacher training in professional activities is growing.

Self-analysis of professional activity is one of the directions for the preschool teachers training. This kind of analysis is important for future teachers' personality formation, it provides an opportunity for professional self-realization, promotes effective professional activities.

The purpose of this article is to present the Russian Arctic region experience in training preschool teachers for self-analysis of professional activities.

The article analyzes the experience of educational organizations (The Northern Arctic Federal University named after M.V. Lomonosov, Arkhangelsk Pedagogical College, Arkhangelsk Region Institute of Open Education) of the Arkhangelsk region which belongs to the Arctic zone. The result of the network interaction of educational organizations is the project work on the topic "Training of specialists in the vocational education system in the North of Russia". Networking of professional educational organizations in the region is not developed enough, it is a promising direction of scientific and pedagogical activity, and it requires further development.

It should be noted that the Arkhangelsk region is one of the largest administrative entities in Russia and occupies 40% of the European North. Arkhangelsk region is one of the leaders of development in the Arctic zone of Russia. The region consists of Nenets Autonomous Okrug, 7 city districts, 19 municipal districts, 24 urban settlements, 179 rural settlements, and the Islands of Novaya Zemlya and Franz-Joseph. Arkhangelsk is the city of the World Arctic with a population of 350 thousand people. It is a developing center of industry, education and science.

At the moment, there are more than 500 preschool organizations of different types in the Arkhangelsk region, including branches and structural units of schools and private kindergartens.

Training of teachers to work in preschool organizations is implemented by vocational education institutions. Higher education of educators in the region is carried out by the Northern Arctic Federal University named after M.V. Lomonosov (NArFU). Secondary vocational education in the specialty of preschool education in our region is presented in four pedagogical colleges: the Arkhangelsk pedagogical College, the Archangelsk industrial-pedagogical College, the Kotlas pedagogical College, the pedagogical College of Kargopol.

Another option to update professional skills in the field of preschool education is retraining. Retraining is organized for employees of kindergartens who do not have professional education and for those who want to get another specialty, in particular, the one of a kindergarten teacher. Professional retraining programs in the Arkhangelsk region are implemented by the Arkhangelsk regional institute of open education and pedagogical colleges.

As it has already been mentioned earlier, the materials of the article are part of a large project launched in January 2018. The primary study of the different-level professional educational organizations (higher, secondary, additional) experience, and the exchange of work experience have shown that in education the progressive methods of educating preschool teachers, which aim at forming the capacity for self-analysis of professional activity, are in demand. The formation of the ability to carry out self-analysis of professional activity is aimed at improving the quality of the training educators and improving their skills. This provision is also relevant because at the moment there is a shortage of qualified personnel in preschool education in the Arctic region.

Ability to self-analysis is formulated as a requirement for educators in the professional standard "Teacher (pedagogical activity in the sphere of preschool, primary general, basic general, secondary general education) (educator, teacher)". In the scientific literature, however, the problem of training teachers for self-analysis of professional activities has not been adequately studied. As a result, there is a contradiction between the theory and practice of educational activity.

The article attempts to solve the problem of forming the capacity of educators for self-analysis, taking into account the experience of the Russian Arctic region educational organizations.

In accordance with the stated problem, the following problems are solved in the article:

1. To reveal the essence of the training preschool teachers problem for self-analysis of professional activity;
2. To describe the specifics of the preschool teacher's training in the Russian Arctic region;
3. To present experience on approbation of methodological recommendations for self-analysis of professional activity during the students' educational process.

2   Literature Review

The works of psychologists, educators, sociologists are devoted to the problem of studying professionalism and preparing a person for the realization of professional activity. The training of an individual for professional activity was considered in the works of Abdullina [1], Babansky [2], Domović [3], Zair-Beck [4], Kuzmina [5], Lomakina [6], Pisareva [7], Piskunova [8], Radionova [9], Slastyonin [10], Tryapitsyna [9] and others.

The analysis of these work allows to say that there is no single definition of the concept of "training". The concept of "vocational training" has a multicomponent character, examined from different points of view.

Drobotenko notes that in the psychological and pedagogical literature of Russia the following understanding of vocational training has developed:

- a system of vocational training, the purpose of which is for students to acquire necessary skills to perform a particular job (Bim-Bad);
- the process of transmitting knowledge and skills to learners and the corresponding result as a set of special knowledge, skills, competences, qualities, labor experience that ensure the possibility of successful work in a particular profession (Yurlovskaya);
- the result of training in an educational institution of the vocational education system [11].

In our opinion the most comprehensive definition is Tutolmin's point of view that "vocational training is a system of organizational and pedagogical activities that ensure the formation of a professional orientation, professional knowledge, skills, competences and professional readiness" [12].

Professional pedagogical activity makes certain demands on a person. To determine what personal and professional qualities that prove the effectiveness of professional activity, a preschool education teacher should have, it is necessary to analyze this activity itself. Effective method in this case is self-analysis. Self-analysis of professional pedagogical activity makes it possible to develop a system of organizational measures aimed at improving it.

The results of the problem in the self-assessment formation among teachers - scientists in Serbia and Croatia reflect the specifics of the problem we are considering. Ilic studied the self-assessment of Serbian teachers. In her work she presents the following definition: "Broadly, self-evaluation is a process whereby teachers collect data on their own learning effectiveness and analysis" [13, p. 79]. As a result, there are three types of self-assessment: "self-evaluation in action aimed at understanding mistakes and correcting them in the future; self-assessment in inaction when there is time to stop and think, and then continue activities and self-evaluation in the course of practical activities that are not realized by the teacher but change the activity itself. The active use of self-analysis promotes the development the teachers' practical skills" [13, p. 79].

The scientists of Croatia Drvodelić, Domović give this definition to the notion of self-esteem. "The self-evaluation process can be defined as a process initiated and conducted by educational institution staff in order to describe and evaluate its own functioning" [3, p. 48]. As a result of the research, they found out that preschool teachers who attended the self-evaluation course during their graduate study were themselves more empowered for the process than the other two groups of respondents. They express the greatest willingness to participate actively in the quality assurance teams [3, p. 47].

The results showed that the self-assessment of a teacher's knowledge, skills and abilities in professional activities depends on the time spent on teaching self-analysis. «In other words, those preschool teachers who had training during their graduate study, which means that it was carried out in a more systematic way and that it lasted longer, feel more empowered than those who completed a form of short-term professional development or have not had any kind of training» [3, p. 57].

Scientific research in Russia on the problem self-analysis of pedagogical activity is mainly concerned with the activities of school teachers (Shelekhova [14], Perenkova

[15], Kunakovskaya [16], Metaeva [17]), teachers of vocational education (Kudinov [18], Alenchenkova [18]), but practically do not concern the activity of preschool education teachers.

Shustova considers that readiness for self-analysis goes the way of development from the "theoretical knowledge to their practical application in real life situations conversion" [19, p. 75]. Readiness for self-analysis includes three components:

– the information (the complex of knowledge about the environment and the ways of self-realization);
– the motivational - value (values, motives of activity);
– and the operational (the mastery of self-analysis methods) [19].

In the course of the study, Perenkova identified the levels of development of a teacher's ability to analyze his own activities:

– analysis of activities based on emotions;
– attention to the result and highlighting the pedagogical actions that are significant for the result;
– critical analysis of their own pedagogical activity [15].

Kudinov, Alenchenkova in their research allocated a group of skills that ensure the success of self-analysis of one's own activity. This group of skills includes:

– to analyze their own pedagogical activities in the process of interaction with subjects of the educational process;
– to identify problems and find effective ways to solve them, taking into account their own experience and knowledge, the educational environment;
– to apply different competencies for interaction with social partners;
– to determine the educational trajectory for the trainee, taking into account individual characteristics and opportunities [18].

Berezhnova considers that self-analysis is a necessary condition for the development of professionally significant qualities. She thinks that "professional self-analysis - it is an individually developed algorithm (method) for the design and organization of the formation of professionalism" [20, p. 3].

Thus, professional self-analysis is important for the personality of the future teacher, because it gives him the opportunity of professional self-realization.

In the last decade scientific research projects in the field of studying competencies of educators have been implemented in the world pedagogical science. Let us consider how the problem of the formation of skills to self-analysis of professional activity among educators is presented in the results of such researches.

In the period 2010–2012 specialists from the research group of the University of East London (UK) and the University of Gent in Belgium implemented the research project CoRe (Competence Requirements in Early Childhood Education and Care, European Commission, Directorate-General for Education and Culture). The aim of the study was to systematize at the European level the competencies of specialists working with preschool children. The study was conducted on behalf of the Europe-an Commission (Directorate General of Education and Culture). As a result of close cooperation with the international associations Diversity in Early Childhood Educa-tion and Training (DECET), the International Step by Step Association (ISSA) and Children in Europe (CiE), a detailed description of the competencies of teachers working with children in the EU countries, their comparative analysis, recommendations are given [21].

It is necessary to pay attention to the results of expertise of specialists in the field of education and care in preschool age in Germany. The study was carried out in 2014 under the auspices of the Federal Ministry for Family, Senior Citizens, Women and Youth Affairs of Germany (Bundesministerium für Familie, Senioren, Frauen und Jugend, BMFSFJ). Specialists of Evangelical High School Freiburg presented a review of the necessary competencies of preschool education specialists in accordance with public and professional-political requirements [22].

A study by German scientists (Evangelical High School of Freiburg) provided an overview of the necessary competencies of preschool education specialists [22]. When describing the competence profile of a teacher working with children of preschool age, the authors (Klaus Fröhlich-Gildhoff, Dörte Weltzien, Nicole Kirstein, Stefanie Pietsch, Katharina Rauh) highlighted the competence of the analysis of their pedagogical roles, attitudes, behavior, professional experience, perspectives, etc. [22].

So, we found that self-analysis is a leading component of the professional competence of the teacher.

Let us turn to the content of normative documents of Russia in the field of professional education. The content of the professional competence of a teacher in Russian Federation determines the professional standard "Teacher (pedagogical activity in the field of preschool, primary general, basic general, secondary general education) (educator, teacher)", approved by Order of the Ministry of Labor of Russia No. 544n of October 18, 2013 (hereinafter the professional standard teacher) [23]. In the professional standard for teachers, the pedagogical activity analysis as a necessary competence of the preschool teacher is allocated that further will be considered as the direction of preparation of the preschool teacher.

The results of professional training in Russia (teacher training colleges, institutes, universities) suggest the formation of self-analysis competencies of future specialists. So, for example, for students of pedagogical colleges in the specialty "Preschool education" the following competencies are allocated:

– to organize their own activities, determine methods for solving professional problems, evaluate their effectiveness and quality;
– to analyze the process and results of the organization of various activities and communication;
– to analyze the lessons;
– to evaluate and analyze the results of work with parents, to correct the process of interaction with them [24].

Thus, the analysis of the content of standards of professional education in Russia, suggests that the category of "self-analysis of professional activity" is presented in the list of allocated competencies of teachers of preschool education.

Preparation of the future teacher implies the availability of theoretical training which is accompanied by practical activities directly in educational organizations.

Theoretical training involves acquaintance with the pedagogical experience of many generations, comprehension of its expediency for solving problems of the child personality development.

Pedagogical practice plays a big role in the preparation of future teachers. According to Novikov and Novikov, being expedient activity, practice acts as the complete system of operations and reveals the essence in the following moments: the purpose, the expedient activity, a subject, means, result of practical activity [25].

Kodzhaspirova notes the huge role of production practice for the formation of a teacher as a professional. During the production practice, the future specialist analyzes the teachers and students' professional activities, carries out self-analysis of "their pedagogical attempts".

Initially, the student makes analysis and self-analysis of professional activities guided by the model and only then in the process of practice there is a "standard of analysis and self-analysis " that the teacher will be guided in his future activity [26, p. 52]. Thus, professional self-analysis is a means of self-education, therefore it should become one of the conditions for the qualitative preparation of the future teacher [26].

Averkieva, Shchekina confirm the importance of production practice and reveal its reflexive potential, noting that professional experience is formed only in the process of reflecting the professional difficulties encountered in pedagogical activity [27].

Thus, production practice that is a favorable environment for the formation self-analysis skills of the professional activities of future educators.

3 Methodology

We have analyzed more than 30 sources in Russian, English and German in accordance with the subject, purpose and objectives. As a result, the following was studied and selected: the concept of "self-analysis of professional activity", characterized by the features of training teachers in the Arctic region of Russia, systematized experience in teaching of student's professional activity self-analysis.

The study was conducted in several stages, using the following methods:

1. theoretical (analysis and synthesis of scientific literature on the problem), the study of normative and methodological documents in vocational education;
2. empirical (study, analysis of actual experience of professional teacher education, observation, survey);
3. statistical (counting statistics, identifying links, visualization of the results).

The base of the research is Arkhangelsk Pedagogical College. The study involved 160 2–3-year students and 23 lecturers.

The study was conducted in 2018 and was carried out in three stages.

At the first stage (search-theoretical), a theoretical analysis of psychological, pedagogical literature on the issue of teacher training for professional activity self-analysis was conducted, an analysis of experience the formation of professional competencies of future preschool teachers was carried out.

At the second stage (pilot-experimental), methodical recommendations on production practice for students, cards for self-analysis, approbation of developed materials; a survey of students using recommendations for self-analysis was carried out.

At the third stage (final-generalizing) results are systematized, conclusions are formulated.

4 Results

We have analyzed professional educational programs of education in different levels, studied the experience of professional educational organizations of the Arkhangelsk region for the training of preschool children teachers. Table 1 shows the system of vocational training of educators for kindergartens in the region.

Formation of professional self-analysis is a long, labor-intensive process, inseparable from the general development of the personality of preschool education teachers. Self-analysis is formed in the process of practice. In our opinion, the primary kinds of activities aimed at developing a student's ability of self-analysis are defined as: planning their own teaching activities with children of preschool age, organization of games, joint educational activities with children, and analysis of the process and results of the organization of various activities and communication of children.

By observing and interviewing students, we found that the implementation of systematic pedagogical activity self-analysis in the process of production practice allows students to identify the most effective moments of the event, timely eliminate shortcomings in the organization of pedagogical activity, and improve the level of professional excellence.

The analysis of pedagogical literature and practical educational activities allows to claim that carrying out by students of each event in the course of production practice has complex character and includes the following stages:

- Preparational stage (familiarizing students with the goals, tasks of the event, planning and preparing activities for their further implementation, consulting with teachers - practitioners, paper work);
- The main stage (active pedagogical activities of students, direct implementation of the event);
- The final stage (self-analysis of the event, interview).

Another result of our study was the statement that doing self-analysis during the production practice, the future teacher of preschool education not only realizes the importance of his future profession, but also develops his professional qualities. Professional readiness to performing self-analysis is formed in the course of overcoming professional difficulties and becomes new knowledge. The development of self-analysis skills will be successful if the student has a tool that helps him to analyze his own activities.

Instructions are set out in the form of methodological recommendations on the production practice. Students are offered cards aimed at self-analysis of various types of professional activity:

Table 1. System of training specialists for preschool education in the Arkhangelsk region.

Educational organization	The name of the program	The level of education	Terms of training
The Northern Arctic Federal University named after M.V. Lomonosov	44.03.02 Psychological and pedagogical education. Psychology and pedagogy of preschool education	Higher, bachelor's degree	Full-time education - 4 years Correspondence education - 5 years
	44.03.05 Pedagogical education (with two training profiles) Primary Education and Preschool Education	Higher, bachelor's degree	Full-time education - 5 years
	44.04.02 Psychological and pedagogical education Psychological and pedagogical support of preschool education	Higher, master's degree programme	Full-time education - 2 years
The Humanitarian Institute, a branch of the NArFU, Severodvinsk	44.03.01 Pedagogical education Preschool education	Higher, bachelor's degree	Full-time education - 4 years Correspondence education - 5 years
Arkhangelsk Pedagogical College Kotlas pedagogical College Kargopol Pedagogical College	44.02.01 Preschool education	Secondary vocational education	Full-time education - 3–4 years Correspondence education - 5 years
The Arkhangelsk Regional Institute of Open Education	Program of professional retraining "Preschool education"	This Additional vocational education gives the right to conduct activities in the field of preschool education	768 h
	Program of professional retraining "Theory and practice of preschool education"		532 h
Arkhangelsk Pedagogical College	Professional retraining program	This Additional vocational education gives the right to conduct activities in the field of preschool education	504 h
Kotlas pedagogical College			504 h

- Conducting gaming activities with preschoolers;
- Conducting classes with preschool children;
- Carrying out an event with parents.

The content of the cards is compiled taking into account the modern requirements of preschool pedagogy and World Skills Russia in the competence "Preschool education". These cards were tested in the work with students of the Arkhangelsk Pedagogical College in 2017–2018 years (Tables 2, 3 and 4).

Table 2. Self-analysis card of conducted game with preschool children (in reduction).

Criteria for self-analysis	The results of self-analysis		
	Full correspondence	Partial correspondence	Discrepancy
1. The goals and objectives correspond to the theme of the game and the age of children			
2. The structure of the game are met			
3. The selected methods and techniques correspond to the age characteristics of children			
4. The content of the game corresponds to the age characteristics of children			
5. The organization of the workplace children: children are appropriately placed in the playing space			
6. The Creative approach to the conducting game			
7. Emotionality			
8. The safety rules and sanitary requirements are met			

The cards present the criteria for self-analysis, which we have compiled on the basis of our theoretical literature review and practice of educational activities. For better understanding of students, the criteria are specified. For example, doing the self-analysis of the event with parents according to the "Teacher's Behavior" criterion, the student will assess his behavior, his manner of holding, the tone of the conversation with parents, the presence of contact with parents and free communication. The criterion "Compliance with the structure of classes" implies that the student has met the methodological requirements for organizing classes with preschoolers, there are all the structural components: the introductory, the main, the final parts. After each event, students filled out one of the above self-analysis cards, depending on the type of professional activity they conducted and then made conclusions about the quality of the

Table 3. Self-analysis card of conducted classes with preschool children (in reduction).

Criteria for self-analysis	The results of self-analysis		
	Full correspondence	Partial correspondence	Discrepancy
1. The goals and objectives correspond to the theme of the classes and the age of children			
2. The structure of the classes are met			
3. Logical sequence and interrelation of structural stages			
4. The selected methods and techniques correspond to the age characteristics of children			
5. The content of the classes corresponds to the age characteristics of children			
6. The organization of the workplace children			
7. The creative approach to the conducting classes			
8. Emotionality			
9. The safety rules and sanitary requirements are met			

Table 4. Self-analysis card of the event with parents (in reduction).

Criteria for self-analysis	The results of self-analysis		
	Full correspondence	Partial correspondence	Discrepancy
1. Event organisation			
2. The behavior of the teacher			
3. A appeal to parents			
4. Pedagogical literacy of the teacher in the presentation of the topic			
5. Visualization			
6. Interest and active participation of parents			
7. The General tone of the event			

event. The implementation of self-analysis allows students to gain the necessary knowledge about their own abilities, in the long term, to correct it.

In order to determine the effectiveness of using methodological recommendations for students' self-analysis, students were interviewed. In the survey, 160 students of 2–3 courses at the Arkhangelsk Pedagogical College, using methodological recommendations for self-examination during the production practice took part. During the survey there was used a compiled scale of opinions on students' self-assessment. A comparative survey analysis of the second and third years students is presented in Table 5.

Table 5. Opinion scale on self-analysis of 2–3 courses students of specialty "Preschool education".

Instruction: choose one of the three proposed statements that corresponds to your opinion	Percentage	
1st statement	2 course	3 course
I do not like to analyze my activities, because I think it is unnecessary	79%	24%
I like to analyze other teachers rather than analyze my own activities	46%	56%
I evaluate every event I conducted	15%	62%
2nd statement	2 course	3 course
I get new knowledge when I carry out the analysis of professional activity	12%	25%
By analyzing my own activities, I find ways to overcome professional difficulties	33%	42%
The analysis of the conducted event helps me to correct the process of communication and education of preschool children	78%	92%
3rd statement	2 course	3 course
Self-analysis of the event is important for me	14%	76%
Self-analysis has a positive impact on the development of my professional qualities	36%	35%
Self-analysis in the work of preschool education teachers will lead to positive results in the functioning of the preschool educational organization	70%	89%

The results presented in the table show that the importance of self-analysis of professional activity among students increases. The main mechanism of self-development is the rethinking of the content of one's own activity. Self-analysis allows the student to understand the merits and demerits in his professional activities. We see how the student's attitude toward self-analysis is changing, in the second year many students did not realize self-analysis importance for professional development, but a small part of the students noted that the analysis of the event helped to correct pedagogical activities, while believing that they do not receive professional knowledge in the course of self-analysis. Still, second-year students have knowledge of the importance of selfesteem for the teacher activities, received in theoretical studies, so many students noted its importance for the functioning of the preschool educational organization. As an example, Fig. 1 observes the student's answers.

A survey of third-year students showed the following, students understand the importance of self- assessment that's why they evaluate every event held with children and parents. Carrying out self-analysis students find ways to overcome professional

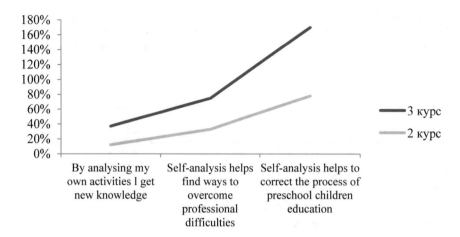

Fig. 1. The survey results of students on the importance of their own professional activities self-analysis.

difficulties and receive new necessary knowledge for the correction of professional activity. Students note the importance of self-analysis for the development of professional qualities and, in general, to improve the quality of preschool education.

5 Conclusion

The problem of self-analysis ability formation of educators in the professional pedagogical activities is relevant at the present stage of the education development which is reflected in international and domestic pedagogical studies and regulatory documents.

In the Arkhangelsk region of Russia, belonging to the Arctic region, a multilevel system of training of educators has been created, there are different training options (university, colleges, institutions of additional professional education). The analysis of educational standards and programs made it possible to learn that regardless of the training option, the training of educators pays attention to the formation of the ability to self-analyze their own professional activities.

The skills of self-analysis are formed in the process of training a professional teacher. A prerequisite is the participation of future educators in pedagogical activities, such an opportunity is provided during the production practice.

In order to successfully carry out self-analysis of professional activities, the authors developed methodological recommendations in the form of self-analysis cards for pedagogical activities with children and parents. Approbation of these guidelines was conducted with students of the Arkhangelsk Pedagogical College in the production practice frame-work, their effectiveness was proved. The majority of students of the last training courses note the importance of conducting self-analysis for their further professional development. The results obtained by us confirm the prospects for further study of the problem and the search for new areas of activities in the process of future teachers training and their production practice conducting in preschool institutions of the Arctic region.

Acknowledgements. The authors of the article express special gratitude to the translator Ekaterina Istomina and the consultants Feng Liu and Inga Zashikhina for presenting the materials of the article in English.

References

1. Abdulina, O.: Lichnost' studenta v processe professional'noj podgotovki [The student personality in the professional training process]. Vysshee obrazovanie v Rossii [Higher education in Russia] **3**, 165–170 (1993). (in Russian)
2. Babanskij, Yu.: Izbrannye pedagogicheskie trudy [Selected pedagogical works]. http://elib. gnpbu.ru/textpage/download/html/?book=babanskiy_izbrannye-pedagogicheskie-trudy_ 1989&bookhl. Accessed 4 Aug 2018. (in Russian)
3. Drvodelić, M., Domović, V.: Preschool teachers' attitudes towards the self-evaluation of preschool institutions. Croatian J. Educ. **18**, 47–60 (2016)
4. Zair-Bek, E.: Novye strategii professional'noj orientacii i podderzhki programm "Obuchenie dlya kar'ery"[New strategies for career guidance and support "The learning for career" programs]. CHelovek i obrazovanie [Human and education] **2**(35), 89–93 (2013). (in Russian)
5. Kuz'mina, N.: Formirovanie pedagogicheskih sposobnostej [Pedagogical abilities formation]. Izdatel'stvo Leningradskogo gosudarstvennogo instituta, Leningrad (1961). (in Russian)
6. Lomakina, I.: Politika kooordinacii ES: tendencii i osobennosti razvitiya [EU coordination Policy: trends and features of development]. Vestnik kul'tury i iskusstv [Vestnik of culture and arts] **1**(21), 98–103 (2010). (in Russian)
7. Pisareva, S.: Metodologiya ocenki kachestva dissertacionnyh issledovanij po pedagogike [Methodology for assessing the quality of dissertation research in pedagogy]. Ph.D. thesis. Russian State Pedagogical University named after A. I. Herzen, Saint-Petersburg [Online] (2005). http://www.dslib.net/obw-pedagogika/metodologija-ocenki-kachestva-dissertacionnyh-issledovanij-po-pedagogike.html. Accessed 20 Aug 2018. (in Russian)
8. Piskunova, E.: Podgotovka uchitelya k obespecheniyu sovremennogo kachestva obrazovaniya dlya vsekh: opyt Rossii: Rekomendacii po rezul'tatam nauchnyh issledovanij [The teacher Training to ensure the modern quality of education for all: the experience of Russia: Recommendations on the results of scientific research]. Publishing house of the Russian State Pedagogical University named after A. I. Herzen, Saint-Petersburg] (2007). (in Russian)
9. Radinova, N., Tryapicyna, A.: Perspektivy razvitiya pedagogicheskogo obrazovaniya: kompetentnostnyj podhod [Prospects for the development of the pedagogical education: competence-based approach]. CHelovek i obrazovanie [Human and education] **4–5**, 7–14 (2006). (in Russian)
10. Slastenin, V.: Professional'noe samorazvitie uchitelya [Professional self-development of the teacher], http://cyberleninka.ru/article/n/professionalnoe-samorazvitie-uchitelya. Accessed 22 Aug 2018. (in Russian)

11. Drobotenko, Yu.: Aspektny`j analiz ponyatiya professional`noj podgotovki v pedagogich-eskom vuze [Aspects analysis of the concept of professional training at pedagogical university]. Sovremenny`e issledovaniya social`ny`x problem [Modern Research of Social Problems] **12**(56), 53–72 (2015). (in Russian)

12. Tutolmin, A.: Terminologicheskij glossarij professional'no-tvorcheskoj pedagogiki [Glossary of terms on vocational and creative pedagogy]. https://novainfo.ru/article/3298. Accessed 14 Aug 2018. (in Russian)

13. Ilic, T.: Geography teachers attitudes toward self-evaluation: the case of Serbia. Eur. J. Geogr. **5**(4), 78–86 (2014)

14. SHelekhova L.: Diagnosticheskie priznaki refleksivnoj uchebnoj deyatel'nosti [Diagnostic indications of the reflective educational activity]. Vestnik Adygejskogo gosudarstvennogo universiteta. Seriya 3: Pedagogika i psihologiya [Vestnik of the Adyghe State University. Series 3: Pedagogy and psychology] **1**, 89–94 (2011). (in Russian)

15. Perenkova, E.: Razvitie sposobnosti uchitelya k samoanalizu professional'noj deyatel'nosti v processe metodicheskoj raboty [Development of the teacher's ability to professional activity self-analysis in the process of methodical work]. Ph.D. thesis. Academy of advanced training and retraining of employees of education Russian Federation Ministry of Education, Moscow] [Online] (2003). http://www.dissercat.com/content/razvitie-sposobnosti-uchitelya-k-samoanalizu-professionalnoi-deyatelnosti-v-protsesse-metodi. Accessed 2 Sep 2018. (in Russian)

16. Kunakovskaya, L.: Refleksivnaya kul'tura pedagoga [The reflexive culture of the teacher]. Publishing and printing center of Voronezh state University, Voronezh (2011). (in Russian)

17. Metaeva, V.: Refleksiya: formirovanie novogo tipa myshleniya cheloveka [Reflection: the formation of a new human thinking type]. CHelovek i obrazovanie [Human and education] **6**, 32–34 (2006). (in Russian)

18. Kudinov, V., Alenchenkova, A.: Metodicheskie aspekty razvitiya refleksivnoj kul'tury pedagogov v processe dopolnitel'nogo professional'nogo obrazovaniya [Methodical aspects of the teachers reflexive culture development in the additional professional education process]. Uchenye zapiski. EHlektronnyj nauchnyj zhurnal Kurskogo gosudarstvennogo universiteta [Scientists notes. Electronic scientific journal of Kursk state university] 2(34), (2015). https://cyberleninka.ru/article/v/metodicheskie-aspekty-razvitiya-refleksivnoy-kultury-pedagogov-v-protsesse-dopolnitelnogo-professionalnogo-obrazovaniya. Accessed 30 Aug 2018. (in Russian)

19. Shustova, I.: Rol` pedagogicheskoj podderzhki v samoopredelenii starsheklassnikov. [The role of pedagogical support in self-determination of senior schoolchildren]. Otechestvennaya i zarubezhnaya pedagogika [Foreign and Domestic Pedagogy] **5**(20), 74–88 (2014)

20. Berezhnova, L.: Formirovanie navy`kov professional`nogo samoanaliza u studenta [Skills development of the student professional self-analysis]. Armiya i obshhestvo [Army and society] **4**(41), 97–106 (2014). (in Russian)

21. CoRe Competence Requirements in Early Childhood Education and Care. Final Report. – London: University of East London, Cass School of Education,Ghent: University of Ghent, Department for Social Welfare Studies (2011)

22. Fröhlich-Gildhoff, K., Weltzien, D.: Expertise Kompetenzen früh-/kindheitspädagogischer Fachkräfte im Spannungsfeld von normativen Vorgaben und Praxis: erstellt im Kontext der AG Fachkräftegewinnung für die Kindertagesbetreuung in Koordination des BMFSFJ. BMFSFJ, Berlin (2014)

23. Professional`ny`j standart pedagoga [Professional Standards "Educator"]. https://минобр-науки.рф/. Accessed 14 Aug 2018. (in Russian)

24. Federal`ny`j gosudarstvenny`j obrazovatel`ny`j standart srednego professional`nogo obra-zovaniya po special`nosti 44.02.01 doshkol`noe obrazovanie [Federal state educational standard of secondary vocational education on a specialty 44.02.01 pre-SCHOOL EDUCATION]. http://classinform.ru/fgos/44.02.01-doshkolnoe-obrazovanie.html. Accessed 14 Aug 2018. (in Russian)
25. Novikov, A., Novikov, D.: Metodologiya: slovar` sistemy` osnovny`x ponyatij [Methodology: dictionary of system concepts]. Librokom, Moscow (2013). (in Russian)
26. Kodzhaspirova, G., Borikova, L., Bostandzhieva, N.: Pedagogicheskaya praktika v nachal`noj shkole [Pedagogical practice in an elementary school]. 2nd edn. Publishing center "Academy", Moscow (2000). (in Russian)
27. Averkieva, G., Shhekina, S.: Pedagogicheskaya praktika kak refleksivnaya sreda formirovaniya professional`nogo opy`ta budushhego uchitelya [Pedagogical practice as a reflexive sphere of the future teacher professional experience formations]. Vestnik Severnogo (Arkticheskogo) federal`nogo universiteta. Seriya: Gumanitarny`e i social`ny`e nauki [Bulletin of the Northern (Arctic) Federal University. Series: Humanities and social Sciences] **4**, 141–149 (2014). (in Russian)

Self-directed Learning in Pre-service Teacher Education

Irina A. Kazachikhina[(✉)] [ID]

Novosibirsk State Technical University, Novosibirsk 630073
Russian Federation
kazachixina@corp.nstu.ru

Abstract. Many recent studies have focused on various aspects of self-directed learning (SDL). However, still little attention has been paid to a holistic situation as the place for developing SDL. It is important for teachers to be aware of the learning situation which transforms students from dependent to self-directed. In this study we focus on students' perception of their learning progress in pre-service foreign language (FL) teaching course. The aim of this study was to examine what elements the students perceive as contributing to their successful learning and to discover whether and how these elements change through the course. The situational analysis (SA) allowed covering elements of students' successful learning, and explore a learning situation of 25 bachelor program students of the teaching profession at the beginning and end of the FL teaching course. Data were collected through students' written reflective practice. As the qualitative study showed, learning situations comprise different elements perceived by students as contributing to their success in learning. In terms of the situational theory, they can be categorized as human and non-human factors, discourse structures of individuals or/and collective human factors, discursive construction of non-human factors, explicit/implicit learning strategies and learning outcomes. After a semester through the course designed to develop SDL skills, the students remained either dependent, or became interested or involved, whereas toward the end of the course the students became either self-directed or involved.

Keywords: Self-directed learning · Pre-service teacher education · Reflective practice · Situational analysis

1 Introduction

Over the past three decades, the shift of focus from teaching to learning has led educators and researchers to revise concepts and teaching practices in higher education. However, this fundamental change would have remained an idea unless it had become an integral part of teacher education. For students who become teachers vs. agents of change in teacher education themselves in future, it is crucial to experience the refocused paradigm and become aware of a broader meaning of 'education' which does not necessarily equal to 'training' [7, 19]. As Freeman has put it 'within teacher education were housed the allied processes of teacher training and teacher development' [6]. If they happen simultaneously, either teacher-facilitated or student-driven or both,

© Springer Nature Switzerland AG 2019
Z. Anikina (Ed.): GGSSH 2019, AISC 907, pp. 292–302, 2019.
https://doi.org/10.1007/978-3-030-11473-2_31

students equip themselves with personal and professional competences for lifelong learning [13]. Nevertheless, learning of this kind takes place at the student's initiative [18] and active engagement which means taking responsibility for setting learning goals, choosing the best ways to learn, evaluating outcomes, and developing new learning strategies. Clearly, the need of individuals for continuous new knowledge and skill acquisition has caused education researchers and practitioners to create new meaning of learning which is known as 'self-directed learning'.

Many studies have been carried out to explore SDL from teaching and learning perspectives. However, clear differentiation between the components of SDL makes it difficult to present a holistic situation which learners perceive as successful and educators as self-directed.

Being inspired by Kalendaa and Vávrováa [15, 16] experience and results of exploring learning context not limited to the classroom situation only, but as it naturally exists, we applied the situational theory [2, 3] to enhance our understanding of SDL in teacher education. The aim of this study was to examine what elements students perceive as contributing to their successful learning, and to discover whether and how these elements change over the time.

2 Theoretical Framework

Providing favorable conditions for SDL and e-learning [23] in a formal setting meets the needs of students called 'the Net Generation (N-Gen)' or 'digital natives'. It is important to consider how students think, and whether they think the way that differs from the way most teachers do, as they have grown in a non-digital environment. Tapscott claims students' change in thinking and learning caused by interactivity, the first part can be regarded as a characteristic of teachers thinking (unless they are digital natives themselves): 'from linear to hypermedia learning, from instruction to construction and discovery, from teacher-centered to learner-centered education, from absorbing material to learning how to navigate and how to learn, from schooling to lifelong learning, from one-size-fits-all to customized learning, from learning as torture to learning as fun, and from the teacher as transmitter to the teacher as facilitator' [27]. Clearly, the opposite characteristics of those who teach and who learn cannot be ignored. The discrepancy resulted in the explosion of research worldwide in the last three decades. Thus, responsibility of the teacher in SDL mainly refers to *cultivating* self-directed learners, which means, as Gibbson explains [8], '[t]he teacher shifts from recitation to provocation, from telling to asking, and from instruction to guidance, teaching students to think and find out for themselves.' The study is based on the understanding of SDL as 'the degree of choice that learners have within an instructional situation' [9].

Currently, to cultivate self-directed learners, teachers are expected to provide favorable learning conditions, such as open access to the required resources for learning purposes, space for directing learning process and communication and instruments helping students to reflect on their success through learning. Since the importance of the proper social and psychological environmental factors for cultivating self-directed learners has been proved [11, 25] a digital environment is considered as one of the main factors of a digital era [22]. The other two factors can be provided by the instructional design of the learning environment, where '[a]ctivities and assignments become the vehicles by and through which learning occurs' [28].

Regardless of the theories supported and developed, most researchers would agree with Hiemstra about the fact ranking first in the list of the known features of SDL that 'individual learners can become empowered to take increasingly more responsibility for various decisions associated with the learning endeavor' [12]. The factor proves psychological foundation [1] of the theory of self-directed learning based on a 'person-centered approach' developed by Rogers [24] within the humanistic approach.

3   Educational Context

The study explored a situation of SDL among students in pre-service teacher education in Novosibirsk State Technical University. The participants of the study were students of Bachelor program in Linguistics, majoring in Foreign Language and Culture Teaching. According to the Federal State Educational Standard for Higher Education for Linguistics,[1] degree holders are required to possess professional competencies in the fields of foreign languages, theory of linguistics, theory of intercultural communication, foreign language teaching, and translation. Among general cultural competencies they are expected to have developed:

'ability to apply methods and means of cognition, training and self-control for intellectual self-development, raising the cultural level, professional competence, health care, moral and physical self-improvement' (General Cultural Competence (GCC) 8); 'readiness for constant self-development, improvement of the qualification and skills, critical self-evaluation of strengths and weaknesses, and identifying ways and choosing appropriate means of self-development' (GCC 11); 'ability to understand the social significance of future profession, and possess high motivation to perform professional activities' (GCC 12).

Though the terms are vague, they relate to self-directed learning. To form the competencies program curriculum provides conditions for active learning using ICT and LMS. Students attend lectures, workshops, and they are also expected to work independently (so-called 'self-study').

The study took place in the context of 4 semesters allocated for Foreign Language Teaching course. The semester assessment approach consists of four elements: (1) a theory-based talk on one of the themes studied, (2) a course book material evaluation task and material development task, and (3) a portfolio.

---

[1] Linguistika (bakalavr) [the Federal State Educational Standard for Higher Education for Linguistics (Bachelor Degree)]. Documents of the Ministry of Education and Science of the Russian Federation (in Russian) http://fgosvo.ru/uploadfiles/fgosvob/450302_Lingvistika.pdf, last accessed 2018/08/24.

The study addressed SDL issues through written reflection assignment as a part of the student's portfolio at the end of each semester on the FL teaching course. They keep portfolios through the course collecting evidence of learning themes studied. The collection comprises reviews, essays, teaching material evaluation notes, lesson plans, teaching material developed reflection reports, and other evidence of students' choice within a loose framework. It manifests the student learning experience, thus, illustrating their awareness of learning process and professional growth. In addition, portfolio stimulates reflection [26]. Written reflection reports (500–600 words) contained 'self-reflection on how learning took place and/or what was learned', 'critical review of learner past experience', 'learning strategies', 'what was learnt' [20], and contributed to students' successful performance.

4 Methods

The research problem determines an epistemological approach to the study [5]. For data analysis, we applied the situational theory [2, 3] adapted to the exploration of learning situations [15, 16]. To explore the elements that contributed to student progress through the course, we conducted a qualitative research using written reflection reports as an instrument.

The reflective reports did not contain any guidelines or questions which could limit students' thinking. The only stress was on 'your' success. First, it was done to make students feel more comfortable and relaxed to talk about their successful performance. It should be clarified that Russians might feel awkward when admitting their success openly. They would rather talk about their problems and failures. As the students were Russian, it was necessary to exclude this culturally-determined behavior. Second, 'your' was also stressed to show respect for students' personal efforts during the semester.

Participants of the study were the third and fourth-year students of 3 cohorts in 2015 (N = 10), 2016 (N = 3), and 2017 (N = 12). 2 male and 13 female participants ranged in age from 18 to 21. The first portion of data was collected from the third-year students after a semester of studying the course. The other portion of data was received when they submitted their reflection reports at the end of the course.

Applying the mapping approach, first, I identified the key elements within the data collected with written reflection instrument, and obtained the first version of two situational maps (see Figs. 1 and 2). Second, the raw data was categorized and the second versions of the maps were created (see Tables 1 and 2). Third, using relational analysis [16], I narrowed down the focus conceptualizing data and obtained final versions of the maps (see Figs. 3 and 4) which showed the elements students perceive as contributing to their successful learning through the course. Finally, to discover how elements change through the course I compared situational maps of the situations 1 and 2.

theme #; my active work in class; cooperative activities; e-course with all materials; awareness of the ELT course value; my being prepared for the lesson; compiling my portfolio; peer-review task; new activities; teacher encouraged me; I wasn't afraid of giving my opinion in front of the group; inspiring content and activities; the way we listen to each other; I used extra material; I could say what I really thought about education issues; self-assessment procedure; I completed all assignments; all students were interested; role-play 'Teacher and Student'; all components of the course were new and interesting; learned new things about education in other countries; making mind-maps to remember; the lessons seemed to be useful in future; the way we explored themes together; clear instructions; I always felt confident; interesting themes; opinions of my groupmates; I attended all classes; we checked the tasks and corrected them; doing tests in a self-checking mode first; receiving feedback; I used resources I found appropriate; friendly atmosphere;

Fig. 1. Messy situational map 1.

microteaching; peer-evaluation; I revised the theory (my notes and extra) for the exam; advantages of self-evaluation; my confidence; support from the teacher; work in small group; microteaching; peer-observation; being autonomous; working with my groupmates as with colleagues; I realize my strength and weaknesses as a teacher; material evaluation tasks were most valuable; followed my learning style preparing for the lessons; peer teaching; learned more about my myself; my development material skills; I enjoyed doing tasks; I used knowledge and skills on teaching reading when I taught pupils at school; I realized I had to learn more about classroom management; new theme about teaching and learning strategies; I liked working independently; I seemed to understand what my teachers at school did right and wrong; I'm proud of my mini-textbook; applied theory learned to lesson planning and teaching; I managed to write an article; peer-evaluation helped me to understand how I did the same tasks; I was always prepared and I was glad to be engaged in class activities; value of sharing experience with groupmates; I always knew what to do; e-course on ELT; appreciate my teacher's criticism; improved analytical ability; exam preparation with my friend; fun; mutual understanding between teacher and students; I studied hard; my collection in portfolio; a list of ELT resources in English; observing university teachers' best practices;

Fig. 2. Messy situational map 2.

Table 1. Ordered situational map 1.

Individual human factors – I; we; all; teacher; groupmates; students; each other; together;

Collective human factors – group;

Discourse structure of individuals or/and collective human factors – teacher encouraged me; all students were interested; the way we listen to each other; the way we explored themes together; we checked the tasks and corrected them; opinions of my groupmates; receiving feedback; making mind-maps to remember; friendly atmosphere;

Discursive construction of non-human actors – theme #; interesting themes; e-course with all materials; inspiring content and activities; all components of the course were new and interesting; the lessons seemed to be useful in future; new activities

Explicit / implicit learning strategies – clear instructions; doing tests in self-checking mode at first; cooperative activities; peer-review task; role-play 'Teacher and Student'; used extra material; self-assessment procedure; used resources I found appropriate; I attended all classes; my being prepared for the lesson; compiling my portfolio; my active work in class;

Learning outcomes – awareness of the ELT course value; I could say what I really thought about education issues; completed all assignments; I always felt confident; I wasn't afraid of giving my opinion in front of the group; learned new things about education in other countries;

Table 2. Ordered situational map 2.

Individual human factors – I; we; all; teacher; friend; colleagues; groupmates; students; university teachers;
Collective human factors – group; small groups;
Non-human factors – at school; university;
Discourse structure of individuals or/and collective human factors – mutual understanding between teacher and students; fun; exam preparation with my friend; appreciate my teacher's criticism; value of sharing experience with groupmates; I was glad to be involved in class activities; I seemed to understand what my teachers at school did right and wrong; I liked working independently; work in small groups; support from the teacher; working with my groupmates as with colleagues; observing university teachers' best practices; peer teaching;
Discursive construction of non-human actors – material evaluation tasks were most valuable; new theme about teaching and learning strategies; a list of ELT resources in English; e-course on ELT;
Explicit/implicit learning strategies – I studied hard; I always knew what to do; I was always prepared; being autonomous; I realized I had to learn more about classroom management; advantages of self-evaluation; self-evaluation; I revised the theory (my notes and extra) for the exam; followed my learning style preparing for the lessons; I enjoyed doing tasks;
Learning outcomes – my collection in portfolio; improved analytical ability; peer-evaluation helped me understand how I did the same tasks; applied theory learned to lesson planning and teaching; I'm proud of my mini-textbook; I used knowledge and skills on teaching reading when I taught pupils at school; my development material skills; learned more about my myself; my confidence; micro-teaching; I realized my strength and weaknesses as a teacher; I managed to write an article;

Degree of dependence

theme #; **my active** work in class; **cooperative** activities; **e-course with all materials**; **awareness** of the ELT course **value**; **my being prepared** for the lesson; compiling **my portfolio**; **peer-review** task; **new activities**; teacher **encouraged me**; **I wasn't afraid of giving my opinion in front of the group**; **inspiring** content and activities; **the way we listen to each other**; I used **extra** material; **I could say what I really thought** about education issues; **self-assessment** procedure; **I completed all** assignments; all students were **interested**; role-play '**Teacher and Student**'; all components of the course were **new and interesting**; **learned new** things about education in other countries; **making** mind-maps **to remember**; the lessons seemed to be useful in future; **the way** we **explored** themes **together**; clear instructions; I always **felt confident**; **interesting** themes; **opinions of** my groupmates; **I attended all** classes; we **checked** the tasks **and corrected** them; doing tests in a **self-checking** mode first; **receiving feedback**; I used resources **I found appropriate**; **friendly** atmosphere;

Fig. 3. Rational analysis of Situation 1.

Degree of responsibility

micro**teaching**; **peer-evaluation**; **I revised** the theory (**my** notes **and extra**) for the exam; **advantages** of self-evaluation; **my confidence**; **support** from the teacher; work in small group; microteaching; **peer-observation**; **being autonomous**; **working with my group mates as with colleagues**; **I realize my strength and weaknesses as a teacher**; material **evaluation** tasks were **most valuable; followed my learning style** preparing for the lessons; **peer teaching; learned more about my myself; my** development material **skills; I enjoyed doing** tasks; **I used knowledge and skills on teaching reading when I taught pupils at school**; I **realized I had to learn more about** classroom management; **new** theme **about teaching and learning strategies; I liked** working **independently;** I seemed to **understand** what my teachers at school did **right and wrong;** I'm **proud of my** mini-**textbook; applied theory learned to** lesson planning and teaching; **I managed to write an article; peer-evaluation helped me to understand how I did** the same tasks; I was **always prepared** and I was **glad to be engaged** in class activities; **value of sharing experience with group mates;** I always **knew what to do; e-course** on ELT; **appreciate my teacher's criticism; improved analytical ability;** exam **preparation with my friend; fun; mutual understanding** between teacher and students; **I studied hard; my collection** in portfolio; a list of ELT **resources** in English; **observing university teachers' best practices;**

Fig. 4. Rational analysis of Situation 2.

5 Results

The SA allowed exploring all elements that students in teacher education perceive as contributing to their success doing a new course in FL Teaching. To extract data from reflective reports for that purpose is sure not a simple procedure for a researcher. For students, reflection also always remains hard and challenging, and requires expressing thoughts in some abstract terms [14]. Nevertheless, reflective reports and essays are used by the researchers as instruments to reveal learners' personal theories about teaching and learning.

The next step was to identify key words and statements which corresponded to the focus or focuses of student's reflection. Thus, 34 elements were identified in the reports at the end of the first semester, and 40 elements at the end of the final semester of the course. Those were given in Messy situational map 1 and Messy situational map 2 respectively (see Figs. 1 and 2).

The identified elements of the reflective reports after Semester 1 of the course and at the end of the course were grouped into 6 categories in Ordered situational map 1 (see Table 1), and into 7 categories in Ordered situational map 2 (see Table 2). The elements referred to other important for students, individuals, collective entities, material elements, important activities that took place within the learning situation, valuable relationships and attitudes, individual achievements, and learning strategies individuals used to progress in learning either confidently or without being aware of them.

According to Clarke's research and projects, messy maps prove to be valuable sources of relational analysis [4]. In the present study messy maps helped to illustrate degree of students' responsibility for their learning at the beginning and the end of the FL teaching course. Every element of each messy map was analyzed in relation to the process of becoming a self-directed learner [9]. The bold elements in Figs. 3 and 4 reflected a degree of students' responsibility for learning.

6   Discussion

A complicated nature of teaching/learning situation with numerous elements and connections between them was quite evident even at the stage of presenting data in the form of messy maps. It might be suggested to focus on learning strategies students develop during FL teaching course rather than on the whole situation at once. But messy maps revealed that other individuals engaged in the situation, their actions, attitudes and learning outcomes could not be discriminated. The complexity of learning as a cognitive, metacognitive and social phenomenon in general and of this case in particular, explains the existence of numerous approaches to the SDL theory. Even focusing only on the situation, excluding a broader educational context, could have distorted the results of the study.

The university where the research was carried out organizes its curriculum combining traditional didactic teaching and various forms of direct instruction with active and ICT-based learning, and moves toward the competence-based education. It seems that university teachers often accuse students for their inability to control their own learning, as they complain 'they often encounter passive, dependent, and grade-driven students […] yet they fail to see the role of their own courses in helping students cultivate lifelong skills.' [21]. In other words, the learning environment for the first- and second-year students of bachelor program in Linguistics require more SDL-oriented syllabi, assignments and, therefore, better learning conditions for students to develop SDL strategies and skills.

To strengthen students' willingness to take their learning process and outcomes, FL teaching course was designed. The syllabus, learning environment, activities and assessment procedures were designed to provide students with choices about their learning directions.

As Fig. 3 shows, after a semester of studying ELT course the students demonstrated that they did not belong to dependent learners, however, we cannot refer them to self-directed. Though the aim of the research is not to measure the readiness of students in the teaching profession to SDL, we should remember about Guglielmino (1977) self-report questionnaire designed to measure an individual's level of readiness to manage their own learning [10]. Since then that valuable instrument of SDL assessment has been widely used in academic and professional context worldwide. But, clearly, the present study benefits from SA and relational analysis which give a holistic view of the multifaceted learning situation.

If we compare the evidence of taking control over learning in Figs. 3 and 4, and relate them to the categories of 'dependent – interested – involved – self-directed' [9], we find that at the beginning of doing FL teaching course a few students can be identified as dependent learners, but most are interested and involved. There is still no evidence of SDL. Hopefully, toward the end of the course we find that most students can be considered as self-directed, a few as involved and probably as interested.

The ordered situational maps provide three main dimensions for two learning situations. The social dimension is manifested through types of individual and collective human and non-human factors, discourses of high priority to the students in two opposing points of time. Ordered mapping showed social implications of learning process, strategies and outcomes.

7 Conclusion

The results of a sequential analysis showed a potential of applying SA for exploring multifaceted teaching/learning process of pre-service teacher education due to the ability to cover elements of the real situation and provide them visually. In the research, based on the situational theory, it should be noted that marking boundaries of the studied situation can determine the angle of its observation [17].

The research was focused on students' perception of their learning progress.

The research showed that acting in the same learning environment students can be influenced by different factors full of different personal meanings and values.

The factors belong to the categories below:

- Human factors, presenting participants of the learning situation, perceived as important to students (e.g. a group mate perceived as 'friend', not all group mates; 'university teachers' as other teachers who provided best models for students),
- Non-human factors presenting some space for a complex meaningful discourse (e.g. 'at school' perceived as professionally significant structure; 'university' perceived as a particular educational institution with highly-professional teaching staff),
- Discourse structures of individuals or/and collective human factors representing relationships, activities, interactions, perception of the mentioned above (e.g. 'I was glad to be involved in class activities' indicated some appreciation of providing an opportunity to learn'; 'fun' as learning as pleasure not torture),
- Discursive construction of non-human actors representing material construct of high value for learning (e.g. 'material evaluation tasks were most valuable' as understanding a value of the professional activity; 'e-course on ELT' as an opportunity to learn autonomously),
- Explicit/implicit learning strategies representing unconscious and conscious ways of learning (e.g. 'I studied hard'; 'self-evaluation'),
- Learning outcomes representing evidence of students' academic performance, personal and professional development (e.g. 'learned more about myself'; 'used knowledge and skills on teaching reading when I taught pupils at school').

FL teaching course designed to develop SDL skills provided students with favorable conditions for self-directed learning. Its instructional design was focused on learner-centered methods. After a semester of studies there were identified three types of students [9] such as dependent, interested and involved, whereas toward the end of the course the students became either self-directed or involved.

Limitations of the study should be noted. First, I relied only on students' reflective reports. Second, the new data will definitely reflect the new situation under study. However, applying the situational theory to pedagogical research could contribute to the theory of SDL. If this study helps my colleagues to study learning situations in their educational context, I believe it will have served a useful purpose.

References

1. Avvo, B.V., Akhayan, A.A., Zair-Bek, E.S., Komarov, V.A., Gorokhovatskaya, N.V., Feofilova, T.G., Fedorova, N.M., Sosunova, NYu.: Instructional Strategies and Learning Technologies in the Implementation of The Competence-Based Approach in the Light Humanitarian Technology in Pedagogical Education: Teaching Guidelines. Izd-vo RGPU im. A.I. Herzena, Saint-Peterburg (2008)
2. Clarke, A.E.: Situational analysis. Gr. Theor. After Postmod. Turn Symb. Interact. **26**(4), 553–576 (2003)
3. Clarke, A.E.: Situational Analysis: Grounded Theory After Postmodern Turn. Sage, London (2005)
4. Clarke, A.E., Friese, C., Washburn, R. (eds.): Situational Analysis in Practice. Mapping Research with Grounded Theory. Left Coast Press Inc., London (2015)
5. Creswell, J.W.: Research Design: Qualitative, Quantitative, and Mixed Methods Approaches, 3rd edn. Sage Publications, Thousand Oaks, California (2009)
6. Freeman, D.: Foreword. In: Richards, J.C. (ed.) Beyond Training: Perspectives on Language Education, pp. vii–xi. Cambridge University Press, Cambridge (1998)
7. Freeman, D.: Observing teachers: three approaches to in-service training and development. TESOL QUATERLY **16**(1), 21–28 (1982)
8. Gibbons, M.: The Self-Directed Learning Handbook: Challenging Adolescent Students to Excel. Jossey-Bass, Hoboken, NJ (2003)
9. Grow, G.O.: Teaching learners to be self-directed. Adult Educ. Q. **41**(3), 125–149 (1991)
10. Guglielmino, L.M.: Development of the Self-Directed Learning Readiness Scale. Doctoral dissertation, University of Georgia (1977)
11. Hiemstra, R. (ed.): Creating Environments for Effective Adult Learning (New Directions for Adult and Continuing Education, 50). Jossey-Bass Publishers, San Francisco, CA (1991)
12. Hiemstra, R.: Self-directed learning. In: Husen, T., Postlethwaite, T.N. (eds.) The International Encyclopedia of Education, 2nd edn, pp. 9–19. Pergamon Press, Oxford (1994)
13. Isaeva, T.E.: Competence of Students and Teachers in Higher Education: formation and evaluation. Rost.gos.un-t putei soobshcheniya, Rostov/D (2010)
14. James, P.: Teachers in Action: Tasks for In-service Language Teacher Education and Development. Cambridge University Press, Cambridge (2001)
15. Kalenda, J., Vávrová, S.: Mapping of self-regulated learning of adults. In: 12th International Conference Efficiency and Responsibility in Education, pp. 239–249. Czech University of Life Sciences Prague, Prague, Czech Republic, EU (2015)
16. Kalendaa, J., Vávrováa, S.: Self-regulated learning in students of helping professions. Proced. Soc. Behav. Sci. **217**, 282–292 (2016)
17. Kacperczyk, A.: Badacz i jego poszukiwania w świetle „Analizy Sytuacyjnej" Adele E. Clarke [The Researcher and his scientific inquiry in light of the 'Situational Analysis' Adele E. Clarke]. Przegląd Socjologii Jakościowej **III**(2), 5–32 (2007)
18. Knowles, M.: Self-Directed Learning: A Guide for Learners and Teachers. Pearson Learning Group, Cambridge (1975)
19. Larsen-Freeman, D.: Training teachers or educating a teacher? In: Alatis, J., Stern, H.H., Strevens, P. (eds.) Georgetown University Round Table on Languages and Linguistics, pp. 264–274. Georgetown University Press, Washington, D.C. (1983)
20. Lew, M.D.N., Schmidt, H.G.: Self-reflection and academic performance: is there a relationship? Adv. Health Sci. Educ. **16**(4), 529–545 (2011)

21.
Nantz, K., Klaf, S.: Cultivating Self-Directed Learners by Design. Fairfield University. AAC&U General Education & Assessment Meeting. March 1, 2014. Session Materials. https://www.aacu.org/sites/default/files/files/meetings/GE14_CS43.pdf. Accessed 18 Aug 2018

22. Noskova, T.N., Pavlova, T.B., Yakovleva, O.V.: Analysis of domestic and international approaches to the advanced educational practices in the electronic network environment. Integracija obrazovanija [Integration of Education] **20**(4), 456–467 (2016)

23. Penland, J.L.: Constructivist Internet-Blended Learning and Resiliency in Higher Education. Sul Ross State University, USA (2015)

24. Rogers, C.R., Lyon, H.C., Tausch, R.: On Becoming an Effective Teacher—Person-centered Teaching, Psychology, Philosophy, and Dialogues with Carl R. Rogers and Harold Lyon. Routledge, London (2103)

25. Spear, G.E., Mocker, D.W.: The organizing circumstance: environmental determinants in selfdirected learning. Adult Educ. Q. **35**, 1–10 (1984)

26. Tanner, R., Longayroux, D., Beijaard, D., Verloop, N.: Piloting Portfolios: using portfolios in pre-service teacher education. ELT Journal **54**(1), 20–30 (2000)

27. Tapscott, D.: Growing up Digital: The Rise of the Net Generation. McGraw-Hill, New York (1998)

28. Weimer, M.: Learner-Centered Teaching: Five Key Changes to Practice, 2nd edn. Jossey-Bass, San Francisco, CA (2002)

Peer Coaching as a Means of EFL Teachers' Professional Development

Elena A. Melekhina and Irina V. Barabasheva[✉]

Novosibirsk State Technical University,
Novosibirsk 630073, Russian Federation
{melexina, barabashyova}@corp.nstu.ru

Abstract. Professional development course *Peer Coaching in Work Practice of EFL Teachers* has as its purpose to advance the professionalism of English as a Foreign Language (EFL) university teacher by training them to promote frequent, informal, helpful observations by one professional educator to another, to give necessary rationale feedback and coach each other. The course provides some practical guidelines on how to effectively implement the educational strategy of Peer Coaching as one of the most successful professional development technique in teaching EFL at Novosibirsk State Technical University (NSTU). The article presents the findings aimed at understanding the advantages and disadvantages of incorporating peer-coaching into the proposed in-service training course for EFL teachers at NSTU. Data collected over the period of the training demonstrate positive attitude of the trainees towards peer coaching and reveal aspects of the newly adopted educational strategy that are perceived as highly motivational for professional development by the participants.

Keywords: Professional development · Peer coaching · EFL teacher

1 Introduction

Professional development (PD) of an English as a foreign language (EFL) teacher is a continuous and complicated process of life-long learning to become a better professional. The process may take different forms from institutionally or individually initiated projects to conferences and seminars, refresher courses or graduate studies. It might involve various activities aimed at extending teaching skills and obtaining insights into the psychological aspects of the process of foreign language acquisition. The range of these activities includes doing research, extensive reading, classroom observation, case analysis, writing for a journal, blogging, and many other means and strategies for facilitating professional and personal growth.

Many studies on teachers' professional development suggest that schools and universities are expected to provide opportunities and conditions for their teachers' development [5, 6, 8, 10, 11]. As Richards and Farrell [8, p. 11] write:

> From the institutional perspective, professional development activities are intended not merely to improve the performance of teachers but to benefit the school as a whole.<...> Improvement of teaching skills and acquisition of new information, theories, and understanding are not goals in themselves: they are part of the process of institutional development.

© Springer Nature Switzerland AG 2019
Z. Anikina (Ed.): GGSSH 2019, AISC 907, pp. 303–309, 2019.
https://doi.org/10.1007/978-3-030-11473-2_32

Shanks et al. [10], for instance, render the experience of organizing school-university partnership resulted in starting Professional Development School where pre-service teachers implemented action research, which helped them to observe and learn from peers. O'Dwyer and Atli [6] focus on the in-service teacher educator roles and conclude that "in-service teacher development recognizes the need for coaching over time, an understanding and facility with contextual variables, and feedback to skills in action with a broad range of clients". Therefore, they have developed the model of teacher educator's roles including different categories that meet the requirements of a university school of English in Turkey. Tack and Vanderlinde [11] stress the importance of teacher educators' research engagement and the need for initiating the programs of research-oriented professional development at the University of Applied Courses in Flanders. Five professional development models to raise self-efficacy of EFL teachers are applied in Ilam Province Teacher Training Center in Iran [5]. These models comprise in-service training, observation/assessment, development/improvement process, study group, and mentoring, and mostly refer to formal professional development.

With regard to informal professional development, it is a common practice when teachers discuss and find solutions to problems with their peers in informal situations in a relaxed manner. Johnston [2] believes that collaboration with peers is 'the most balanced relationship', which promotes teachers' development through 'shared professional understandings'. Cooperation with peers also allows teachers to take on responsibilities for achieving shared professional goals and governs the process of professional development [8]. Teachers might collaborate both in groups and in pairs. One form of collaboration when two teachers help each other to solve pressing problems of teaching practice is known as peer-coaching.

According to Robbins [9, p. 1]:

> Peer coaching is a confidential process through which two or more professional colleagues work together to reflect on current practices; expand, refine, and build new skills; share ideas; teach one another; conduct classroom research; or solve problems in the workplace.

Joyce and Showers [3] claim that peer-coaching is a process that effectively promotes professional development. They explain that companionship makes the process of 'sharing new practice' more 'pleasurable' than working in isolation. This is in line with Richards J. and Farrell T., who advocate the non-judgmental manner of providing peer feedback as a part of the peer-coaching process and trusting relationship in which teachers become confident in each other [8, p. 167].

The research reported here targets motivational aspect of the collaborative process of in-service professional development of English language teachers based on the principles of peer-coaching.

2   Research Questions

Taking into account that professional development of most language teachers proceeds in various formal and informal collaborative ways, we set the following research questions:

1. What are the advantages and disadvantages of incorporating peer-coaching into the in-service training course for EFL teachers at NSTU?
2. What aspects of peer-coaching are perceived as most motivational for professional development by EFL teachers?

2.1 Educational Context

The participants involved in the research were a group of EFL teachers engaged in the in-service training course *Peer Coaching in Work Practice of EFL Teachers* at the Faculty of Teacher Re-training and Professional Development of Novosibirsk State Technical University (NSTU) in 2017. The group consisted of 18 NSTU teachers with different work experience in teaching English to university students, as presented in Table 1.

Table 1. Percentage of trainees according to work experience.

Work experience	0–3 years	4–10 years	11–15 years	16–20 years	>20 years
Trainees	11.11%	16.66%	33.33%	11.11%	27.77%

The trainees' decision to sign up for the course was explained by the need for developing professional and personal competencies in order to meet the changing requirements of the Russian Federation educational system and fulfill the labor functions of the professional educational standard «Teacher of vocational training, professional education and additional professional education». The course lasted one semester.

2.2 Methodology

We employed an approach of mixed methods research including surveys at the beginning and at the end of the course, semi-structured interviews, round-table discussions, and reflective essays.

2.3 Procedure

The in-service training course *Peer Coaching in Work Practice of EFL Teachers* was designed by a group of teachers from the Foreign Languages Department for the Humanities of NSTU.

The objectives of the course were as follows:

- to expand knowledge about the current approaches to professional development
- to increase awareness of contemporary educational technologies
- to engage the staff in peer coaching process by taking on the roles and responsibilities of both the coach and the coachee
- to enhance the skills of self-reflection and self-evaluation
- to establish collaborative relationships with colleagues

In the beginning of the course, the trainees reflected on their own experience in pedagogical activity as members of the university faculty with the purpose to speak on and assess the professional achievements, difficulties and challenges they faced while working. First in pairs, then as a whole group, they considered the most successful professional practices and discussed the problems of drastic changes in the educational context of modern universities.

Further on, based on the reflection and self-reflection, the trainees worked out the ideal model of the language teacher excellence they would like to pursue. The model represented a set of personal and professional qualities including language proficiency, knowledge and skills required for a successful career of an EFL university teacher.

As soon as the 'model' was composed, the trainers presented an overview of a wide range of educational strategies and activities for language teachers' professional development including *Team teaching, Teaching portfolios, Case analysis, Teacher support groups, Workshops, Self-monitoring, etc.* [8]. The focus was on the opportunities provided by those educational strategies in addressing the participants' immediate professional needs as well as in facilitating the process of achieving their short- and long-term goals within EFL teaching. Afterwards, the trainees defined the role of those educational strategies in advancing language teachers' professional development in general and their professional competencies in particular.

At the next step, the participants were introduced into the Peer Coaching strategy (its aims and goals, steps and stages, forms of activity, roles and responsibilities of the performers, etc.) and were familiarized with its specific types such as *Technical Coaching, Collegial Coaching, Challenge Coaching, Team Coaching,* and *Cognitive Coaching* [4]. Having chosen the most suitable Peer Coaching *type* to match their urgent professional needs, the trainees practised one of the educational strategies and activities mentioned above in terms of the EFL teachers' professional development.

The next stage of the training process was micro-teaching with one trainee acting as a coach, another – as a coachee, and the rest of the participants assuming the roles of students in the classroom. The participants followed the Peer Coaching guidelines worked out by the course developers. The trainees adhered strictly to the rules while observing the three main phases crucial for the successful implementation of Peer Coaching in pedagogical practice: *Peer Watching* (observation and note-taking without any comments from the coaches), *Peer Feedback* (presentation of the collected data and facts by the coaches), *Peer Coaching* (true peer coaching between the coaches and coachees accompanied by presentation of ideas and subsequent discussions on how to improve the EFL teaching process) [1]. The practice required trainees to stay focused on regular self-reflection and self-evaluation, and to produce objective feedback on the course activities both orally and in the written form.

3   Results and Discussion

A survey based on the 'Motivation of professional activity' method proposed by Zamfir in the modification of Rean [7] was conducted at the very beginning (pre-implementation period) and at the very end (post-implementation period) of the course. The trainees' level of professional motivation as a crucial aspect for professional development of teachers revealed the correlation between the existing extrinsic and intrinsic motives, which provide for EFL teachers' professional promotion, and identified the core motives for their professional performance. When comparing the results of the conducted survey presented in Table 2, a positive change for the ideal motivation complex with **IM > EPM > ENM (4.32 > 4.127 > 3.834)** from the initial poor motivation complex with **IM < EPM < ENM (3.705 < 3.945 < 4.036)** became obvious.

Table 2. Level of extrinsic and intrinsic professional motivation.

Implementation period	Intrinsic motivation (IM)	Extrinsic positive motivation (EPM)	Extrinsic negative motivation (ENM)
pre	3.705	3.945	4.036
post	4.32	4.127	3.834

The results of the post-implementation survey demonstrated a remarkable growth in *Intrinsic motivation* (of more than 14%), which reflected the trainees' satisfaction with an EFL teaching process, their awareness of the newly adopted educational strategies and techniques, achieved aims and goals as well as self-realization and distinction in the professional sphere, etc.

Positive changes in *Extrinsic positive motivation* comprising the respondents' promotion intentions were confirmed by a 4.4% increase.

Extrinsic negative motivation, which lowered slightly (by 5% only), highlighted the trainees' lessened fear of criticism from the administration and colleagues associated with their possible low-performance, and proved the developing culture of collegiality among the participants.

A further survey conducted in pre- and post-implementation periods revealed the percent of trainees who were ready to adopt new educational technologies and use Peer Coaching in the teaching process. The results presented in Table 3 indicate that the majority of respondents experienced difficulties when applying new educational strategies and techniques in the classroom. It was a common problem for most of the teachers irrespective of their EFL teaching experience (about 72%). It is important to note that by the end of the course virtually all the trainees (94.5%) had acknowledged Peer Coaching to be very useful and supportive in acquiring new teaching methods and educational techniques.

In the course of round-table discussions and in final reflective essays the trainees spoke on the advantages and disadvantages of the in-service training course *Peer Coaching in Work Practice of EFL Teachers* focusing on the challenges they faced (e.g. non-evaluating peer- and self-observation) and the results they had finally achieved. In spite of different work experience, all the trainees, both experienced and inexperienced, underlined the efficiency of Peer Coaching process in acquiring new strategies and techniques that were mastered during the course. They stressed its non-threatening character and principles of informal, helpful observation that contributed to

Table 3. Percentage of the trainees ready to use Peer Coaching in EFL teaching.

Implementation period	YES		DO NOT KNOW		NO	
	pre-	post-	pre-	post-	pre-	post-
Are you ready to use educational technologies in teaching?	27.7%	**100%**	55.5%	–	16.8%	–
Is collegiality efficient in teaching?	66.7%	**88.9%**	33.3%	**11.1%**	–	–
Is Peer Coaching helpful in teaching?	61.2%	**94.5%**	27.7%	**5.5%**	11.1%	–
Are you ready to use Peer Coaching in teaching?	38.86%	**83.2%**	44.36%	**16.8%**	16.8%	–

a friendly and relaxed atmosphere in the classroom and helped to cultivate open and fruitful collaboration with other participants. As soon as the positive sides of Peer Coaching strategy were stated, the trainees marked some downsides they experienced while doing the course. Thus, they pointed out insufficient training in mastering the time-management skills that they lacked to organize regular self- and peer-observation.

The course developers revealed a remarkable growth in trainees' professional motivation. It was underlined that in the course of time the atmosphere in the classroom improved significantly, the trainees became less tensed and more open to communication and interaction with their colleagues, and much more ready to advance their knowledge and master new skills.

4 Conclusion

The study has explored the motivational potential of peer-coaching for professional development of EFL teachers. The results indicate that collaboration with peers makes teachers feel less anxious and more motivated to share their positive and negative experience without constraints. The changing roles of a coach/coachee allow trainees to gain a better insight into the process of teaching and teacher-student interaction. Reflection and peer-feedback provide trainees with a means of building up their professional competence and actively enhancing their teaching practice. Due to its universal nature, the proposed professional development course can be recommended to university teaching staff of other specializations.

References

1. Gottesman, B.: Peer Coaching for Educators, 2nd edn. The Scarecrow Press, Inc., Lanham, Maryland, and London (2000)
2. Johnston, B.: Collaborative teacher development. In: Burns, A., Richards, J.C. (eds.) The Cambridge Guide to Second Language Teacher Education, pp. 241–249. Cambridge University Press, New York (2009)

3. Joyce, B., Showers, B.: The coaching of teaching. Educ. Leadersh. **40**(1), 4–10 (1982)
4. Joyce, B., Showers, B.: The evolution of peer coaching. Educ. Leadersh. **53**(6), 12–16 (1996)
5. Karimi, M.N.: The effects of professional development initiatives on EFL teachers' degree of self efficacy. Aust. J. Teach. Educ. **36**(6), 50–62 (2011)
6. O'Dwyer, J.B., Atli, H.H.: A study of in-service teacher educator roles, with implications for a curriculum for their professional development. Eur. J. Teach. Educ. **38**(1), 4–20 (2015)
7. Rean, A., Bordovskaya N.: Pedagogika: uchebnoye posobiye [Pedagogics: Textbook for Colleges]. Piter, St. Peterburg (2007). (in Russian)
8. Richards, J., Farrell, T.: Professional Development for Language Teachers: Strategies for Teacher Learning. Cambridge University Press, New York (2005)
9. Robbins, P.: How to Plan and Implement a Peer Coaching Program. Association for Supervision Curriculum Development, Alexandria (1991)
10. Shanks, J., Miller, L., Rosendale, S.: Action research in a professional development school setting to support teacher candidate self-efficacy. SRATE J. **21**(2), 26–32 (2012)
11. Tack, H., Vanderlinde, R.: Teacher educators' professional development: towards a typology of teacher educators' researcherly disposition. Br. J. Educ. Stud. **62**(3), 297–315 (2014)

Cultural Studies

"The Rivers of France" by J. M. W. Turner in the Intermedial Perspective

Irina V. Novitskaya[1(✉)] ⓘ, Irina A. Poplavskaya[1] ⓘ,
Lyudmila A. Khodanen[2] ⓘ, and Victoria V. Vorobeva[3] ⓘ

[1] Tomsk State University, Tomsk 634050, Russian Federation
irno2012@yandex.ru, poplavskaj@rambler.ru
[2] Kemerovo State Institute of Culture, Kemerovo 650056, Russian Federation
hodanen@yandex.ru
[3] Tomsk Polytechnic University, Tomsk 634050, Russian Federation
victoriavorobeva@mail.ru

Abstract. The article deals with intermedial features of the book edition of engravings "The Rivers of France" (1837) based on W. Turner's watercolours. It examines how the interplay of perceptual modes contributes to the construction of a multimodal narrative based on a journey and transmission of aesthetic and emotional load produced by the affordances of its textual and pictorial formats. It also investigates how principles of "poetic painting" philosophy are instantiated in the composition and thematic content of the engravings. The article describes the analyzed material, summarizes its discourse clues, identifies narrative stimuli in the engravings, and interprets them with regard to the underlying concepts of intermediality and poetic impact of the pictorial art. Such an approach results in identification of the central semiotic object in the engravings – the river – that represents a visual metaphor allowing for multiple interpretations. Underlying W. Turner's conceptual domain, the river symbolizes both life dynamics and its stability.

Keywords: Intermediality · Narrative · Turner · Engraving · Poetic painting

1 Introduction

The Research Library of National Research Tomsk State University owns a book edition of engravings titled "The Rivers of France" that was published in London in 1837 [18]. Our interest in this book is accounted for by several reasons, one of which is the fact that it is an integral part of a family library that used to belong to Russian philanthropist and statesman Earl G. A. Stroganov (1770–1857) and later was presented to the first Tomsk university at the time of its establishment. The Stroganovs' family library has always been regarded as a gem of the university Research Library collection and is known to contain more than 24,000 volumes in major European languages, which is indicative of its former owner's multifaceted personality [12].

Another factor that drew our attention towards this book is its artistic narrativity that is grounded in an interplay of perceptual modes. In "The Rivers of France", the authors experiment with combining two narrative modes – pictorial and literary – by

© Springer Nature Switzerland AG 2019
Z. Anikina (Ed.): GGSSH 2019, AISC 907, pp. 313–324, 2019.
https://doi.org/10.1007/978-3-030-11473-2_33

alternating between "showing" and "telling". In terms of literary studies, such multi-sensory experiment is known as a multimodal approach in literature. At the beginning of the XIX century, the appearance of multimodal texts was prompted by technical advancements in printing during the time [4, p. 424].

The produced aesthetic and literary effects have also been critically evaluated by researchers who have concluded that editions like that exemplify what one calls "poetic painting", an innovative approach in pictorial art of those times [20]. The present study considers a series of engravings known as "The Rivers of France" (1837) produced after the drawings by famous British watercolorist and painter Joseph Mallord William Turner (1775–1851), R. A. (full member of the Royal Academy). The key premise here is the assumption that pictorial arts in general and paintings in particular are the media with a narrative potential. Whether or not a viewer "hears or reads", a narrated story is largely dependent on the active involvement of the viewers in dynamic communication on the one hand and on the ability of the narrator and his narrated world to trigger narration clues and make the narrated world "appeal" to viewers. The poetic aspect of such a narrative complies with the theory of "poetic painting" introduced by Turner's teacher, Joshua Reynolds (1723–1792), the first president of the Royal Academy of Arts [20].

The present paper focuses on how the interplay of perceptual modes contributes to transmission of aesthetic and emotional load produced by a multimodal narrative as recorded in textual and pictorial formats. So to analyze a narrative, jointly constructed by J.M.W. Turner and his engravers, as well as the author of the verbal text, L. Ritchie, it is necessary to outline the key operational concepts of the present study (Sect. 2), to present the analyzed material, summarize its discourse clues, identify narrative stimuli of the engravings, describe and interpret them with regard to the underlying concepts of intermediality and poetic impact of the pictorial art (Sect. 3).

2 Key Concepts

A preliminary note concerns the status of pictorial images in narrative studies, the primary context of the approach developed in this article. Within the purview of the present article, the questions also arise as to *How does the pictorial narrative fit into the domain of intermedia studies? Can we define it as a media convergence if pictorial images are accompanied by written texts, or it is a case of multimodal communication? To what extent do such "texts" allow multiple interpretations?*

Although culture has always had a visual component, it is only recently that visual images have come to the foreground of research in the humanities and in social sciences owing it to a developing theory of multimodal communication [4, 9, 11]. This "visual/pictorial turn" has led to a reconceptualization of the term "text" with a view to various modes of communication and to the introduction of a new term "multimodal text" helpful to denote "texts that use a variety of signs, such as image, language, and sound" [4, p. 426; 15, p. 28]. Within the new paradigm, a visual image is understood as a special kind of text with some framing potential which requires an interpretation within the type of discourse it evolved in [15, p. 144]. Taken from this perspective, visual images are regarded as still pictures of people's social and cultural beliefs, their ordinary life in the previous periods of history, imprinting those moments or details that are hardly recoverable from other historical sources like books and chronicles. However, to perceive and interpret information encoded by visual images one should rely on the intra- and extratextual clues which channel viewers' attention and shape their frame of thought [17, pp. 161–179]. In addition to it, a certain perspective of "reading" of a visual image is determined by its author's active mediating role, in which he or she is responsible for the multi-layered perspectivization in the communication process [8].

All said above brings to the fore the artistic strategies and techniques that Turner employed in his paintings in compliance with the philosophy of "poetic painting" developed by J. Reynolds [20]. According to this concept, a painter should concentrate on depicting an idealistic world and turn to images of "the highest order" like views of cities, palaces, castles and landscapes. It is also a requirement of this art form to reiterate the themes well known from the world history and literature like "Snow Storm, Hannibal and his Army Crossing the Alps" (1812), or "Ulysses Deriding Polythemus – Homer's "Odyssey" (1829). This system of thought also required painters to use gradation as an artistic technique in presenting objects or shades of colours as well as to employ elements of the "décor", for example, time of day or night, light, weather conditions to create semantic expressiveness and richness. Another feature of the approach involved utilizing associative symbols like rivers, seas, cliffs, ruins of buildings, bridges, and clouds as expressive visual metaphors allowing for multi-layered interpretations. One of such symbols – the river seen as a representation of time, fate, memory, fertility, and mirror – functions as a key image in Turner's paintings that is central to the narrative in "The Rivers of France".

3 Turner's Pictorial Narrative "The Rivers of France"

3.1 A Description of the Edition

A collection of engravings "The Rivers of France", published in London in 1837, features various sites of France at the beginning of the XIX century.

The book contains 57 engravings depicting scenes along two major rivers in France: the Seine and the Loire. These engravings were first published in three successive volumes of travel stories titled "Wanderings by the Loire" [14] and "Wanderings by the Seine" [13], also known under the title "Turner's Annual Tour", which were reminiscent of Turner's journeys to France in the period 1821–1832. The sketches that Turner had made while touring the Loire in 1826 and the Seine over a decade resulted in the drawings prepared for engravers. However, in the "Wanderings…" the engravings after Turner's watercolours were mere illustrations to the text [6, p. 222]. As Leitch Ritchie (1800–1865), the author of the text of the "Wanderings…", put it in the first volume, he was determined to try to describe the rivers as he [Turner] saw them "… with his eyes and imagination – intromitting as little as possible with the descriptions of the artist – and to beguile the tedium which may be supposed to attend a lengthened description, by romantic narratives illustrative of its history or scenery, or the manners and habits of the people" [14, Advertisement]. It is a well-established fact that the two men were on friendly terms but did not travel together [7, p. 170; 19,

p. 116], so one cannot but admire how well Ritchie's descriptions of the sites and scenery contributed to an enhanced perception of Turner's travel sketches which were later turned into watercolours and engravings.

In the edition of 1837, the focus seems to have been shifted from the text to the images because each engraving is accompanied by a short passage written in two languages (French and English), which seemed to refer to the story that had once been told. The subjects in the engravings were scenes along the Loire River, which runs for 1000 km from its origins in the southeastern mountains of France, north to Orleans and then west to the Bay of Biscay, as well as of the busy Seine River, which passes through the center of Paris and exits into the English Channel. According to Ruskin, one of Turner's most loyal admirers, the engravings in "The Rivers of France" are "the best … ever done from Turner except a few vignettes to Rogers' Poems" [16, p. 332, V. 38.].

The book edition opens with an engraving in the frontispiece that depicts the Italian Boulevard in Paris. As we browse further through the book, we cannot stop admiring the wonderful river views (*Between Quilleboeuf and Villequier, La Chaise de Gargantua near Duclair, Scene on the Loire, Confluence of the Seine and Marne, Quilleboeuf*), and become overwhelmed at the sight of grand cathedrals and castles (*Orleans, Palace at Blois, Tancarville, Rouen Cathedral, Saint Julien's, Saint Florent*), comprehend the rhythm and every minor detail illustrative of city dwellers' daily life (*Canal of the Loire and Cher, Riez, near Saumur, Chateau de Nantes, Havre, Harfleur, Saint Denis*). The collection of images ends with engravings featuring all the views of Paris – Paris, from the Barrière de Passy, The Marchè-aux-Fleurs and the Pont-au-Change, Pont Neuf, Hòtel de Ville and Pont D'Arcole and others. Conceived as a part of an ambitious plan to portray the great rivers of Europe, the book of engravings "The Rivers of France" was pitched at a new middle class readers and turned out to be Turner's most celebrated series, coffee table-top best sellers [2]. "The subjects on the Seine are on the whole the most wonderful work he ever did, and the most admirable in artistic qualities; while those on the Loire, less elaborate, are more majestic and pensive" [16, p. 449, V. 13].

As Turner's biographers point out, it was not without reason that the painter turned to such an artistic form as engravings to disseminate his works [7, p. 36, 170]. His interest in engravings revealed itself in the 1790 s. The painter was quick enough to realize commercial benefit of this artistic genre, which was in fashion at that time, and soon had engravings based on his drawings published in illustrated magazines and books. The first engravings after his topographical drawings appeared in the *Copper-Plate Magazine* (1794–1798) and the *Pocket Magazine* (1795–1796). What is more, it was the flourishing market for landscape and antiquarian topography that provided his first real income, and little by little Turner was able to become financially independent and to rent an accommodation in a luxurious neighborhood in 1799.

The process of producing the engravings began with Turner painting a watercolor design. Professional engravers then rendered the figures, landscapes, boats and buildings using hatchings, cross-hatchings and marks. "He etched with his own hand the foundation outline on the copper plate, which was then handed over to the professional engravers in mezzotint, who worked under his supervision, and a hard taskmaster he proved" [7, p. 66]. It was customary for Turner to improve views during the engraving process. As a result, views could be considerably altered, as in Troyes, for instance. "The blocked in pinkish L-shaped mass center right in the watercolor design becomes a cathedral. The blue-gray flow of the river becomes a playground for ducks. The structures on both sides of the poplars are more distinct. And does not the bluish night sky appear happier lit up by a crescent moon!" [The J.M.W. Turner Museum, London, https://jmwturnermuseum.org/product/troyes/].

The painter was known to be a perfectionist as far as his drawings were concerned, so he kept improving his works even when they were exhibited to the public. In addition to it, he was always determined to provide viewers with as many details in the titles of his painting as it was necessary to be understood adequately and fully. It seemed the painter was "chewing up" the precise meaning of his paintings in the titles, like in the case with Turner's famous painting with a ship caught in a snowstorm. Its full title says: "Snow Storm – Steam-Boat off a Harbour's Mouth Making Signals in Shallow Water, and going by the Lead. The Author was in this Storm on the Night the "Ariel" left Harwich" (1842). This Turner's recurrent practice is consistent with the obligations incurred by the "poetic painting" concept. In order to contextualize his paintings in a wider cultural discourse the painter included quotes from poems by John Milton (1608–1674), James Thomson (1700–1748), Lord Byron (1788–1824) and his own unfinished poem "The Fallacies of Hope" as accompaniments in the titles [20]. Turner's forays into poetry enhanced the narratives of his landscape paintings by means of creating a synesthetic intermedial text allowing for paintings to "visualize" poetry and for poetry to enhance impression by giving way to emerging associations.

There was a permanent group of engravers, who worked under Turner's supervision on "The Rivers of France", some of whom were members of the Birmingham group [6, p. 222]. In the 1837 edition, we can enjoy works made by twelve engravers. All engravings for this series were done on steel plates, which were capable of producing far more perfect prints than the traditional copper plates.

3.2   A Description of the Multimodal Text

The book of engravings "The Rivers of France" presents itself as a narrative recorded in textual and pictorial formats featuring some French regions.

In 1832, commissioned the second time by the publisher Charles Heath, for whom he had just produced a series of Loire valley compositions, to provide illustrations for a popular publication project, Turner set out to tour the coast, the countryside, and the urban areas of northern France. That leisurely cruise up the Seine from Le Havre to Troyes is thought to have lasted approximately from 17 August till 23 October [5, pp. 171–183] and resulted in two sketchbooks "Seine and Paris" and "Paris and Environs" that are attributed to that tour [3, p. 394]. The tour was planned for the painter to be able to focus mostly on Paris and its satellites, and also on the picturesque banks of the river meandering through Normandy. Likewise, Turner's previous tour along the Loire river, across the region from Lyons to Brittany, also known as the Valley of the Loire or the Garden of France, had given the artist ample opportunity to catch the "feeling" of the location, to produce snapshots of eternal moments in everyday life with genuine precision and admiration. So, in the edition of 1837, there is a constellation of images featuring various places like Nantes, Havre, Orleans, Blois, Amboise, Tours, Saumur, Rouen and others each narrating a visual history of the area.

At the time of Turner's last tour, Normandy was recovering from the losses and devastation incurred by the Revolution in 1789 and the Napoleonic wars, economy and culture in the region were gaining prosperity owing it to agriculture and developing industries. Much of Normandy displayed its rural character comprising grasslands, picturesque riverside and rural landscapes allowing the artist to do hundreds of sketches. The landscape views were impressive, inspiring and appealing to literary enthusiasts to produce poetic descriptions of the area like the one written by Henry Wadsworth Longfellow (1807–1882) in his poem "Oliver Basselin", first published in 1858.

The region of the Loire valley was also bound to withstand the longstanding effects of the Revolution and the XIX century wars in both materialistic and psychological sense. Numerous ruins continued reminding people of the failures and damages that the people of France had survived and were the source of bitterness ingrained in their psyche. Nevertheless, the beauty and charm of the region's finest cities – Blois, Tours, Orleans, Amboise – and their suburbs, all steeped in history, character and culture, could not leave anyone unmoved. As a result, inspiring images of the region were captured in the works of art by numerous artists like George Clarkson Stanfield (1828–1878) in his "Saumur, by the Loire Valley", or by Gustave Courbet (1819–1877) in "La source de la Loire", or by Jean-Jacques Delusse (1758–1833) in "Vue of Les Rosiers-sur-Loire" and others. It is natural to assume, that the scenery Turner was able to observe resonated with his "system of thought", as Ruskin put it, and was expertly displayed in the painter's romantic landscapes. So let us take a closer look at the visual basis of the narrative jointly created by Turner and Ritchie.

At the first glance, it is peacefulness and magnificence of the nature that the scenes in the engravings seem to be bringing to the foreground for the viewers' eyes to catch. Owing to the narrative stimuli in the engravings – a panoramic and aerial perspective of the river, vast meadows and hills, castles, overhanging mountains, dozens of people and the sky with its atmospheric effects – a "story" that is conjured up in the imagination of viewers is all about tranquility, harmony and beauty, aesthetic and pacifying powers of the nature as well as balance in every aspect of life.

The genre of landscape, in which Turner was second to none, rendered its conventions to best suit the artist's thirst for freedom. It was while working on his Liber Studiorum (1807–1819) when Turner presented a great variety and range of landscape. That publication was issued in parts consisting of five plates each and covering all the styles of landscape composition, including Pastoral, Marine, Mountainous, Historical, Architectural, and Epic Pastoral.

In "The Rivers of France" one can hardly find any historical landscape scenes although every depicted site is known to have a historic background which is occasionally referred to in Ritchie's commentaries to the engravings. "The earth is covered with ruins, piled themselves upon the ruins of an earlier age; the atmosphere is thick with the shadows of history; our ear is filled with the hum of perished nations." [13, p. 58]. Similarly, in the book of engravings, we find that:

a Roman fortress, we are assured by some writers, existed on the spot; while others bring down the origin of the chateau [Chantoceau] to the year 992. In all events, it was celebrated in the beginning of the thirteenth century as a strong hold of a robber, who constructed the antique archers resembling a bridge seen in the plate, for the purpose of levying contributions more easily on the passing mariner. Inhabiting himself a castle perched on the summit of an imposing mass of mountains, that rose majestically several hundred feet above the water's edge, from which it seemed to glare around, with a jealous and threatening aspect, upon the whole valley of the Loire; his vassals lay watching below in the shadow of the bridge for their expected prey. This freebooter was attacked and dislodged by the famous Pierre de Dreux, Duke of Brittany, who took the castle by assault after a siege in which many of the defenders were slain, for which service he received in gift from Louis VIII, the domains of Chantoceau and Montfaucon [18, p. 6].

Thus, this verbal description of a ruined chateau contributes to a visually presented history by appealing to viewers' imaginative abilities.

The kinds of landscape that are more abundantly exemplified in the book of engravings are those regarded as pastoral, marine and mountainous. However, it should be noted that the scenes on the French rivers are presented in such compositions that are more of a mixed nature, for instance, a combination of a sea or river view with a vast countryside or a hillside. What is of great significance, as far as such landscape views are concerned, is recurrent perspectives (either downwards or upwards) and key types of subjects endowed with symbolic meaning. So, it is habitual to see people, a waterbody, a building and a bridge which are functioning as narrative stimuli in the intermodal text produced by Turner and Ritchie.

The thematic subject of the edition – the river – is depicted in a series of engravings representing the marine/river type of landscape. Both rivers in the edition, the Loire and the Seine, are seen rolling their placid waters either in broad daylight or in the night with the moon illuminating them. In the engravings, we can see a vast body of water occupying a third part or more of the painting surface. For the most part, we observe a river whose water surface is flat, quiet and mirror-like with reflections of surrounding objects, like in the views *Scene on the Loire, Tancarville (front view) or Chateau de Tancarville, Caudebec, Chateau de la Mailleraie, Confluence of the Seine and Marne, Coteaux de Mauves, Honfleur, Chateau Chantoceau*, etc. It is noteworthy that the painter is persistent in showing us the river as the center of life: there are always people in the paintings going about their daily businesses (*Saumur, Saint Denis, Reid near Saumur, Tours, looking backward; Vernon, Blois, Bridge of Meulin, Bridges of St. Cloud &Sevres, Mantes, Montjean, Rouen, Rouen looking up the river*) or just socializing and relaxing (*The Lanterne of St. Cloud, Bridge of Meulin, Troyes, View of the Seine, Jumièges*). The river with its traffic, which is rather heavy at times (*Tours, Amboise, Beaugency, Between Clairmont and Mauves, Canal of the Loire and Cher, Nantes*), takes center stage as a source of subsistence for common dwellers, tradesmen, businessmen (*Rouen, Rouen (looking up the river), The Marchè-aux-Fleurs and the Pont-au-Change, Havre, Chateau Gaillard*), and as a place where the old technology meets the new (*Between Quilleboeuf and Villequier, Havre with the Tower of Fransis I*). However, seeking balance in presenting life as it appeared around him, Turner hints at the might of the nature and at the devastation that high waters of rivers or seas can inflict on people and their possessions. In a couple of engravings, one can "read" a message about unpredictability of the nature, anticipation of its risks, and probable ensuing damage that people will have to cope with in their lives. For example, the gestures of the boaters in *La Chaise de Gargantua near Duclair* suggest that they are aware of the fact that the storm is going to hit them soon, there is also a thunderbolt illuminating a huge hill, a cloudy sky and an overall gloomy colour of the scene. There is an implication of a confrontation that the people will have to face in no time. Ritchie's text partially contributes to this effect: in his passage about *Quillenboeuf*, he mentions that it used to be a fortified port and fortress located in an important position which was once attacked by the troops of the Duke de Mayenne, but managed to turn them in retreat.

In quite a number of landscape images, the painter presents a stunning aerial view of a river elegantly bending or gracefully meandering through the valley into the sky (*Rouen (looking down the river), Saint Germain, Bridges of St. Cloud &Sevres, Tancarville (front view), Chateau Gaillard,* etc.). The river is known to encode a symbolic meaning of life, so it is rather essential that Turner provides us with no narrative clues triggering the idea of life's beginning (a spring gradually transforming into a river) or life's end (a river flowing into a sea) in the depicted scenes. Viewers are expected to understand Turner's underlying idea, which he eloquently expresses without any words: life is going on, life is in its full swing, and one should accept it with all its beauty, turns and surprises. Ritchie echoes this attitude in his commentary: "He who can gaze from such an elevation on a picture like this without an inflation of the breast, a tingling of the blood, a perceptible waxing of the principle of animal life throughout his frame, a disposition to *shout* as he was wont in the brave joy of boyhood, – let him descend at once into the valley" [18, p.10]. To enhance the sensory perception of visual images the writer tends to describe those sites by means of superlatives or emotionally-coloured vocabulary like *the scene of mingled grandeur, unconceivably fine, fine, picturesque country, the beauty of the landscape increases to a degree of magnificence, the delicious view, a superb/magnificent view, the most remarkable site.*

Another object that enriches Turner's visual narrative is a bridge that is depicted in a series of engravings representing what one can classify as a kind of an architectural landscape. A bridge per se is a beautiful architectural construction which adds an aesthetic value to the visualized scene. As a symbol, it is believed to encode the concept of transition and change of every kind. It can be a means of crossing a river, or crossing from one country and world to another. It can even symbolize a critical juncture in a person's life, a link between life and death, a change from old to new. In the book of engravings, the bridges that come center stage are *Pont de L'arche (the Bridge of Arch), the Bridge of Meulan, the Bridges of St. Cloud and Sevres, the Pont Neuf (The Bridge of Nine Pillars), the Pont-au-Change (The Exchange Bridge),* a bridge in Tours which some travelers compare to the Waterloo bridge in London, a bridge across the Loire in Beaugency with its 39 arches, bridges in Saumur, in Rouen, Vernon, and Mantes, as well as *the Suspension bridge* in Paris. In one of the engravings, Turner managed to depict three famous bridges in Paris together: *the Pont-au-Change,* "so called when the Exchange was established by Louis-le-Jeune", *the Pont Neuf* (inaugurated by Henry IV in 1607), the oldest standing bridge over the Seine, called so to distinguish it from older bridges in the city, and a small portion of the *Pont-Notre-Dame* [18, p. 19].

The edition also presents another variant of architectural landscape in which the focus is on a prominent construction – either a cathedral (*Orleans, Saint Julien's*), or a chateau (*Chateau de Nantes, Chateau Gaillard, Chateau of Amboise, Graville,*

Jumièges, Lillebonne (Chateau), or a castle (*Montjean, Tar*), or a palace (*Blois palace*). This type of landscape can involve a city view as well (*Beaugency, Troyes, Rouen*). Thanks to the mastery with which the painter depicts architectural objects embedded in a landscape scene, viewers' perception is guided towards some social and ethical values helpful to reconstruct a cultural and historical model of the world with its temporal and spatial dimensions specific to that period. Rather than bring to light the utilitarian functions alone of the architectural objects seen in the engravings, the painter links them to their conceptual and artistic functions. Such an approach is in line with the idea of a transient, historical and unique nature of each culture, of the history as a continuous process of qualitative change and development. Turner's romantic style turns architecture into a symbolic means of construing his personal view on the world, of comprehending the place that each person occupies in the world, of conveying the concept of the nation as a driving force of any change.

It is essential that people are an indispensable element of the composition in every visual story in the engravings, they enliven the otherwise unemotional scenes and are there to indicate that people have created those architectural buildings to meet their own needs and purposes. Castles, chateaus and mansions are for people to live in, cathedrals and churches are for their religious spiritual practices, streets, squares, bridges and functional buildings as attributes of cities are designed and planned for people to go about their other purposes and interests. On the one hand, it is through architectural designs that people can realize their creative urge, on the other – these constructions can serve as a source of admiration and aesthetic pleasure. What is more, an architectural object can be "read" like a text of special kind whose function is to present history of some particular period via the visual mode. Any architectural object is thought to be a "real mental image object ... that place[s] itself directly in our existential experience and consciousness", it is a dialogue between our thought and feeling [1, p. 97]. Thus, the artistic dimension of any work of art, including architectural, lies in the domain of a viewer's consciousness. Only a person experiencing architecture can evaluate how it confines and frames human existence in the world. So, viewers' experiences of some place or object, evoked by unconscious emotions and images through sensory perception, allow reconstructing those layers of meaning that the "once lived" people inhabiting it filled it with. One can say that architecture embodies our memories, so it goes down to the viewer's natural imagination to "read" the memories enclosed in a building.

In "The Rivers of France", images of architectural objects present memories of two core layers: those of a distant past and those contemporary with the time when the authors lived. Take a look at the happiest scene in the book of engravings – *the Boulevard in Paris* – that opens the edition. Apart from the fact that it provides us with a snapshot of people's social practices at the time of the painter's visit to the French capital city, this image gives us some clues helpful to comprehend how the real world used to be constructed within a certain social and historical context. In other words, this visual image can be regarded as some code or a text with encrypted information, readily susceptible to multiple interpretations of viewers. In summation, heterogeneous elements of the urban or rural environment that people live in construct a "visual lifestyle", i.e. a visual representation of the matrix of interaction among social groups, socially imposed system of values and standards of behavior. Taken from this perspective, all engravings in the edition collectively represent the lifestyle of the residents inhabiting towns and villages along the two rivers.

A much more subtle level of narrative refers to the memories of the past that may be evoked by certain objects in the images, and on the whole, can be numerous and individually ascribed. Let us consider one of the frequently observed objects in the engravings – that of a ruin – that is endowed with the potential to conjure up memories of some tragic historical events or wars, or decline. Images of ruins can be seen in a series of engravings like *Palace at Blois, Montjean, Lillebonne (the Chateau and Tower), Jumièges, Chateau Gaillard, Saint Germain*. Ritchie's poetic description puts readers in the picture:

> Always ruins, however, – still ruins! On approaching nearer, we perceive only a mass of roofless walls, and broken turrets, – wild-flowers in the windows and nettles in the hall – ivy instead of tapestry and carpets of the long grass that grows upon graves. It is the once-famous abbey of Jumièges, whose remains just stand like a monument to itself. The annexed view is taken from a different point; but the idea it conveys of the mouldering edifice is excellent. The human figures in the piece add to the effect; they seem hastening away from the spot sacred to solitude and desolation [18, p. 11].

However, such upsetting scenes do not take center stage in the images presented in the book: the ruins are never shown in a close-up in the engravings, they are either on top of a hill, or at a distance and are always lit by the sunlight or moonlight. Their presence in the landscape scenes seems to be an indication to the human memories that are, for the most part, contained in people's minds, legends and stories.

> This chateau is a vast and massy edifice now falling into ruin. Its architecture presents a specimen of the taste of several different ages, beginning earlier that the fourteenth century, an ending with the seventeenth, when its western façade (generally supposed to be the most beautiful) was constructed by Gaston D'Orleans after the designs of Mansard. In the year 1575 the states general were assembled here by Henry III, and afterwards in 1588; the latter epoch was signalized by the murder within the walls of the Chateau of the Duke de Guise and his brother the Cardinal. The apartment is still shewn where the bloody deed was committed [18, p. 1].

One can only enhance memories associated with this palace, for instance, with the fact that it was once occupied by King Louis XII, King of France from 1498 to 1515, who was born in Blois in 1462, or that Marie de Medici, wife of King Henry IV, lived in the chateau from 1617 to 1619 having been exiled from the court. These links to the royalty and the court perceive notions of politics as a theatre, and a palace as a space of power.

Thus, Turner and Ritchie's multimodal text presented in "The Rivers of France" is a narrative that unfolds in two complementary modes – pictorial and textual. Taken individually, each of the modes provides its affordances to ensure the intended message is delivered. The pictorial mode functions as primary in this edition and exploits its conventions to present landscape scenes illustrative of daily life, social and religious practices, hardships and pleasures of residents. The textual mode complements the visual story by providing timely and insightful comments on the depicted objects and locations. Used in combination, the two modes prove efficient in encoding a message grounded in the principal symbolic object of the book – the river. "Images enhance the authenticity of narratives" [10], so the engravings in "The Rivers of France" manage to instantiate all the symbolic notions associated with the concept of river.

4   Conclusion

Our analysis of the book of engravings "The Rivers of France" sheds new light on the way semiotic modes interact in the communication and progression of narratives. The present research on word/image combination has the merit of bringing to the fore nineteenth century recurring practices of constructing intermodal narratives based on a journey. It focuses on the way the two perceptual modes were integrated in a narrative about a topographical and symbolic object – the river, which is prone to multiple interpretations owing to the plurality of its meanings.

As it can be "read", the rivers under study – the Seine and the Loire – form an expressive visual metaphor conjuring up its symbolic notion of a path. On these grounds, the river is associated with spatial and temporal progression of the French nation, with the development of France's history and culture. The rivers function as a specific communicative space linking the country's various regions, the capital with its environs. Along with that, the rivers encode an idea of a boundary in the literal and figurative senses, a boundary between urban and rural scenes, between present, past and future of this nation. The element that is functionally linked to the notion of a path is a bridge, which in the context of the book is indicative of an emergent concept of transition: from war to peace, from turbulence to steadiness, etc. Drawing on these assumptions, one can make inferences on the duality of the symbolic meaning of the river as the authors of the book present it. Whilst the river with its currents is perceived as an embodiment of progression, time, life, and continuously changing history, the motionless banks of the river symbolize stability, sustainability and salvation.

All said above allows for the identification of such properties of "The Rivers of France" as intermediality, narrativity, and semiotic polysemy.

References

1. Auret, H.: Toward the poetic in architecture. S. Afr. J. Art Hist. **25**(2), 97–111 (2010)
2. Brown, D.B.: Joseph Mallord William Turner 1775–1851, artist biography. In: Brown, D.B. (ed.) J.M.W. Turner: Sketchbooks, Drawings and Watercolours. Tate Research Publication, December 2012. https://www.tate.org.uk/art/research-publications/jmw-turner/joseph-mallord-william-turner-1775-1851-r1141041. Accessed 1 Aug 2018
3. Finley, G.E.: A 'New Route' in 1822 Turner's colour and optics. J. Warbg. Court. Inst. **XXXVI**, 385–390 (1973)
4. Gibbons, A.: Multimodal literature and experimentation. In: Bray, J., Gibbson, A., McHale, B. (eds.) The Routledge Companion to Experimental Literature, pp. 420–434. Routledge, London, New York (2012)
5. Herrmann, L.: Turner Prints: The Engraved Work of JMW Turner. Phaidon, Oxford (1990)
6. Hind, A.M.: A History of Engraving and Etching. Courier Corporation, New York (2011)
7. Hind, C.L.: Turner's Golden Visions. TC & EC Jack, London, Edinburgh (1910)
8. Igl, N., Zeman, S. (eds.): Perspectives on Narrativity and Narrative Perspectivization, vol. 21. John Benjamins Publishing Company, Amsterdam, Philadelphia (2016)
9. Kress, G., van Leeuwen, T.V.: Multimodal discourse: the modes and media of contemporary communication. Lang. Soc. **33**(1), 115–118 (2001)

10. Nørgaard, N.: Modality, commitment, truth value and reality claims across modes in multimodal novels. J. Lit. Theory **4**(1), 63–80 (2010)
11. Pignagnoli, V.: Paratextual interferences: patterns and reconfigurations for literary narrative in the digital age. Amst. Int. Electron. J. Cult. Narrat. **7**(8), 102–119 (2016)
12. Poplavskaya, I.A.: Problemy izuchenija biblioteki Stroganovykh v Tomske: knigi frantsuzskikh pisatelej XIX v. [Research on the Stroganovs' collection in Tomsk: books of French writers of the XIX century]. Tomsk. State Univ. J. Philol. **4**(20), 87–97 (2012). (in Russian)
13. Ritchie, L.: Wanderings by the Seine. Proprietor, London (1834)
14. Ritchie, L.: Wanderings by the Loire. Longman, Rees, Orme, Brown, Green, and Longman, London (1833)
15. Ryan, M.L., Thon, J.N. (eds.): Storyworlds Across Media: Toward a Media-Conscious Narratology. University of Nebraska Press, Lincoln and London (2014)
16. Ruskin, J.: The Works of John Ruskin. Edward T. Cook and Alexander Wedderburn (eds.). Library Edition, vol. 39. George Allen, London (1903–1912)
17. Schöttler, T.: Pictorial narrativity. Transcending intrinsically incomplete representation. In: Ign, N., Zeman, S. (eds.) Perspectives on Narrativity and Narrative Perspectivization, vol. 21, pp. 161–182. John Benjamins Publishing Company, Amsterdam, Philadelphia (2016)
18. Turner, J.M.W., Ritchie, L.: The Rivers of France. From drawings by J. M. W. Turner. R. A. Proprietor, London (1837)
19. Thornbury, W.: The Life of JMW Turner, RA: Founded on Letters and Papers Furnished by his Friends and Fellow Academicians. Chatto & Windus, London (1877)
20. Mosejchenko, A.E.: William Turner. Transl. from Engl. Bertelsmann Media Moskau AO, Moscow (2011)

Word of the Year as a Cultural Concept in Media Discourse

Marina Yu. Ryabova🆔 and Tatyana S. Sergeichik$^{(\boxtimes)}$🆔

Kemerovo State University, Kemerovo 650000, Russia
mriabova@inbox.ru, lalli8@mail.ru

Abstract. The article deals with the analysis of a linguocultural concept "word of the year" in the discourse of media communication; it studies the correlation of the concept with significant social events in the culture of Germany, the USA, the UK and Russia. The discussion that follows will explore the important historic events, the process of priorities shift, and emergence of new cultural values influencing on the language development, reflecting in the appearance of new lexemes and linguistic concepts encoded in the forms of Words of the Year. We observe that globalization in the modern world is the basis of mutual influence of different national cultures. The most obvious role is that of the English language, functioning in the USA and UK. The influence results in that modern English becomes a powerful donor for many other languages. The availability of mobile and Internet technologies, popularity of World Wide Web, especially among the youth, stimulate the process of lexemes penetration into various languages. The use of social sites as information grounds by media communication agencies contribute to forming of the common global information space and also dynamic global language discourse on the basis of the English language primarily.

Keywords: Linguocultural concept · "Word of the year" · Media discourse · Communication · Lexemes · Cultures

1 Introduction

In this article we discuss the specifics of functioning of the linguocultural concept Word of the Year in the space of media communication, as a manifestation of its connection with socially significant events in different national cultures (Russian, German and English), as well as dynamics of changing the relevance of sociocultural themes that served as the basis for these concepts generation in the form of certain lexemes.

It should be mentioned that there are comparatively few studies devoted to analysis of Words of the Year in linguistic cultural anthropology at present. The explanation of this fact may be due to several reasons. The first reason is evidently a comparatively small actual time of Internet access in a global scale, which provides users with access to the information they are interested in. For example, the first "Russian word of the year" dates back to 2007 – the time when it became possible to talk about a massive Internet access of Russian users which confirms the first thesis. On the other hand, it is important to take into account that the Internet allows the implementation of special

Z. Anikina (Ed.): GGSSH 2019, AISC 907, pp. 325–332, 2019.
https://doi.org/10.1007/978-3-030-11473-2_34

data computing technologies to process high frequency words, as well as an adequate representation of statistical data. Secondly, sometimes it is assumed that the word of the year rating is a sort of seasonal entertainment. However, one should recognize the importance of such linguistic ratings, since the choice of a word of the year is based on processing a huge number of media discourse publications, on a statistical analysis of the lexicon use, conducted by complex methods and methodology of corpus linguistics, and expert assessment of every sample's relevance for a calendar year period. Thus, the goal of this study is to trace the process of formation and functioning of the Word of the Year concept in various linguistic cultures. The choice of methodology is defined by the above mentioned research problems. To obtain reliable and clear results, the research methods employed here are as follows: discourse analysis of media discourse, linguistic description, contextual analysis, and content analysis.

2 Definition of a Cultural Concept. The Notion of the Linguistic Cultural Concept Word of the Year

This study assumes a linguocultural approach to definition of the concept, according to which the concept is understood as an element of the national linguoculture and mentality, a specific individual and group mode of perception and worldview, defined by a combination of cognitive and behavioral stereotypes and attitudes in their connection with national values and features of this culture [7]. In accordance with this understanding, the concept is a phenomenon that contains the experience of world comprehension by an individual or an amount of knowledge and ideas about the surrounding reality accumulated by society [10]. Concepts, as a reflection of cultural values, can be expressed in words and images, or in material objects. In comparison with a text that develops in space and time, the concept is always a minimization [9]. Accordingly, the phenomenon of Word of the Year as a reflection and result of the linguistic representation of certain cultural value characteristics in a given society at a particular moment of time, is understood as a cultural concept, or rather, a linguocultural one, taking into account its linguistic nature.

3 Word of the Year Analysis in Different Linguocultures

It is assumed that a Word of the Year is the word (or phrase) characterized by the greatest frequency and occurrence in media discourse during a calendar year, which, according to the results of expert rating evaluation, is defined as the most linguistically resonant. Experts (scholars, publishers, leaders of organizations and societies) are usually initiators of these ratings, together with linguists and researchers from other fields of science. Different aspects of Word of the Year's problems are considered by scholars. Thus, Nikolaeva studies a Word of the Year as a linguocultural concept. She states that Words of the Year "mark not only the most important political, economic and socio-cultural events and phenomena of a particular year, but also that they fix new meaningful elements in a linguocultural picture of the world in a moving, constantly changing semantic matrix of the national language" [6, p. 156].

Issers analyzes the practice of defining the Russian Word of the Year as a socio-linguistic phenomenon. She insists that Word of the Year not only helps to understand through language overt, but also covert social changes, adding that "these words fixing historically significant, often tragic, events of the present in the Russian discursive space inevitably fall into a sphere of playing interpretations with time, acquiring an ironic or sarcastic potential" [2, p. 25], which, in turn, reflects the socio-political situation in a society. Issers adds that Words of the Year "reflect not only the specifics of the public dialogue in Russia, but also its subjects' creative potential" [2, p. 25]. However, the view about "playing interpretation" and "creative potentials" in a society, in our opinion, is highly controversial, and somewhat curious and subjective.

According to Melnik, the definition of the Word of the Year allows to make the language analysis of a whole calendar year processing a very hot linguistic material, to summarize its verbal and conceptual content, to see the priorities in social values and the prospects of its further development [5].

3.1   Word of the Year Analysis in the German Linguoculture

Observing the choice of the German Word of the Year, Bogdanova remarks that this practice is a way to draw attention and interest to studying the German language [1]. Actually, institutions that make suggestions of a Word of the Year are the National Language Training Centers or government sponsored language societies. For example, the Association for the German Language in Germany (Gesellschaftfür deutsche Sprache) is aimed at stimulating the interest to German, encouraging its study and usage in the world, thereby ensuring its relevance and significance; the other institution – the American Dialect Society operates in the USA. Both organizations carry out this language research starting from 1971 (from 1977 on a regular basis) and 1991 respectively.

At the same time, in many cases it is important to mention an underlying financial reason for choosing a Word of the Year. This relates to such distinguished commercial publishers as Oxford University Press, Merriam-Webster, Collins, Macquarie Dictionary, which started the practice of choosing a Word of the Year for a period of 2004–2006. With the development of their global branches, the will to draw attention to themselves and their work by rating words is evident as they deal with words by definition.

During last forty years in Germany there is a strong tendency toward domination of political terms in choosing a Word of the Year. There are such examples as: *Gesundheitsreform* (health care reform, 1988); *die neuen Bundesländer* (the new lands in Germany, 1990); *Politikverdrossenheit* (apathy towards politics, 1992); *Superwahljahr* (the year of super elections, 1994); *Reformstau* (stagnation in reforms, 1997); *Bundeskanzlerin* (woman-chancellor, 2005) and many others.

Since 2008 the year associated with the global financial recession and 'crisis', the 2008 Word of the Year reflects the fact that an economic focus typically dominates in political affairs in all world developed countries. So, the crisis determined the buzzwords in the following ten year period in Germany as well.

For instance, in 2011 the problems of political economy prevailed in the context of *'Stresstest'* (the 'stress testing'), which the German banks, nuclear power stations and the country's strategic sites experienced like patients, examined for the operating life and modes of failure.

Another example is the 2012 Word of the Year – *'rettungsroutine'* (routine salvation). It denotes a multitude of German government's measures to save the economy in the long European crisis. The neologism *'rettungsroutine'* shows the power of economic influence on politics, as well as emphasizes the quotidian politicians' fatigue to take decisions in allocating multibillion euro tranches.

The abbreviation *'GroKo'* (from *'GroßeKoalition'*, i.e. big coalition) – the 2013 Word of the Year – as a paronym for *'crocodile'* expresses an ironic attitude and mockery over a big federal governmental coalition in Germany which gave no chance to other candidates during the 2013 election.

Year of 2017 was marked by the word with political connotation – *'Jamaika-Aus'* (Jamaica Out), reflecting the difficulties during formation of the new government after the Bundestag elections in September 2017. The meaning of the word *'Jamaika-Aus'* refers to the failure in forming 'Jamaica Coalition' of the Social Democratic Party of Germany (SPD), political alliance of two German political parties – the Christian Democratic Union of Germany (CDU) and Christian Social Union in Bavaria (CSU), Alliance 90/The Greens. The color symbols of these political parties – black, yellow and green respectively – are the colors of Jamaican flag. The addition of *'Aus'* in spoken German refers to unsuccessful negotiations. So, the word *'Jamaica Out'* is a political neologism denoting the ongoing failure to form a governing majority in the German Parliament.

In general, among all popular German words selected for rating, the top ten lists, especially for the last three decades, principally include the vocabulary connected with the society life and development. The most prominent for German people was the event of the country's joining to the European Union, and the word *'Reisefreiheit'* (freedom to move, 1989) reflects this fact, denoting the right of abode and freedom from immigration control. These words include: *'Sozialabbau'* (social welfare cuts, 1993); *'Hartz IV'* (unemployment benefits – euphemism for socially deprived, named after the fourth stage of Peter Hartz' reforms of the German labour market and job agencies, 2004); *'Abwrackprämie'* (scrappage program – to promote the replacement of old vehicles with modern vehicles, 2009); *'Wutbürger'* (angry citizens, 2010); *'Flüchtlinge'* (refugee, 2015) and others.

Above all, German people are traditionally concerned with ecological problems which influenced the choice of the Word of the Year several times: Umweltauto (eco-friendly vehicle, 1984), Tschernobyl (Chernobyl, 1986), Klimakatastrophe (climatic catastrophe, 2007) and others [3].

3.2   Word of the Year Analysis in the English Linguoculture

The action "English Word of the Year" started in 1991 in the USA by the initiative of the American Dialect Society. Later, a word of the year was chosen according to the version of various lexicographic publishing houses (since 2004 – according to the version of the Oxford Dictionary USA Word of the Year, since 2006 – according to the version of the Merriam-Webster publishing house). Since 2000, Global Language Monitor publishes a top list of words, phrases and names of the year. Since 2004, the list of Words of the Year is published by American lexicologist Grant Barret in the New York Times. During the whole period of choosing the Word of the Year in the United States, the inexhaustible interest of Americans in information technologies is observed [4]. Many American researchers can rightly be called pioneers in the field of information technology: due to US research groups, the original concepts of network packages which became the basis for creation of a World Wide Internet Web emerged. Subsequently, development of the world's most popular social networks – Facebook, Twitter, Instagram – has changed the lives not only of Americans, but of all people in the world.

In 1993 the words *"information superhighway"* were defined as the Word of the Year, in 1995 it was the word *"web"*, and in 1998 – *"e-mail"* and *"e-commerce"* (electronic commerce using the Internet), in 1999 – the abbreviation *"Y2 K"*.

For some time – from 2000 to 2008 – the rating of the Words of the Year in the United States has noticeably shifted to the level of political economy issues: *chad* (small pieces of paper or cardboard produced in punching paper tape or data cards - reflecting the scandal during the 2000 presidential election, which resulted in termination of using the voting ballots with perforation), *red state/blue state/purple state* (red/blue/violet states - republican states/states supporting democrats/vacillating states during 2004 presidential elections in the USA), *subprime* (high-risk - mortgage lending, 2007), *bailout* (saving companies from the financial crisis, 2008), etc.

Since 2009, when the word *"twit"* was coined (the ability to encapsulate human thought in 140 characters), a series of the words of the year related to Internet technologies continued to appear: *hashtag* (hashtag is the keyword indicated by the sign # – 2012), *#BlackLivesMatter*(#Lives of the Blacks matter – a movement that became massive in 2014 during street demonstrations caused by the murder of blacks, 2014), *fake news* (fake news is Donald Trump' expression who accused the media and Internet publications in lying, 2017).

In the UK the choice of "UK Word of the Year" is published by the Oxford University Press (Oxford University.com). In 2017, according to the data of this publishing house, the 2017 Word of the Year was *"youthquake"* (a significant cultural, political or social change arising from the actions or influence of young people (Oxforddictionaries.com). The word *"youthquake"* is a blend created by analogy with the word *"earthquake"* with two root bases - *youth* and *quake*. The new word *"youthquake"* repeats the prototypical base rhythmically, intonationally and even phonetically. Thus, a ludic element (playing mode) of communication emerges that helps to refer to a known pattern (*"earthquake"* as a metaphor of political upheaval, activity), and a new metaphorical meaning appeared - political or social change, an event caused by the youth, e. g: *On 18 April, Prime Minister Theresa May, leader of the Conservatives, called a snap election triggering seven weeks of intense political campaigning. After the British public went to the polls on 8 June, headlines emerged of an unexpected insurgence of young voters. So despite higher engagement figures among the baby boomer generation and despite Labour ultimately ending up with fewer seats than the Conservatives in the House of Commons, many commentators declared that 'It was the young wot "won" it for Jeremy Corbyn', and dubbed their collective actions a 'youthquake'* [11].

The history of the word *'youthquake'* dates back to 1965 when Diana Vreeland, editor-in-chief of "The Vogue", declared the Year of the Youthquake. In an editorial in the Vogue US January edition that year, she wrote: "The year's in its youth, the youth in its year. ... More dreamers. More doers. Here. Now. Youthquake 1965." Vreeland coined *"youthquake"* to describe the youth-led fashion and music movement of the swinging sixties, which saw baby boomers reject the traditional values of their parents. In 2017 the UK was the heart of youthquake, with the 'London look of boutique street-style individualism taking the high fashion houses of Paris, Milan and New York by storm to inform a new mass-produced, ready-to-wear fashion directive world-wide' [11].

3.3   Word of the Year Analysis in the Russian Linguoculture

Russian Word of the Year has been chosen since 2007 to continue the existing world practice upon an initiative of Mikhail Epstein, Russian-American linguist, cultural researcher and critic. The project is commonly referred to as sociolinguistic, because it involves a significant number of informants and experts as well as an extensive discussion of ratings in mass media. From 2009 to 2016, the buzzwords have been selected along with the experts by means of people Internet voting through the system *imhonet.ru*.

The first Russian word of the year – *'glamur'* (glamour) – was chosen in 2007. The buzzword reflected the orientation of then Russian society toward consumerism, entertainment, and, in a certain sense, idle way of life [8].

The year of 2008, marked with a global financial collapse, however, sharply changed the people's interest vector to political and economic direction. Thus, socio-political problems became matters of their deep concern which is reflected in the semantics of the Words of the Year: *'crisis'* (2008), *'reset'* (restart in the relationship between the USA and Russia after reaching agreements, signed by President Obama and the Russian Prime Minister Dmitry Medvedev (2009), *'police'* (after Russian police reform (2011), *#крымнаш* (#Crimea is ours – a neologism hashtag, Twitter-style cliché – expressing the reaction of the Russian people to the admission of Crimea into the rule of the Russian Federation (2014).

The 2013 Word of the Year *'Gosdura'* was triggered by a speaking mistake of Vladimir Pozner, a well-known journalist, the ex-president of the Russian Television Academy. Famous for his commitment to the principles of liberalism and democracy, V. Pozner mispronounced the word Gosduma (the State Duma in Russia) as *Gosdura* which sounded like an insulting word *'fool'* for the lower house of the Federal Assembly of the country. The Council of Experts of the Russian Word of the Year, in their turn, selected the lapse *'Gosdura'* taking into account several laws passed by the State Duma which were considered unreasonable or absurd (namely, the anti-adoption law known as "the DimaYakovlev's law"; "The Law on Sanctions for Individuals Violating Fundamental Human Rights and Freedoms of the Russian Federation"; the antigay propaganda law named as "Protecting Children from the Information containing a Denial of Traditional Family Values"; 148 article of the Russian Criminal Code declaring that "public actions clearly defying the laws of society and committed with the purpose of insulting religious beliefs" are criminal acts). The choice of the

Russian 2013 Word of the Year, evidently, was not made by the public opinion, since these laws restrict actions of the specific minor groups and a very small part of Russian citizens; it rather presents the divergent viewpoint and individual emotional reaction of the rating experts to these laws.

The Russian Word of the Year chosen in 2017, *'renovation'*, denotes the Moscow government program for mass displacement of citizens from shabby houses built in a period 1957–1968s. The choice of the 2017 Word of the Year can hardly be called reasonable, because this renovation program concerns only some smaller amount of Moscow citizens who are less than 1.5 million people (while the population in Moscow is 12 million), whereas the total Russian population is more than 146 million.

On the other hand, the choice of the words for 2015 and 2016 years – *'refugee'* and *'brexit'* respectively, appears reasonable enough. At first glance, the problems associated with a large scale population migration from Asia and Africa to Europe, as well as the United Kingdom's withdrawal from the European Union, were not directly relevant to Russian citizens, though they were of great interest to them. People care about global problems due to their full media coverage and wide availability of Internet information sources. So, the choice of the buzzwords *'refugee'* and *'brexit'* as 2015 and 2016 Words of the Year is assumed to be rather objective in terms of their frequency and semantic significance.

4 Conclusion

It therefore seems that the choice of the Word of the Year is usually determined by mass media publications under the influence of big money and politics, especially these days. This statement is proved by the fact that motivation to make up Words of the Year ratings is demonstrated, as a rule, by the economies seeking to dominate in the modern political space, such as the USA, Great Britain, Germany, and recently Russia joined these countries. It is worthy to note that the concepts of the Words of the Year are sometimes identical in the analyzed linguocultures, which is the evidence of similar interest vectors and common linguistic presentation of certain cultural values. This is the reason of the coincident choice of the same word of the year in different countries. The examples are: *'refugee'* as 2015 Word of the Year in Russia and Germany; *'postfaktisch'* or *'post-truth'* as 2016 Word of the Year in Germany and Great Britain, and, of course, *'fake news'* as 2017 Word of the Year most often used in mass media of the USA, Great Britain and Russia.

On the other hand, it should be said however that there is no any availability of word ratings and the lack of popular words selection in African cultures, such as: Egypt, Tunisia or the Republic of South Africa (RSA), for example. These countries have a number of the national language varieties, as well as mass media. Meanwhile, English is an official language and/or used as lingua franca in some African countries, though there is no motivation to choose the Word of the Year, which is explained by the interests and priorities that are different from the ones of the world leading countries.

It is remarkable that there is no research or any significant publications on the Word of the Year in China which can be explained by the interest closure of the country, their language difficulty, an ancient culture, and what seems to be more significant, their unwillingness to confront openly the competitor countries in geopolitical issues.

References

1. Bogdanova, Yu.Z.: "Slova goda" v Germanii kak sredstvo povysheniya interesa k izucheniyu nemeckogo yazyka ["Words of the year" as a means of promoting interest to learning the German language]. Forum molodyh uchenyh **4**(4), 173–175 (2016). (in Russian)
2. Issers, O.S.: Ot ser'eznogo – do smeshnogo: igrovoj potencial rossijskogo slova goda [From serious stuff – to funny things: language pun potential of the Russian "Word of the year"]. Politicheskaya lingvistika **4**, 25–31 (2015). (in Russian)
3. Kobenko, Y.V., Kostomarov, P.I., Meremkulova. T.I., Poendaeva, D.S.: Standard German hybridization in the context of invasive borrowing. In: Filchenko, A., Anikina, Z. (eds.) Linguistic and Cultural Studies: Traditions and Innovations. LKTI 2017. Advances in Intelligent Systems and Computing, vol. 677, pp. 275–285. Springer, Cham (2018)
4. Martseva, T.A., Snisar, A.Y., Kobenko, Y.V., Girfanova, K.A.: Neologisms in American electronic mass media. In: Filchenko, A., Anikina, Z. (eds.) Linguistic and Cultural Studies: Traditions and Innovations. LKTI 2017. Advances in Intelligent Systems and Computing, vol. 677, pp. 266–274. Springer, Cham (2018)
5. Melnik, Yu.A.: Obzor social'no-lingvisticheskih proektov «Slovo goda» [Survey of social and linguistic projects "The Word of the Year"]. In: Proceedings of 2nd International Scientific Conference on the Problems of Russian Higher Education Modernization, pp. 63–66. Izdatel'stvo Ippolitova, Moscow (2016). (in Russian)
6. Nikolaeva, E.V.: "Slova goda" kak lingvokul'turnye koncepty ["Words of the Year" as linguocultural concepts]. Philol. Sci. Issues Theory Pract. **10**(1), 154–157 (2017). (in Russian)
7. Popova, Z.D., Sternin, I.A.: Kognitivnaya Lingvistika [Cognitive Linguistics]. Vostok-Zapad, Moscow, AST (2007). (in Russian)
8. Ryabova, MYu.: Glamur kak kul'turnyj koncept i filosofiya povsednevnosti [Glamour as a cultural concept and an everyday life philosophy]. Bull. Kemerovo State Univ. **2**(58/1), 215–220 (2014). (in Russian)
9. Stepanov, Yu.S.: Koncepty. Tonkaya Plyonka Civilizacii [Concepts. A Thin Film of Civilization]. Yazyki slavyanskoj kul'tury, Moscow (2007). (in Russian)
10. Volodina, N.V.: Koncepty, Universalii, Stereotipy v sfere literaturovedeniya [Concepts, Universalia, Stereotypes in the Sphere of Literary Criticism]. Flinta, Moscow (2010)
11. Word of the Year 2017 is… The official announcement of the Oxford dictionaries word of the year 2017. Oxford dictionaries. https://en.oxforddictionaries.com/word-of-the-year/word-of-the-year-2017. Accessed 10 Aug 2018

Linguistic Studies

Emotive Metaphors in Professional Jargons

Mariia I. Andreeva[✉], Olga Yu. Makarova[ORCID],
Daria V. Gorbunova[ORCID], and Marina V. Lukina[ORCID]

Kazan State Medical University, Kazan 420012, Russian Federation
lafruta@mail.ru,
{mrs.makarova,darya.gorda,loukin-v}@yandex.ru

Abstract. The work investigates semantic structure and features of nominations of emotion fear – emotives – in professional and social jargons. Military
and navy jargons are found to be productive. The research was performed in
four stages. Stage one reveals that fear is primarily manifested through secondary emotions. At the second stage it is determined that the number of
descriptive emotives, as opposed to direct ones, prevails. The lexico-semantic
group 'Fear' is proved to comprise four subgroups of descriptive emotives,
namely, 'emotion', 'person in a state of emotion', 'personal actions in a state of
emotion', and 'causation of emotional state'. The dominant nominations constitute subgroup two – 'person in a state of emotion', whereas 'emotion' and
'personal actions' are poorly nominated by the emotives in question. Stage III
proves emotives to follow four metaphorical models formed according to
mapping of source and target domains. In particular, person in a state of emotion
is rather compared with personal actions than with a disease. Models resting on
comparisons with artifacts and nature are also determined. Stage IV implied
comparison of the obtained results. Applied complex methodology and research
stages provide detailed study. The study shows that in linguistics 'fear' is a
person-oriented lexico-semantic group with core nominations of a person in
professional and social jargons.

Keywords: Emotion · Fear · Emotive · Semantics · Metaphor · Models ·
Professional jargon

1 Introduction

The nominations of emotions are frequently researched in modern linguistics, in particular by Apresyan [3], Pennenbaker [18], Wierzbicka [27, 28]. Present research
focuses on semantic and structural parameters of emotives, i.e. nominations of emotions, originating from professional and social jargons. The latter were researched, in
particular, by Caballero [6], Low [12] and Malyuga [13]. The research aims to structure
the meanings of the emotives [4], determine and specify metaphorical ways of their
formation [11, 15]. The work is relevant and significant as it contributes to the theory of
metaphor and semantics. A deep insight into semantic structure of emotive units in
question provides better understanding of professional and social language cultures and
applied nomination strategies.

Z. Anikina (Ed.): GGSSH 2019, AISC 907, pp. 335–342, 2019.
https://doi.org/10.1007/978-3-030-11473-2_35

1.1   Literature Review

There are numerous studies of emotions' nominations held in modern linguistics [3, 17, 21, 22, 24, 28]. Emotives are defined as linguistic units with an emotive component – seme in their meaning [21]. Emotiveness is viewed as a linguistic equivalent of emotionality. The latter, however, is also studied within linguistics, in stylistic emotional marking of dialects in particular [30].

The three dimensional model of emotions developed in the theories of Plutchik and Kellerman [19] classifies emotions into primary and secondary. The former is also called basic and is no further subdivided into separate emotions. The scholars differentiate between eight primary emotions: anger, fear, joy, surprise, sadness, anticipation, trust and disgust [19]. A primary emotion can be a constituent of a more complex one. For example, love is a combination of joy and trust. Fear as a primary emotion comprises a number of secondary ones (see below, Stage I of the analysis).

The nominations of emotions fall into two categories: direct and descriptive [3, 21]. Direct or 'basic' nominations (term of Apresyan), are not considered as emotive as they only nominate an emotion, but not a person in a state of emotion. According to classification of Babenko [4], descriptive or metaphoric emotives (that describe personal emotional state), on the contrary, verbalize a personal emotional characteristics and their achievement (see below lexico-semantic group 'Action in a state of emotion'), personal actions and emotional impact (causation) [4]. The material of the research suggests that descriptive nominations are implemented by figurative means – metaphors.

The work studies professional emotive metaphors. Solnyshkina et al. [23] define units of professional jargons as substitutes of terms used in low-register communication [23]. Korovushkin [10] states that units of social jargons are applied by particular social group and are characterised by vernacular parameters [10]. The prior research of metaphors in professional jargons was implemented by scholars on the lexis of architecture and design [6], wine experts [12], science [29], and business [13].

To classify professional emotive metaphors we followed theories of Lakoff and Johnson [11]. The terms 'source' and 'target' domains are used to correlate literal and figurative meanings of the emotive units under study. Metaphorical mapping is achieved when target domain is structured and nominated according to the resemblance to source domain [11].

2   Materials and Methods

The research material comprised 83 linguistic units collected from dictionaries of professional jargons [1, 9, 14, 20, 25, 26], namely, military (17) (hereinafter the number of analysed linguistic units is indicated in brackets), navy (17), aviation (2), medicine (3), hunting (1) business (5), photography (3) and sport (1). The units under study verbalize fear and its secondary emotions. Social jargon is represented by 34 nominations of fear. It is to be highlighted that present research does not make comparisons between professional and social emotives and regards them as one lexico-semantic group. Moreover, 34 contexts with units in question were collected from language corpora – British National Corpus [5] and Corpus of Contemporary American English [7]. The contexts prove emotives to function in discourses.

The research is based on the following methods:

- continuous sampling method was used to collect emotives from printed and electronic dictionaries;
- it is to be mentioned that material collection was combined with the method of semantic or componential analysis, as the units in question were collected in accordance with emotive seme present in their meanings. Further the componential analysis was used to determine core and differential emotive semes [24] as well as to differentiate between direct and descriptive emotives. The core seme is 'generic integral seme conveying mutual categorical characteristics of the specific type units' [2]. Differential semes indicate differences in the meanings of units under study [24].
    - the differential semes were determined through analysis of dictionaries definitions [16];
    - description of the obtained results followed (see research Stages II and III);
    - elements of statistical method were used to achieve quantitative results;
    - the obtained data were compared according to every parameter determined at Stage IV.

3   Stages of Analysis

The research consisted of four stages. *The first stage (I)* involved material collection – linguistic units which nominate fear (83) – from the dictionaries of professional and social jargons (see References).

According to the obtained data, we distinguish between the following secondary emotions of fear: anxiety (20), nervousness (15), worry (12), cowardice (9), fright (7), horror (3), panic (3), vigilance (2), dread (1), and danger (1). Moreover, 10 units under study nominated absence of fear, namely, tranquility (8) and courage (2). The above-mentioned nominations constitute a lexico-semantic group 'Fear'. In terms of professional jargons' distribution, units of navy (17) and military (17) origin prevail.

Semantic components of the units under study were determined at *Stage II*. According to the emotive semantic primitive's (or seme's) position in the meaning of the units under study, we differentiate between direct and descriptive nominations of emotions and person in a state of emotions. The lexical-semantic group 'Fear' demonstrates the following statistics:

- direct nomination –13 units (for example, **wuss,** *soc. a coward,* **heebie-jeebies,** *soc. nervousness*) (hereinafter the unit is marked by professional or social origin, which is shortened. See Materials and methods);
- descriptive nomination – 70 units (for example, **cop out,** *soc. to be a coward,* **to get the wind up,** *mil., to be frightened*).

The quantitative analysis suggests that emotions and people in a state of emotions are verbalized by figurative means in professional and social jargons (84% of the units).

The research proceeded with the componential analysis which provided data for semantic classification of the descriptive emotive units. The authors developed a code system of semes, which were marked as following: Em (Emotion), Pers (Person), Ac (Action), Caus (Causation). Further the obtained semes were combined. Every combination corresponded to one of the following lexical-semantic subgroups:

(1) emotion (Em) – 10 units: **care,** *busin. worry,* **willies,** *mil., nervousness, worry,* **flap,** *mil., anxiety.* For example, '*There was no sense in letting a silly thing like an unsolicited horoscope put her **in a flap**'* [5];

(2) person in a state of emotion (Pers Em) – 35 units: **uptight,** *soc., nervous,* **chilled,** *soc.* **calm, arsed,** *soc. worried.* For example, '*Calm the hell down, don't **be so uptight** about it*' [20];

(3) personal actions in a state of emotion (Ac Pers Em) – 6 units: **to look out for squalls,** *navy, to be vigilant;*

(4) causation of emotional state (Caus Pers Em) – 19 units: **to catch sb. flat(-)footed,** *mil., to frighten,* **hairy,** *soc., frightening.* For example, '*She's going to be a scratchy, **hairy** traveller, she complains bitterly*' [8].

As a person-oriented notion, emotion is primarily nominated in the subgroup (2) 'Person in a state of emotion'. The statement proves the research to be anthropocentric.

Stage III involved classification of descriptive metaphorical nominations of emotions. Metaphors rested on the comparison of a person in state of emotion with (1) person (his actions, physiology, and characteristics), (2) artifact (its actions and characteristics), (3) nature (animals and forces of nature), and (4) disease. Table 1 below represents metaphorical models formed according to corresponding source and target domains [11].

Table 1. Metaphorical models of emotive units, nominating 'Fear' in professional and social jargons.

Source domain	Target domain	Number of units
Person	Person in a state of emotion	32
Artifact		21
Nature		15
Disease		2

Model 1 'Person → person in a state of emotion' is represented by 16 units. In particular the target domains of the units nominate personal state, namely, **uptight,** *soc. nervous, anxious,* **arsed,** *soc. nervous,* **hairy,** *soc., scaring,* **shook up,** *soc., nervous.*

Being frightened is compared with physiological characteristics and processes of a person, namely, **to be in a cold sweat,** *soc., to be shocked and frightened,* **cack**

it/oneself, *soc., to be horrified,* **brown trouser moment,** *soc., scaring,* **tongue-tied,** *soc. unable to express oneself due to fear.*

The following texts prove the units under study to be applied to, when person described, faces the person of an opposite gender or when a person is placed in a potentially dangerous environment. For example, *'They talked easily. Benny stumbled from time to time, and became* **tongue-tied** *when she looked at the handsome boy sitting beside them'* [5], *"I* **cacked myself** *when I looked over the edge of the cliff at the sea 200ft below."* [1].

The meaning of a descriptive emotive **Dutch (man's) courage,** *navy, courage of a drunk person* is motivated by alcohol consumption by soldiers in wartime in order to warm up and be brave. For example, *'We all know drink gives us all* **Dutch courage** *and if ever bravery medals should be awarded then give it to a drunk'* [8].

Model 1.1. 'Personal action → person in a state of emotion' comprises 16 emotives. According to the idiom **take it on the chin,** *to be courageous* sportsman's courage is compared with withstanding the hardest blow in boxing. For example, *'It was disappointing, but there are a lot of players down south and very few contracts so I* **took it on the chin.'** [5].

Military jargon provides emotives which nominate body organs and parts in source domain. For example, **have the guts,** *to be courageous.* The emotive is applied in the following passage: *'Who* **has the guts** *to take the fair decision, the tough decision, to offer to put an extra penny on income tax to pay for education?'* [5].

Flat feet in male are regarded as a physical contraindication to military service. The condition described motivates figurative meaning of frightening a person in the emotive **to catch sb. flat (-) footed,** *soc., to frighten.* For example, *'The rise of regional conflicts has* **caught** *military planners* **flat-footed'** [8].

Nomination of *vigilance* is motivated by professional military action **placing somebody on the look out for,** *soc., to make somebody vigilant.*

Nine emotives form Model 1.2. 'Personal action performed with instrument → person in a state of emotion'. The core seme of the emotives which constitute the model is 'action'.

We would like to emphasize that nominated instrument or artifact is, primarily, profession-related (military, navy and medical). The nervousness, fright, anxiety and tranquility of a person are compared with performed manipulations. Being nervous is associated with firing a gun in a specific manner in the emotive **to be light on the trigger,** *mil., to be nervous.*

The integral differential semes 'blockage' and 'stop' (of movement (source domain) and of fear (target domain)) are found in the idiom **choke (the) luff,** *navy, to calm down.* The same state is achieved by taking a 'medication' or applying ice (as to the traumatized area), namely, **put some ice on it/take a chill pill,** *med., to calm down.* For example, *'The police officer told Jack* **to take a chill pill** *and answer the questions'* [20], *'Stop behaving like that. Just* **put some ice on it'** [20].

Model 2. 'Artifact → person in a state of emotion' contains 21 emotive units of professional and social jargons under study.

A person is compared with mechanism which is wound, in particular, in the idiom **to have the wind up,** *navy, to be scared, to panic.* Navy origin of an idiom **to show smb. white feather,** *navy, to be show one's cowardice* is represented by the following description. When leaving the enemy at full speed, ships' steam boilers function excessively and white steam is discharged and appears above the ship which resembles a feather.

Person is compared with ship and its' location in the following units: **to be in deep water(s),** *navy, to be anxious,* **to be in smooth water,** *navy, to be calm, to* **seek the shore,** *navy, to look for tranquility.*

Metaphorical phrase **a (big/great) girl's/girls' blouse/blouses,** *mil., a coward, weak sensitive man* rests on comparison of masculine behaviour with that of a woman (differential semes 'cowardice', 'sensitivity' and 'weakness') achieved by nomination of a feminine item of clothing. For example, *'No matter how a lad feels, it's just not the done thing to display his emotions—he might be accused of being **a big girl's blouse**'* [5].

Model 3. 'Nature → person in a state of emotion' includes 15 units under study. Person is compared with forces of nature in five units under study, in particular, **port after stormy seas,** *navy, tranquility,* **keep a weather (-) eye/to look out for squalls,** *navy, to be vigilant.* For example, *'It's hardly surprising that Federal Reserve policymakers now **keep a weather eye on** the stock market'* [5].

Animalistic metaphors are represented by six emotives in question. The comparison rests on the characteristics of an animal. The professional emotive of hunting **buck fever,** *hun., nervousness* compares a hunter's condition with that of an animal being hunted. The following text of a professional discourse describes hunter's experience. For example, *'To prepare myself for the **buck fever**, I shoot my bow then sprint to my target'* [5].

The motivation of the word **albatross,** *navy, a source of anxiety* rests on the marine superstition. Traditionally, in marine professions to shoot an albatross is considered to be a sign of bad luck [5].

Model 4. 'Disease → person in a state of emotion' is poorly represented in the research material and verbalizes two units of professional and social jargons, namely, **collywobbles,** *soc. nervousness,* **bug out fever**, *mil., nervousness.* The nominated state can be applied to verbalize conflict issues as in the following example, *'Yet what is conflict in its essence? It's easy to recognise its surface symptoms: the anger, the panic, the shakes and the **collywobbles**'* [5].

The obtained results were compared at *Stage IV*. The lexical-semantic group 'Fear' is verbalized by 12 secondary emotions in the material under study. The major part of the researched emotives has professional origin (49 units) as compared to units of social jargons (34 units). The nominations of fear are more productive in navy and military jargons (34%). Direct nomination of emotions is found to be less frequent in the research material, unlike descriptive one (84%).

Four lexical-semantic subgroups formed according to the semantic/componential analysis reveal personal actions in the state of fear which are not frequently nominated. Supposedly, this may be predetermined by psychological factors.

The analysis of nomination strategies suggests that being descriptive, metaphorical emotives are represented in four source domains: person, artifact, nature and disease. Dominant motivation of fear by person and his/her actions prove emotion, fear in particular, to be person-oriented which is manifested by professional and social language cultures under study. Comparisons with animals and natural forces are found, however, not frequent. Frightened person is also compared with inanimate object – artifact, ship or gun in particular. The range of artifacts used in source domain is profession-related.

4   Conclusion

Regarded as a lexical-semantic group, 'Fear' presents complex structure manifested by semantic subgroups in professional and social jargons. The former prevail in number (49%). Determined semantic features are classified in four subgroups, basing on core semes 'emotion', 'person', 'action', and 'causation'. 'Person' is found to be the most common semantic component of studied emotives (50%). Person-oriented lexical-semantic group 'Fear' is, primarily, manifested by descriptive emotives – metaphoric nominations. Thus, research material and obtained results contribute to the general theory of metaphor. The correlation of source and target domains shows four models based on comparisons with: (1) person, (2) artifact, (3) nature, and (4) disease. Person, his/her actions and characteristics are used as basic source domains for descriptive nominations of professional and social emotives included into the lexical-semantic group 'Fear'.

The set of applied methods and stages of work allows detailed and complex semantic study of the professional and social emotives, it can be applied to further research works focused on lexical studies.

References

1. A dictionary of slang. http://www.peevish.co.uk/slang/l.htm. Accessed 14 Aug 2018
2. Andreeva, M.I., Solnyshkina, M.I.: Idiomatic meaning of idiom "halcyon days" in institutional discourse: a contextual analysis. J. Lang. Lit. **1**, 306–310 (2015)
3. Apresyan, YuD: Obraz cheloveka po danným yazyka: popytka sistemnogo opisaniya [Human image in language data: an attempt of systemic description]. Voprosy yazykoznaniya **1**, 37–65 (1995). (in Russian)
4. Babenko, L.G.: Leksicheskie sredstva oboznacheniya emotsiy v russkom yazyke [Lexical means of nominations of emotions in Russian language]. Izd-vo Ural un-ta, Sverdlovsk (1989). (in Russian)
5. British National Corpora (BNC). https://corpus.byu.edu/bnc. Accessed 14 Aug 2018
6. Caballero R.: Re-Viewing space: figurative language in architects´ Assessment of Built Space Mouten de Gruiter. Berlin/New York (2006)
7. Corpus of Contemporary American English (COCA). https://corpus.byu.edu/COCA. Accessed 14 Aug 2018
8. English Oxford living dictionaries. https://en.oxforddictionaries.com/. Accessed 14 Aug 2018
9. Jackspeak: A guide to British Navy slang and usage. https://books.google.ru/books?id=aYrpCgAAQBAJ&printsec=frontcover&hl=de&source=gbs_book_other_versions_r&redir_esc=y#v=onepage&q&f=false. Accessed 14 Aug 2018
10. Korovushkin, V.P.: Nestandartnaya leksika v angliyskom i russkom podyazykah [Nonstandard lexis in English and Russian sublanguages]. Vestnik orenburgskogo gosudarstvennogo universiteta **4**, 53–59 (2003). (in Russian)
11. Lakoff, G., Johnson, M.: Metaphors we live by. University of Chicago Press, Chicago (2004)
12. Low, G., Deignan, A., Cameron, L., Todd, Z.: Researching and applying metaphor in the real world. John Benjamins Publishing Company, Amsterdam (2009)
13. Malyuga, E., Orlova, S.: Linguistic pragmatics of intercultural professional and business communication. Springer International Publishing, AG (2018)

14. Mezhdunarodny musykal'ny klub [International music club]. http://mmk-forum.com/showthread.php?t=29428. Accessed 14 Aug 2018. (in Russian)
15. Moskvin, V.P.: Russkaya metaphora [Russian metaphor], 200 pages. LKI Publishing, Moscow (2012). (in Russian)
16. Myagkova, E.Yu.: Emotsional'no-chuvstvenny component slova [Emotional component of a word]. PhD thesis. Institute of Language Studies RAS (2000). (in Russian)
17. Pavlenko, A.: Bilingual minds: emotional experience, expression and representation clevedon. Multilingual Matters, Clevendon, GBR (2006)
18. Pennebaker, J.W.: Writing about emotional experiences as a therapeutic process. Psychol. Sci. **8**(3), 162–166 (1997)
19. Plutchik, R.: Psychology of individual differences with special reference to emotions. Annu. NY Acad. Sci. **134**(2), 776–781 (1966)
20. Rebrina, L.N.: Slovar' molodezhnogo slenga (na materiale angliyskogo, nemetskogo, frantsuzskogo i russkogo yazykov) [A dictionary of teen slang: English, German, French, Russian]. Volgograd State University Publishing, Volgograd (2017). (in Russian)
21. Shakhovsky, V.I.: Kategorizatsiya emotsiy v leksiko-semanticheskoy sisteme yazyka [The categorization of emotions of lexico-semantic system]. Voronezh State University Publishing, Voronezh (1984). (in Russian)
22. Siroka, D.: A linguistic picture of the world and expression of emotions through the prism of expressive lexis. J. Educ. Cult. Soc. **2**, 297–308 (2013)
23. Solnyshkina, M.I., Kalinkina, T.E., Ziganshina, ChR: Konventsii professional'noy kommunikatsii [The conventions of professional communication]. Phylology Cult. **3**, 138–145 (2015). (in Russian)
24. Sternin, I.A.: Problemy analiza structury znacheniya slova [Issues of a word meaning analysis]. Voronezh State University Publishing, Voronezh (1979). (in Russian)
25. Sudzilovsky, G.A.: Sleng – chto eto takoe? Anglo-russiky slovar' voennogo slenga [What is slang?]. Voenizdat, Moscow (1973). (in Russian)
26. Teen slang. http://teenslang.su/. Accessed 14 Aug 2018
27. Wierzbicka, A.: Emotions across languages and cultures: diversity and universals. Cambridge University Press, Cambridge (1999)
28. Wierzbicka, A.: Language and metalanguage: key issues in emotion research. Emot. Rev. **1**(1), 3–14 (2009)
29. Zeidler, P.: Models and metaphors as research tools in science. Zweigniederlassung, Zurich (2013)
30. Zelenkova, A.V., Makarova, OYu., Gorbunova, D.V.: Obuchenie chteniyu hudozhestvennoy literatury na yazyke originala [Teaching of fiction reading in original]. Gumanizatsiya obrazovaniya **3**, 65–71 (2018). (in Russian)

Transformative Resources
of the Terminological Internationalization (on
the Material of German and English)

Vladimir V. Elkin[1] , Elena N. Melnikova[1] ,
and Anna M. Klyoster[2(✉)]

[1] Pyatigorsk State University, Kalinina Pr. 9,
357532 Pyatigorsk, Russian Federation
{evvvve,e.n.melnikova}@mail.ru
[2] Omsk State Technical University, Mira Pr. 11,
644050 Omsk, Russian Federation
annaklyoster@mail.ru

Abstract. The article on the vocabulary material of German and English presents the experience of studying the terminological internationalization, characterizing the current state of a large number of languages and their interaction which is most actively manifested in such lexical areas as scientific and technical terminology, social and political vocabulary. It discusses the landmarks of replacing German with English as the universal language of science and the universal means of communication in the global arena. It is suggested that the present epoch, though still characterized with the dominance of English as lingua franca, can be called the epoch of internationalization directed at the active intermediation and integration of national languages and cultures of all parts of the world under the auspices of the English language. The research results display the process of terminological internationalization as a consequence of the rapprochement of the nations on the basis of socio-economic, political, scientific and cultural ties resulting from globalization. A special emphasis in the article is placed on the terminological base, formed as a result of the processes of internationalization of the modern German and English languages vocabulary. The process of assimilation goes so deep that the foreign language origin of the terms is not felt by native speakers, on the contrary, a loanword is perceived as a generally accepted international nomination of an object or phenomenon. The obtained data suggest that English and German are moving toward each other, rather than becoming more different.

Keywords: Terminological internationalization · Globalization ·
Language identity · Borrowing · Loanword · Classification of loanwords ·
Compound word · Assimilation

1 Introduction

Globalization, being an integral part of the modern social life development, has a certain impact on many spheres of life, including the language development which, of course, is subject to the intensification of international cooperation. Terminological

© Springer Nature Switzerland AG 2019
Z. Anikina (Ed.): GGSSH 2019, AISC 907, pp. 343–356, 2019.
https://doi.org/10.1007/978-3-030-11473-2_36

internationalization characterizes a current state of a large number of languages and their interaction which is most actively manifested in such lexical areas as scientific and scientific-technical terminology, social and political vocabulary.

Observing the latest achievements in science and technology of the 20th and 21st centuries, we note that English dominates in the realm of science. Nowadays if scientists need to coin new terms, they are most likely to do it in English. And if they are planning to publish a new discovery or the results of their research, as in our case, it is most definitely in English. However, it was not always so.

In different periods of human history and in different spheres of scientific knowledge in the world and the universe there prevailed different languages. Once, the dominant language of science was Greek, then Latin, French, and German. All of them were universal languages of science and universal means of communication in Europe. However, due to certain objective reasons their status of lingua franca began to fracture and each of them became just one of many languages in which science was done.

Michael D. Gordin, a Professor of the History of Science at Princeton and the author of two books "Scientific Babel", in which he explores the history of language and science [8], points out that English was far from the dominant language of scientific communication in 1900. The story of the 20th century is both the rise of English and the serial collapse of German as the dominant scientific language.

World War I had two major impacts on the system of basically having a third of science published in English, a third in French, and a third in German with Latin still preserved in some spheres. First of all, after World War I, Belgian, French and British scientists organized a boycott of scientists from Germany and Austria, denying their access to conferences and publications in the Western European journals. There formed two scientific communities: the German one functioning in the defeated Germany and Austria, and another one functioning in Western Europe which was mostly English and French. Moreover, it was the moment in history when international organizations were established to govern science. And those newly established organizations began to function in English and French.

The second effect of World War I took place in the United States where the official policy of Isolationism in the 1920s ruined the foreign language education. That resulted in a generation of future scientists who came of age in the 1920s with limited exposure to foreign languages. Besides, that was also the moment when the American scientific establishment reading, writing and speaking only English started to take over dominance in the world. So, due to political, economic, military and social factors it all ended up with a very American-centric, and therefore very English-centric community of science after World War II.

The present epoch, though still characterized with the dominance of English, can be called the epoch of internationalization and globalism, i.e. opposite to Isolationism and directed at the active intermediation and integration of national languages and cultures of all parts of the world under the auspices of the English language.

The Internet site Encyclopedia.com, having more than 100 trusted sources, including encyclopedias, dictionaries, and thesauruses, presents the factual material proving that at various times and in various ways English borrowed new words and still does nowadays from all the world's regions and languages.

Most often the need for nomination of new objects and concepts arises in various fields of science and technology, so among the scientific and technical terms the foreign loans are of particular interest because science and technology cannot develop only in a single country or on a separate continent.

Speaking about the terminological internationalization it is essential not to disregard such two notions as a loanword and a borrowing [7]. According to the online version of the Concise Oxford Companion to the English Language, a loanword is a word taken into one language from another [15]. Such words are, on the analogy of usury and money relations, both 'loans' from Language A to B and 'borrowings' by B from A. Thus, a borrowing, on the one hand, is the process of taking a word or phrase from one language into another, or from one variety of a language into another. And, on the other hand, it is the item so taken.

Borrowing is a major factor of language change, but the term itself is quite misleading and could be called a misnomer: it presumes repayment, whereas there is no quid pro quo between languages. The item borrowed is not necessarily returned, because it never leaves the source language and in any case changes in the transfer. The present situation in linguistics gave birth to a three-word German system employed to discuss the process of lending and assimilation: Gastwort, Fremdwort, and Lehnwort [7].

A Gastwort (guest-word) is an unassimilated borrowing that has kept its pronunciation, orthography, grammar, and meaning, but is not used widely. The term Gastwort itself, with /v/ for the W of Wort, a capital letter because it is a noun, and the alien plural Gastwörter can serve as a good example of the phenomenon under discussion. Such words are usually limited to the terminology of specialists and italicized and glossed when used.

A Fremdwort (foreign-word) denotes an advanced stage. It has been adapted into the native system, with a stable spelling and pronunciation (native or exotic), or a compromise has been made by translating all or a part into a native equivalent. The term Lehnwort for general purposes converted to loanword (also: loan-word, loan word) represents an instance of such an adaptation.

A Lehnwort proper is a word that has become indistinguishable from the rest of the lexicon and is open to normal rules of word use and word formation. It is seldom possible, however, to separate the stages of assimilation quite neatly. Assimilation into a language as widespread as English occurs at three levels: local, national, and international. A word may remain local in some variant of a language, then become national, and then international. Such a process often takes years, leaving many loans drifting uncertainly.

2   Subjects and Methods

The object of this scientific project research is the terminological base formed as a result of the processes of vocabulary internationalization of modern German and English languages. The methods relevant for such an analysis are stipulated by the aim of identifying the transformative resources that are manifested in the considered language systems of the Germanic group of languages and can be presented in the form of the method of continuous sampling from modern dictionaries and research materials, the method of etymological analysis, the method of contextual analysis using authentic text materials, the method of comparative analysis using an online search service (Google Books Ngram Viewer), the sociolinguistic method of correlating the intra-linguistic causes of changes inherent to the languages themselves and the phenomena and factors of non-linguistic reality associated with the history of specific nations, as well as with the general development of the human society, i.e. extralinguistic factors, as well as the descriptive method necessary for identifying the specifics of the use of considered borrowings.

The formation of the modern German language has undergone several stages of linguistic development over the past decade. Thus, the period of the end of World War II was marked with the criticism of the authoritarian language that had arisen during the governing of the national socialists, and the emergence of a new language style borrowed from the United States. During the same period there arose ideological differences in the norms of language use between the Eastern and Western parts of Germany which were primarily conditioned by the consequences of the economic development. The internationalization of the terminological structure of the language became even more noticeable in the period of industrial development of the society and the emergence of services sector, for the service companies that retained their original name in German. The terminological transformations that took place in German at the time of the European Union emergence affected such spheres of life as economy, public life (a large number of borrowings appear in connection with the opportunity to travel), food consumption, as well as goods of the services sector.

Since the 1990s the process of terminological internationalization has been greatly accelerated by the new technologies, free trade and globalization, which in the current macroeconomic environment leaves virtually no chance of preserving the so-called "small languages". English acquires the status of the language for international communication (lingua franca), and the increase of its role and influence on the structure of other languages in linguistics is designated as an intellectual disaster. The active penetration of elements of English into the modern German language, as it is underlined by some linguists [5, 6, 11, 17], can have negative consequences which are seen primarily in the loss of linguistic identity. In this regard, there is a fear that the linguistic structures that were originally formed in the inflectional Roman-Germanic group of languages, based on Latin are beginning either to change or disappear altogether. Although Latin, along with ancient Greek, has long been a source for formation of scientific terminology and socio-political vocabulary, researchers note the loss of the common language experience.

The modern society which is developing in the digital world of mass communication, and which thus received the name "digital natives", is more capable of making an impact upon the formation of the language, as the youth language is actively using the appropriate terminology and graphics to a great extent simplifying and accelerating the process of communication. Such substitutions are ambiguous among those linguists who wonder whether such transformations contribute to development of the languages and their creative renewal, or whether the matter is their decline and semantic impoverishment. Nevertheless, the processes of the terminological internationalization are actively manifested both in modern German and in modern English which undoubtedly have an impact on transformation of vocabulary and grammatical structure of the languages.

3   Results

The universal dictionary of terms of modern German DUDEN 2017 [6] records 5000 new borrowings, some of which are studied in our scientific research. Having considered some historical processes that influenced the development and formation of modern German, it is necessary to identify those areas of the social activity where the internationalization of the terminological repertoire of the language is represented to a greater extent. The notion term not only includes separate lexical units, but also word-combinations and abbreviations. The greatest number of the terms borrowed from English is presented in the sphere of economy, in the so-called language "Business-Deutsch": Abteilungsleiter, Personal-referent, Sachbearbeiter – Manager; Maßstäbe – Benchmarks; Besprechungen – Meetings; Marktwert – Shareholder Value; Tafel – Flipchart; Ausziehbanner – Rollup; Rechnungswesen – Controlling; Erläuterungen – Briefing; Verlagerung von Aufgaben – Outsourcing; Hausmeister – Facility manager; Fensterputzer – Vision Clearance Engineer; Hilfsköchin in der Kantine – Nourishment Production Assistant; Markenpflege – Brand Management; Fähigkeiten der Mitarbeiter – Skills.

The emergence of a large number of borrowed terms in German can be explained by the fact that currently we observe the active internationalization of business and the development of production and economic cooperation among different countries which entails the displacement of native German words and phrases and their replacement with the lexical units of the global English language, serving as a means for international communication.

Having conducted a comparative analysis with the application of the online search service Google Books Ngram Viewer which allows to make diagrams of the frequency of use of language units on the basis of a huge number of printed sources published since the 16th century and collected in the service Google Books, we determined the frequency of use of the following concepts in speech with the meaning/a head of a department/: Personalreferent – 4%, Sachbearbeiter – 15%, Abteilungsleiter – 23% – Manager – 58%. Having considered several pairs of dictionary units by this method, we have obtained the following results: Fähigkeiten der Mitarbeiter – 24%, Skills – 76%; Verlagerung von Aufgaben –11%, Outsourcing – 89%.

Thus, it becomes obvious that the borrowed American-English terms Manager, Skills, Outsourcing are actively used in the vocabulary of modern German, reducing the percentage of the national vocabulary use, and thereby affecting the grammatical structure of German, since the borrowed terminology is used with the grammatical features of the English language.

The extensive internationalization of terms takes place in such spheres of the public life as leisure and food where one can observe the emergence of a large number of the borrowed American-English words due to the spread of the American material culture: Feierabend – After-Work-Party; Bäckerladen – Back-Shop; Imbissbude – Restaurant; belegte Brötchen – Burger, Fast-Food, Henkelmann – Cheeseburger; Stulle – McRibs; Brotdose – BigMac; Kinderportion – Kids Menu.

The process of the language internationalization has also affected large German enterprises which actively use the English equivalents instead of common national terms, denoting the type of their activity. For example, the company Ron Sommer (Deutsche Telekom) actively introduces the following terminology: Telefonate tagsüber – Sunshine-Call; Abendgespräche – Moonshine-Call; Ortsgespräche – Short-Distance-Call; nationales Ferngespräch – German-Call; R-Gespräch – Free-Call; Deutsche Bahn AG now uses only the terms Tickets, Meeting-Points, Service-Points, and the faculties of higher education institutions in Germany are designated by the term Department.

Among the new words of the universal dictionary DUDEN [6], actively used by the native speakers of the German language, a significant part is represented by the borrowed terms from the American-English social media sphere: Selfie, Selfiestick, Tablet, Social Bot, liken, and facebooken. An interesting fact is the use of the verbs 'liken' and 'facebooken' which, having English basis, display in the German language the grammatical features of the German verb: liken – hat geliket. However, the conducted comparative analysis with the application of the online search service Google Books Ngram Viewer proved that the German verb 'gefällt mir' is used in the social networks more often than its English equivalent 'liken' and the ratio is: gefällt mir – 63%, liken – 37%.

Thus, we see that English, going beyond its historical area, has a significant impact on the transformation and development of the modern German language. In the context of the increasing process of globalization, the level of the terminological internationalization is not weakening but rather continues to increase. To the greatest extent, the process of internationalization influences the public thematic groups of the vocabulary which in German, unlike in some other languages, has clearly defined adaptive features such as preservation of the phonetic, grammatical and semantic characteristics. The process of terminological internationalization is also facilitated by the vital need to nominate the new concepts that appear in the context of the development of social activities.

Taking into consideration and summing up the existing data concerning the sorting and classification of the German loanwords in English, we distinguish four major lexical semantic groups based on the factors of the sphere of recurrence [1–4, 9, 12, 13, 16, 19], degree of assimilation in the recipient language, created or intended pragmatic effect:

(1) German terms commonly used in the English everyday language practice;
(2) German terms commonly used in the English academic contexts;
(3) German terms mostly used in English for creating a literary effect;
(4) German terms rarely used or intended for use in English.

In the course of the analysis within the above mentioned major groups, the following thematic groups and subgroups were identified:

1. German terms commonly used in the English everyday language practice. Most of the following examples are easily recognized by the majority of English speakers for they are commonly used in the English everyday colloquial contexts. Some of them still retain German connotations, while others retain none. Not every word is recognizable outside its relevant context. A number of these expressions are used in American English, under the influence of German immigration, but not in British English.

1.1. Food and Drink: Berliner Weisse (German spelling: Berliner Weiße), Biergarten, Braunschweiger, Bratwurst (sometimes abbreviated to brat), Budweiser, Bundt cake, Delicatessen (modern German spelling Delikatessen), Emmentaler (or Emmental), Frankfurter (sometimes abbreviated to frank or Frankfurt), Gummi bear (also found with the Anglicized spelling gummy bear, German spelling: Gummibär), Hamburger, Hasenpfeffer, Hefeweizen, Jagertee, Kipfel (also kipferl), Kinder Surprise (also known as a Kinder Egg), Kirschwasser, Knackwurst, Kohlrabi, Kommissbrot, Lager, Leberwurst, Liptauer, Mozartkugel, Muesli (Swiss German spelling: Müesli, standard German: Müsli), Noodle (from German Nudel), Pilsener (also Pils, Pilsner), Pretzel (Standard German spelling: Breze(l)), Pumpernickel, Quark, Radler, Rollmops, Sauerkraut (sometimes abbreviated to Kraut), Schnapps (German spelling: Schnaps), Seltzer, Stein, Streusel, Strudel (e.g. Apfelstrudel, milk-cream strudel), Wiener, Wurst, Zwieback.

1.2. Sports and Recreation: Abseil (German spelling: sich abseilen, a reflexive verb, to rope (seil) oneself (sich) down (ab)); the term abseiling is used in the UK and Commonwealth countries, roping (down) in various English settings, and rappelling in the US), Blitz (abbreviated from Blitzkrieg), Blitz chess (from German Blitzschach), Forlaufer, Karabiner (modern abbreviation of the older word Karabinerhaken), Kibitz (from German Kiebitz), Luft, Patzer, Rucksack (more commonly called a backpack in American English), Schuss, Sitzfleisch, Turner, Turnverein, Volksmarsch/Volkssport/Volkswanderung; Zeitnot, Zugzwang, Zwischenschach, Zwischenzug.

1.3. Animals: Dachshund, Doberman Pinscher, Hamster, Poodle, Rottweiler, Schnauzer, Siskin (from Sisschen, dialect for Zeisig), Spitz.

1.4. Philosophy: An sich, Anschaung, Dasein, Ding an sich, Geist, Lebensraum, Lebenswelt, Mitsein, Neanderthal (modern German spelling: Neandertal), Schadenfreude, Übermensch, Wanderlust, Weltanschauung (calqued into English as world view), Welträtsel, Wertfreiheit, Wille zur Macht, Zeitgeist.

1.5. Society and Culture: Doppelgänger (also spelled in English as doppelgaenger), Dreck, Dummkopf, Fest, Gastarbeiter, Gemütlichkeit, Gesundheit, Hausfrau (colloquial, American English only), Kaffeeklatsch, kaput (German spelling: kaputt), Kindergarten, Kitsch, Kraut (a derogatory term for a German), Lederhosen, Meister (also as a suffix: –meister), Oktoberfest, Poltergeist, Spiel, uber-/über- (e.g. Übermensch), Ur- (German prefix: e.g. Urtext), verboten (in English this word has an authoritarian connotations), Wiener, Wunderkind.

1.6. Technology: -bahn as a suffix (e.g. Infobahn, after Autobahn), Ersatz, Flak, Kraft, Volkswagen, Zeppelin.

2. German terms commonly used in the English academic contexts. German terms are often found in the English academic disciplines and some specific spheres.

2.1. Academia: Ansatz, Doktorvater, Festschrift, Gedenkschrift, Leitfaden, Methodenstreit, Privatdozent, Professoriat, Wissenschaft.

2.2. Architecture: Angstloch, Bauhaus, Bergfried, Biedermeier, Hügelgrab, Jugendstil, Plattenbau, Sondergotik, Stolperstein.

2.3. Arts: Gesamtkunstwerk, Gestalt.

2.3.1. Music: Affektenlehre, Almglocken, Alphorn, Augenmusik, Ausmulti-
plikation, Blockwerk, Crumhorn (German spelling: Krummhorn),
Fach, Fife (German spelling: Pfeife), Flatterzunge, Flugelhorn (German spelling: Flügelhorn), Glockenspiel, Heldentenor, Hammerklavier
(most commonly used in English to refer to Beethoven's Ham-
merklavier Sonata), Hosenrolle, Kapellmeister, Katzenjammer,
Katzenklavier, Kinderklavier, Konzertmeister, Kuhreihen, Leitmotif
(German spelling: Leitmotiv), Lied, Lieder ohne Worte, Liederhand-
schrift, Liederkranz, Liedermacher, Meistersinger, Mensurstrich,
Minnesang, Ohrwurm, Orgelbewegung, Rauschpfeife, Rückpositiv,
Sängerfest, Schlager, Schuhplattler, Singspiel, Sitzprobe, Strohbass,
Urtext, Volksmusik, Walzer (Waltz), Zukunftsmusik.

2.3.1.1. Genres: Kosmische Musik, Krautrock, Neue Deutsche Härte
(NDH), Neue Deutsche Todeskunst, Neue Deutsche Welle
(NDW), Neue Slowenische Kunst, Romantische Oper,
Schranz.

2.3.1.2. Selected works in classical music: Johann Sebastian Bach's
Das wohltemperierte Klavier, Jesus bleibet meine Freude;
Brahms's Schicksalslied; Kreisler's Liebesleid, Liebesfreud;
Liszt's Liebesträume; Mozart's Eine kleine Nachtmusik, Die
Zauberflöte; Gustav Mahler's Kindertotenlieder; Schubert's
Winterreise; Schumann's Dichterliebe; Richard Strauss's Der
Rosenkavalier, Also sprach Zarathustra, Vier letzte Lieder;
Johann Strauss II's Die Fledermaus, An der schönen blauen
Donau; Richard Wagner's Die Walküre, Götterdämmerung,
both from his opera cycle Der Ring des Nibelungen.

2.3.1.3. Carols: Stille Nacht, O Tannenbaum.

2.3.2. Typography: Fraktur, Schwabacher.

2.4. Biology: Ahnenreihe, Ahnenschwund, Ahnentafel, Anlage, Aufwuchs, Aur-
ochs (Modern German: Auerochse), Bauplan, Edelweiss (German spelling
Edelweiß), Einkorn, Krummholz, Lammergeier or lammergeyer (German
spelling: Lämmergeier, also Bartgeier), Oberhäutchen, Schreckstoff, Spitzen-
körper, Spreite, Unkenreflex, Waldsterben, Zeitgeber, Zugunruhe.

2.5. Chemistry: Bismuth, Darmstadtium, Einsteinium, Knallgas Reaction, Meit-
nerium, Paraffin, Roentgenium, Wolfram.

2.6. Economics: Dollar (from German Thaler through Dutch (Rijks)daalder),
Freigeld, Freiwirtschaft, Heller, K (in economics the letter K (from the German
word Kapital) is used to denote Capital), Lumpenproletariat, Mittelstand, Takt.

2.7. Geography: Hinterland, Inselberg, Knickpoint (German Knickpunkt, from
knicken - to bend sharply, fold, kink), Mitteleuropa, Mittelgebirge, Schlatt
(also Flatt), Steilhang, Thalweg (written Talweg in modern German).

2.8. Geology and Metallurgy: Aufeis, Bergschrund, Cobalt, Dreikanter, Fenster,
Firn, Flysch, Gneiss (German Gneis), Graben, Horst, Karst, Loess (German:
Löss), Nickel; Randkluft, Rille, Shale; Sturzstrom, Urstrom, Urstromtal, Zinc.

2.8.1. Minerals: Feldspar (German Feldspat), Hornblende, Meerschaum, Moldavite (German Moldavit), Quartz (German Quarz), Wolframite (German Wolframit), Zinnwaldite (German Zinnwaldit).

2.9. History, Anthropology and Archaeology: Alltagsgeschichte, Aufklarung, Biedermeier, Chaoskampf, Diktat, Gründerzeit, foreworld (from German Vorwelt), Junker, Kaiser, Kleinstaaterei, Kulturgeschichte, Kulturkampf, Kulturkreis, Kulturkugel, Kunstgeschichte, Kunstforscher, Kunstforschung, Kunsthistoriker, Landflucht, Landnahme, Nordpolitik, Ostflucht, Ostpolitik, Ostalgie, Reichstag, The Third Reich, Urheimat, Urmonotheismus, Urreligion, Völkerschlacht, Völkerwanderung, Weltpolitik.

2.10. Linguistics: Ablaut, Abstandsprache, Aktionsart, Ausbausprache, Dachsprache, Dreimorengesetz, Gleichsetzung or Gleichung, Grenzsignal, Gruppenflexion, Junggrammatiker, Loanword (a calque from German Lehnwort), Mischsprache, Rückumlaut, Sprachbund, Sprachgefühl, Sprachraum, Stammbaumtheorie, Suffixaufnahme, Umlaut, Urheimat, Ursprache, Verschärfung, Wanderwort.

2.11. Literature, Mythology and Folklore: Bildungsroman, foreword (from German Vorwort), Formgeschichte, Knittelvers, Kobold, Künstlerroman, Kunstprosa, Leitwortstil, Märchen, Nachlass, Nihilartikel, nix, Q (abbreviation for Quelle), Quellenforschung, Quellenkritik,… über alles (above all, originally from Deutschland über alles, the first line of Hoffmann von Fallersleben's poem Das Lied der Deutschen (The Song of the Germans)), Sammelband, Valhalla, Vorlage.

2.12. Mathematics and Formal Logic: Ansatz, Eigen- (in composita such as eigenfunction, eigenvector, eigenvalue, eigenform), Grossencharakter (German spelling: Größencharakter), Hauptmodul, Hauptvermutung, Ideal, Möbius band (German: Möbiusband), Positivstellensatz, quadratfrei, Stützgerade, Neben- (in composita, such as Nebentype).

2.13. Medicine: Anwesenheit, Diener, Entgleisen, Gegenhalten, Mitgehen, Mitmachen, Rinderpest, Schnauzkrampf, Sitz Bath, Spinnbarkeit, Verstimmung, Vorbeigehen, Vorbeireden, Wahneinfall, Witzelsucht, Wurgstimme.

2.14. Military Terms: Blitzkrieg, Gestapo, Kriegsspiel, Luftwaffe, Panzer, Panzerfaust, Strafe, U-Boot (abbreviated form of Unterseeboot), Wehrmacht.

2.15. Physical Sciences: Ansatz, Antiblockiersystem, Aufbau principle (German spelling: Aufbauprinzip), Durchmusterung, Farbzentrum, Fusel alcohol (German: Fuselalkohol), Gedanken experiment (German spelling: Gedankenexperiment); Gegenschein, Gemisch, Gerade and its opposite Ungerade, Graupel, Hohlraum, Kugelblitz, Kugelrohr, Mischmetall, Reststrahlen, Schiefspiegler, Schlieren, Sollbruchstelle, Spiegeleisen, Trommel, Umklapp process (German spelling: Umklappprozess), Umpolung, Vierbein, Zwitterion.

2.16. Politics: Antifa (abbreviation for Antifaschistische Aktion), Befehl ist Befehl, Berufsverbot, Kritik, Lumpenproletariat, Machtpolitik, Mitbestimmung, Nazi (abbreviation for Nationalsozialist), Putsch, Realpolitik, Rechtsstaat, Überfremdung.

2.17. Psychology: Angst, Eigengrau, Einstellung effect, Galgenhumor, Ganzfeld effect, Gestalt psychology (German spelling: Gestaltpsychologie),

Gestaltzerfall, Grübelsucht, Haltlose personality disorder, Merkwelt, Lebenslust, Schadenfreude, Sehnsucht, Sorge, Umwelt, Wehmut, Weltschmerz, Wunderkind, Zeitgeber.

2.18. 2.18. Sociology: Gemeinschaft, Gesellschaft, Herrschaft, Kulturkreis, Lebensform, Männerbund, Verstehen, Volksgeist, völkisch, Untermensch, Weltbild, Zeitgeist.

2.19. Theology: Gattung, Heilsgeschichte.

3. German terms mostly used in English for creating a literary effect. There are a few terms recognised by the majority of English speakers, but usually used to bring to mind and memory a deliberate German coloring. E.g., Autobahn (referring particularly to German motorways), Achtung, Frau and Fräulein, Führer (always used in English to denote Hitler or to connote a fascistic leader), Gott mit uns (the motto of the Prussian king was used as a morale slogan amongst soldiers in both World Wars. It was bastardized as "Got mittens" by American and British soldiers, and is usually used nowadays, because of the German defeat in both wars, derisively to mean that wars are not won on religious grounds), Hände hoch, Herr, Ich bin ein Berliner (a famous quotation by John F. Kennedy) [10], Leitmotif, Meister (used as a suffix to mean expert (Maurermeister) or master), Nein, Raus, Ja, Jawohl (a German term that connotes an emphatic yes – "Yes, indeed!" in English. It is often equated to "yes, sir" in Anglo-American military films, since it is also a term typically used as an acknowledgement for military commands in the German military), Schnell!, Wunderbar.

4. German terms rarely used or intended for use in English. To this lexical semantic group we refer, first of all, those lexical units that are registered in English contexts but they have not become recurrent yet: Besserwisser (someone who always knows better), Bockmist (meaning nonsense or rubbish), Geisterfahrer (ghost driver, a wrong-way driver; one who drives in the direction opposite to that prescribed for the given lane), Götterdämmerung (Twilight of the Gods, a disastrous conclusion of events), Ordnung muss sein (There must be order. This proverbial phrase illustrates the importance that German culture places upon order.), Schmutz (smut, dirt, filth), Vorsprung durch Technik (competitive edge through technology, used in an advertising campaign by Audi), Verschlimmbessern (to make something worse in an honest but failed attempt to improve it).

Besides, while analysing the contents of the Internet resources - including such mass media as online newsrooms, blogs, personal and university websites, global cross-platform networks - we have noted a stable tendency for English language native speakers' discontent of self-expression and craving for borrowing a specific group of words from German [14, 17, 18, 20, 21]. The problem is that we are hugely dependent on the language to help us express what we really think and feel. But some languages are better than others at crisply naming important sensations. Although English is rich with linguistic treasures it is missing certain words for sentiments. Quite the contrary, German is particularly adept at describing complex emotions in a single word, providing added nuances, description, or color. Speaking about this type of words, especially those only intended for use in English, we mean that Germans invent long compound words that elegantly grasp and signify thoughts, feelings and emotions that we all know, but that other languages require whole clumsy sentences or paragraphs to express.

Here are some examples of German compound words belonging to the thematic group Expression of Thoughts, Feelings and Emotions and already, though not quite extensively, used in English context: Abgrundanziehung (The pull of the cliff edge.), Allgemeinbildung (everything that any adult capable of living independently can reasonably be expected to know), Backpfeifengesicht (A face that is begging to be slapped.), Bildungsroman (German literature is full of a distinctive kind of novel in which we follow the maturation of the central character, normally in relation to love and work. This genre and the word that crowns it presents a particular answer to the vexed question of what a novel should be for. For a great German tradition, the point of novels is to teach us how to live.), Blaumachen (Literally, blaumachen means blue-making. It is used in German to indicate a day off prompted by lack of motivation - basically, playing hooky. The word originates from the term Blauer Montag, or Blue Monday, which refers to the day of rest for fabric crafters in the Middle Ages, who soaked cloth in indigo on Sunday and dried them the next day, allowing the workers to take time off), Drachenfutter (Dragon food is a gift that one has to offer to one's spouse to appease their fury for a wrong one has committed. If an affair is discovered, one may have to cook up an enormous meal of Drachenfutter.), Erkenntnisspaziergang (Going out, in order to gain deeper insights while walking.), Erklärungsnot (Literally, a distress at not having an explanation. Erklärungsnot is something we feel when we realise we do not have any explanations for the big questions of life. It is a word that defines existential angst as much as shame.), Fahrvergnügen (driving pleasure, introduced in a Volkswagen advertising campaign), Fernweh (The distress of always being in familiar surroundings and the longing to go faraway.), Flughafenbegrussungsfreude (The happiness you feel when someone picks you up at the airport.), Fremdscham (vicarious shame, the shame felt for the behavior of someone else), Frühjahrsmüdigkeit (The feeling of dispiriting about the onset of spring.), Futterneid (The feeling when you are eating with other people and realise that they have ordered something better off the menu that you would be dying to eat yourself. The word recognises that we spend most of our lives feeling we have ordered the wrong thing.), Handschuhschneeballwerfer (a coward willing to criticize and abuse from a safe distance), Kummerspeck (A word that frankly recognises how often, when one is deeply sad, there is simply nothing more consoling to do than to head for the kitchen and eat.), Kopfkino (We are inveterate cineastes in our own minds. We shoot little movies in which we say exactly what we mean and seize the advantage when we can. The word connotes though that very few of us ever know how to be efficient and skilled directors outside our own heads.), Lebensmüde (an occasional longing to give up our hold on existence), Luftschloss (Literally, a castle in the air; a dream that is unattainable – a word suggesting that German culture is deeply indulgent about big dreams but also gently realistic about how hard it can be to bring them off.), Ruinenlust (This word means the delight one can feel at seeing ruins.), Ringrichterscham (The embarrassment you feel when you are next to a couple having a fight.), Schadenfreude (the feeling of enjoyment that comes from seeing or hearing about the troubles of other people), Schnapsidee (An idea you had while drunk.), Sitzfleisch (This word describes a character trait of endurance; literally a capacity to sit and put up with what is boring, arduous or painful over long periods.), Sontagsleere (The melancholy emptiness of Sundays), Traumneustartversuch (Desperately trying to continue the dream you were having just before you woke up.), Treppenwitz (The things you should have said but only occur to you when it is too late. It is easy to be wise after the event; It is too late to lock the stable-door when the horse is stolen.), Torschlusspanik (the fear, usually as one gets older, that time is running out and important opportunities are slipping away), Über Alles (disambiguation), Waldeinsamkeit (the feeling experienced while alone in the woods, connecting with nature), Weltschmerz (A word that acknowledges that we are sometimes sad not about this or that thing, but about the whole basis of existence. The presence of the word indicates a culture that is not falsely cheerful but takes tragedy as a given.), Witzbeharrsamkeit (Telling the same joke over and over again until there is no one left who have not heard it from you.).

4 Conclusion

Summing up the discussion, it is necessary to mention that the results of the research offer interesting data about language contacts, terminological internationalization and language change.

Our data show that technologically advanced cultures represented by languages like German and English enrich their vocabularies with loanwords as new lexical units intended to meet the needs in the particular scientific community. Some of these words, while often becoming the standard terms in that community and moving into the general German language, as well as into the general English language as loans, acquire the status of international terms. These items need not exhibit German phonological features in order to be true German creations within a German text, rather than being artificial scientific and technical terms, confined to the given communication environment; for they observe German spelling, noun capitalization, and inflectional and gender-marking principles differentiating German from other languages. As loans, they provide a window into German scientific influence on English-speaking and other language communities. Similarly, nonscientific but classical-based terms show that ordinary Germans also use the classical elements in much the same way that English speakers use the classical elements naturalized in English to create nontechnical words. And a listing of the semantic spheres denoted by our factual material illustrates their impact on socio-economic, political, scientific and cultural aspects of life.

As a result of studying the terminological systems of German and English with the application of the sociolinguistic method for correlating the intra-linguistic causes of changes inherent to the languages themselves and the phenomena and factors of non-linguistic reality associated with the history of specific nations, as well as with the general development of the human society, i.e. extralinguistic factors, we have identified the causes of the terminological internationalization.

The process of terminological internationalization is a consequence of the rapprochement of nations on the basis of socio-economic, political, scientific and cultural ties, resulting from globalization. In most cases, words are borrowed as a means of denoting new objects, concepts and processes. Loanwords can substitute original lexical units and represent secondary nominations of already known objects and phenomena. This process takes place if a borrowed word is used for a slightly different characteristic of an object, if it is an international term. Otherwise, loanwords with the help of only one lexical unit are capable of nominating a complex concept expressed in the recipient language by a whole phrase or even a sentence.

Many loanwords undergo significant phonetic, grammatical and even semantic changes, adapting accordingly to the phonetic, grammatical and semantic laws of the borrowing language. The process of assimilation can be so deep that the foreign language origin of such terms is not felt; on the contrary, a loanword is perceived as a generally accepted international nomination of an object or phenomenon and is detected only by means of a detailed etymological analysis.

And if previously there existed (especially in German) the tendency to borrowing terms by various kinds of translational transformations, nowadays both in German and in English we observe the tendency to bringing over terms as whole lexical units. So, we record different fashions about how people comprehend and feel about the productive capacity of their own language versus borrowing a term wholesale from another. Our data suggest that English and German are moving toward each other, rather than becoming more different. The old historical process by which French and Spanish, German and English, etc. become mutually unintelligible over the centuries may now be ended, if not moving toward reversal.

References

1. Ayto, J.: Dictionary of Word Origins: Histories of More Than 8,000 English-Language Words, Arcade (1993)
2. Bahlow, H.: Deutsches Namenslexikon: Familien- und Vornamen nach Ursprung und Sinn erklärt, Gondrom Verlag, Bindlach (1991). (in German)
3. Bliss, A.: Dictionary of Foreign Words and Phrases in Current English. Warner Books, New York (1992)
4. Bryson, B.: The Mother Tongue: English and How it Got that Way. Penguin Books, London (1990)
5. Duckworth, D.: Der Einfluss des Englischen auf den deutschen Wortschatz seit 1945. In: Fremdwort-Diskussion, pp. 213–325. Wilhelm Fink Verlag, München (1999). (in German)
6. Duden - Die deutsche Rechtschreibung, 27. Verlag/Bibliographisches Institut GmbH, Berlin (2017). https://orange.handelsblatt.com/wp-content/uploads/2017/08/duden-neue-woerter-auszug-auflage-27.pdf. Accessed 12 Aug 2018
7. Encyclopedia.com. Borrowing. https://www.encyclopedia.com/literature-and-arts/language-linguistics-and-literary-terms/language-and-linguistics/borrowing. Accessed 17 Aug 2018
8. Gordin, M.D.: Scientific Babel: How Science Was Done Before and After Global English. University of Chicago Press, Chicago (2015)
9. Hughes, G.: A History of English Words. Blackwell Publishers Ltd., Oxford (2000)
10. Kennedy, J. F.: Ich bin ein Berliner (I am a 'Berliner'), delivered 26 June 1963, West Berlin. http://www.americanrhetoric.com/speeches/jfkberliner.html. Accessed 2 Aug 2018
11. Klyoster, A.M.: Specifika semanticheskih svjazej nemeckih terminov inzhenernoj psihologii [Specificity of semantic links of German terms of engineering psychology]. Omskij nauchnyj vestnik 4(99), 114–117 (2011). (in Russian)

12. Knapp, R.: Dictionary of Germanisms. http://www.humanlanguages.com/germanenglish/. Accessed 12 Aug 2018
13. Knapp, R.D.: German English Words: A Popular Dictionary of German Words Used in English. Lulu Press, New York (2005)
14. Livni, E.: Seven German words that English-speakers need to make sense of 2016 and prepare for 2017. https://qz.com/851953/seven-german-words-that-english-speakers-need-to-make-sense-of-2016-and-prepare-for-2017. Accessed 14 Aug 2018. (in German)
15. McArthur, T.: Concise Oxford Companion to the English Language. http://www.oxfordreference.com/abstract/10.1093/acref/9780192800619.001.0001/acref-9780192800619-e-737?rskey=QvMWBB&result=749. Accessed 10 Aug 2018
16. Pfeffer, J.A., Cannon, G.H.: German Loanwords in English: An Historical Dictionary. Cambridge University Press, Cambridge (1994)
17. Schröder, K.: Zur Problematik von Sprache und Identität in Westeuropa. Eine Analyse aus sprachenpolitischer Perspektive. In: Sociolinguistica. Internationales Jahrbuch für Europäische Soziolinguistik, pp. 56–66. Walter de Gruyter, Berlin/Boston (2015)
18. Smith, E.: High German Loanwords in English. http://germanic.eu/High-German-loanwords-in-English.htm. Accessed 12 Aug 2018
19. Tolzmann, H.D.: The German-American Experience. Humanity Books, New York (2000)
20. World Heritage Encyclopedia. http://community.worldheritage.org/articles/List_of_German_expressions_in_English. Accessed 8 Aug 2018
21. Zheltukhina, M.R., Biryukova, E.V., Gerasimova, S.A., Repina, E.A., Klyoster, A.M., Komleva, L.A.: Modern media advertising: effective directions of influence in business and political communication. Man in India 97(14), 207–215 (2017)

Intercategorial Relations in the Structure of Verbalized Perceptual Situation (Based on the German Language Material)

Galina Galich[1] and Natalia Shnyakina[2(✉)]

[1] Dostoevsky Omsk State University, Mira 17a,
644050 Omsk, Russian Federation
galich2004@rambler.ru
[2] Omsk State Pedagogical University, Naberezhnaja Tukhachevskogo 14,
644099 Omsk, Russian Federation
zeral@list.ru

Abstract. The article attempts to analyze the intercategorial relations which reflect the significant connection between the elements of the perceptual situation and allow structuring the cognitive activity within the perception of the world. This approach gives the opportunity to get closer to the problem of describing the categorization mechanisms by means of thinking and language. The situation is analyzed at the level of the language/text fragment, in which all of its obligatory components/actants can be explicated. They are connected to the categories of ordinary mentality. The language exponents of these actants are in grammatical-lexical and semantic relations which assume the formation of identity vectors between the cognitive content of the fragment and the denotative situation. The focus is on the relationship between the cognitive components of the reflected situation. As it turns out, the sense of intercategorial interaction is closely cor-related with the categorial-grammatical relations in German language: subject-object, attributive, circumstantial, pronominative. Thus, the connections between the linguistic units shed light on the existence and significance of cognitive intercategorial relations as the main components of categorization.

Keywords: Categorial net · Cognitive situation · Perceptual situation · Categorization

1 Introduction

The study of linguistic phenomena from the standpoint of cognitive linguistics makes it possible to see the specifics of the categorial division of the world by human thought. The language fragment is a material which allows identifying significant categorial features that have different manifestation and are associated with the global categorial net.

This article attempts to describe the categorial structure of perceptual situation objectified in the German language. The paper concentrates on the nature of intercategorial relations reflected in the language fragment. Achieving this, the object involves the following issues: firstly, the role of categorization in the cognition; secondly, what

© Springer Nature Switzerland AG 2019
Z. Anikina (Ed.): GGSSH 2019, AISC 907, pp. 357–366, 2019.
https://doi.org/10.1007/978-3-030-11473-2_37

categories of ordinary mentality can be revealed on the basis of the language fragments, describing the perceptual situation; thirdly, what kind of relations unites the revealed categories in the perceptual situation objectified by language.

2 Categories and Categorization

Categorization is a fundamental operation of human mind which contributes to organizing and to systematizing of information about the surrounding reality. Taking into account the general psychological mechanisms, individual and collective knowledge of people, all the objects and phenomena in the world can be combined into the classes called categories. Language being a product of human thought activity helps to understand the logic of information processing: it allows us to realize "how the world is seen and understood by the human mind, how it is represented and categorized by mentality" [17, p. 57], therefore, the study of the categorization, types of categories and principles of their construction functioning in thought and language are very important for understanding this process in general and the nature of category in particular.

"Category" and "categorization" have a significant place in philosophy [19–21, 24, 31, 33], psychology [4, 22, 23, 26, 28, 34] and cognitive linguistics [8, 12, 14, 16, 18, 30, 32, 35]. In the philosophy the following definition is given to "categorization": "these are cognitive processes that provide recognition and revealing in objects, events the "prototype" examples of concepts (categories) [9]. In psychology categorization is understood as the "mental process of assigning a single object, event, and experience to a certain class which may include verbal and non-verbal meanings, symbols, sensory standards, social stereotypes, behavior stereotypes, etc." [27]. The category is understood as "some rule, according to which we assign objects to one class as equivalent to each other" [4, p. 27]. In cognitive linguistics categorization is defined as "assignment of a fact, object, process etc. to a certain category of experience and their acceptance as a member of this category, but in a broader sense - the process of categories formation, the division of external and internal world of a person in accordance with the essential characteristics of their functioning and being ..." [16, p. 42].

A significant element of categorization and its basis is the previous experience. All cognitive processes, as Pankrats writes, citing Mack Shane and Pylishin, are associated with "throwing" of some categorial net on the world" [25, p. 13]. This categorial net is a system of cognitive coordinates, according to which every member of the language community lives and interacts with the world. The realization of a new object or event is finding its place in the system of norms and categories, accepted both in society and in the individual understanding of reality. Classification of the cognizable objects in different taxonomic groups presupposes the activation of existing knowledge, correlation of new information with the existing groups (categories). The categorial net is a kind of prism for experience processing which provides the unity of text creation and interaction within the same conditions of being. All the native speakers have a set of categories that are understandable to every member of the language collective; it provides the understanding between people, based on encoding and decoding information.

In this paper the understanding of categorization is based on the opinion of Kubryakova who speaks about the world's division by language [17, p. 97]. It means that the categorization is understood as a peculiar way of creating taxonomy of the world, taking into account the verbal division of the conceptual space accepted in the language community. The category is a large format of knowledge, reflecting a person's ideas about reality. On the one hand, the category has a common sense inherent in all its members; on the other hand, every element that belongs to the category has its own specific form of explication. The category, therefore, has an expression form and the content form. These aspects can be studied on the basis of language fragments that describe the situation of cognition.

3 Language Fragment as a Result of the Categorial Division of the World

In the paper the language fragment is analyzed as objectified knowledge – as a complex multidimensional and multilevel structure that reflects the common factors of language and mental processes.

In linguistics the attempts to describe the categorial structure of the sentence have already been made. The tradition of revealing the deepest components in language was set up by Fillmore who explained the syntactic structures through the manifestation of the deep cases (agentive, instrumentalis, dative, factitive, locative, and objective). These manifestations are some ideas about things: "who did something", "with whom something happened", "what has undergone any change" [5, p. 405–406]. The idea to decompose the sentence into separate categories was reflected in other works [1–3, 6, 7, 10, 11, 13, 15, 29]. For example, the activity model developed by Kubryakova includes the following: agent (source), operation, patient (object), instrument (tool), goal (result), spatial and temporal coordinates of activity, intentions of the subject in the form of description or evaluation [15, p. 443]. Similarly, on the basis of verbalized structure Furs separates the deep semantic components: agent, action, object of influence, state of the object, instrument, result of action, existence of the object, temporal and spatial characteristics [6, p. 6]. In some works a deep structure of the perceptual situation is studied. For example, Bondarko distinguishes the following components: "the subject of perception/perceptor/observer, observation, observability, object of perception/observed situation, perceptual events …" [3, p. 279]. In Khomyakova's cognitive-information situation there are the subject (experimenter), the object of perception (source of influence), the address, the information actant, and the means of perceptual influence [11, p. 247].

The analyzed theories make it possible to establish that, based on a language fragment, many researchers mention the following components in their categories lists: subject, object, cognitive action, feature, instrument, result, space and time. It is very important to identify the relationships between these components.

4   Intercategorial Relations in the Structure of Verbalized Perceptual Situation

Intercategorial relations are understood as logical relations between the revealed semantic elements of the language fragment. Its awareness is possible not only based on the semantics of the respective components of the sentence, but also by deictic words that create the perspective of the whole fragment and profile it. These logical correlations have the form of elementary propositions that reflect the following basic types of relationships, identified as a form of everyday knowledge of the native speaker: subject-object, attributive, and circumstantial.

Subject-object relations form a kind of "categorial basis" of the fragment and reflect a propositional connection between the subject and the object through the cognitive action binding them.

- The category "subject" is the idea of a person interacting with an object; it is expressed by a noun or pronoun designating a human or an animal.
- Zu Beginn erhalten die Kinder Zeit, das Bild in aller Ruhe zu betrachten.
- Hinter der Kasse kommt nach einigen Minuten ein junger Jagdhund hervorgekrochen, der uns vorsichtig beschnuppert und ziemlich rasch mit Moritz Freundschaft schließt.

As an "object of knowledge" is understood anything of the surrounding reality, falling into the focus of attention of the subject and having perceptual features perceived by the subject.

- Außerdem habe er Stimmen gehört, die ihn zu der Tat aufgefordert hätten.
- Er musterte die winzigen Hände.

The process of cognition has a strategic nature. Through the language objectifications in the context it provides information about the perceptual channel.

- In den letzten fünfzehn Jahren habe ich das Land fast nur aus der Ferne gesehen.
- Wenn er von der Arbeit nach Hause kommt, hört er gern Musik.
- Dieses Wort nicht in deinen Mund, sagte Arlecq bewegt und roch an seinen Händen, die in vierundzwanzig Stunden alles ergriffen, gefühlt, betastet hatten, nur nicht Isabel.
- Das ganze Gras der Insel war essbar, es schmeckte leicht bitter und nussig und passte gut zu gebratenen Kartoffeln.

In reality, presence of all three components of subject-object relations is obligated: such full sentences are often used to describe perceptual situations. For example: Sie hätten Rauchgase eingeatmet und Verbrennungen erlitten. In the above sentence the categorial node "subject" is expressed by the pronoun "sie"; object is represented by the word "Rauchgase"; the cognitive action – by the verb "einatmen".

Analysis of examples shows that the subject-object relation can also be explicated in a shorter form in the language: in one- and two-component constructions. Depending on the syntactic structure of the reflected situation, the subject is represented either as an active person making cognitive action, or as a passive participant, present "behind the scenes". In this case, the language fragment does not directly contain the categorial component "subject".

In one-component construction the subject-object relations are expressed through objectification of the category "object".

- Das Zimmer ihres Vaters riecht noch immer nach Tabak, obgleich seit einiger Zeit niemand mehr dort geraucht hat.
- Die Farbe, mit der ihre Kleider rot oder blau bemalt sind, blättert schon ab.

In a two-component construction, alongside with the category "object", there is also the "cognitive action" component which indicates the way of getting perceptual knowledge.

- Auch der badische Rotwein schmeckt köstlich, die ersten beiden Flaschen sind im Nu ausgetrunken, eine dritte wird für uns geöffnet.
- Durch die verschlossene Tür hindurch riecht es nach Tabak.

The next type of relationship which is present in the verbalized situation of cognition is the **attributive relation** that expresses characteristics about categorial coordinates: subject, object, instrument, feature or about the perceptual action and its variants. It is possible to show eventual forms of realization of attributive relations by the examples:

1. Subject - feature of subject: Die kleine Debo sprang mit Stöcken durch Pfützen und Bäche, atmete vergnügt den Duft von Anemonen, Eichen und Pferdedung ein, ging angeln und reiten.
2. Object - feature of object: Schau einmal, sagt die junge Frau, die vom Garten her ins Zimmer getreten ist, und weist mir ihren Arm, innen an der Beuge ihres Ellenbogens sehe ich einen kleinen ovalen Leberfleck.
3. Instrument – feature of instrument: Bei all diesen Aufgaben kommen ihnen zwei Begabungen zugute: die sehr feine Nase und das aufmerksame Gespür für ihr Gegenüber.
4. Feature – feature of feature: Bald hingen endlose Läufer und großflächige Teppiche im Garten und verwandelten die Landschaft in ein Haus jenseits des Hauses, während das Dröhnen der Staubsauger und Bohnerwachsmaschinen und die Gerüche von Scheuermitteln und Terpentin die Geräusche und Gerüche des Gartens überlagerten.

Circumstantial relations have a complex nature: connecting various categorial nodes, they can also express different types of relations: total-partitive, consecutive, local and temporal.

Total-partitive relations express the connection between the part and the whole and can be traced between the nodes "subject" and "instrument". In many contexts describing the perceptual situation, the idea is expressed that the instrument belongs to the subject of cognition and that its function is significant for life activity.

- Meine Nase ist also das Flexibelste an mir: Sie darf die schönsten und muss die schlimmsten Düfte stundenlang riechen.

- Seine Flügel zerrissen die Luft so scharf, dass es in meinen Ohren schmerzte.
- Mund auf, sagt er und schiebt mir einen Löffel mit einer bitteren Flüssigkeit in den Mund.

Consecutive relations are a consequence of the interaction between subject and object; they both have external and internal manifestations.

Impulses that influence the state and behavior of the subject generate an external (physiological or emotional) reaction, expressed in the categorial coordinate "result", connected with the subject, object and/or cognitive action.

- Ich habe wieder den Gestank der Festzelte gerochen und den Kater gespürt.
- Mir war oft schlecht, wogegen nur Essen zu helfen schien, und so knabberte ich abwechselnd Nüsse, Salzstangen und Schokolade, trank Limonade dazu und wurde bald schon so schwerfällig, als stünde ich kurz vor der Niederkunft.

The internal reaction is the result of comparing the new perceptual information with the past experience of a person: it includes descriptive and evaluative reactions. Description is the result of intellectual action, associated with determining the object's real attributes. Evaluation is based on individual preferences and personal character-istics. Based on the examples, it can be seen that the consecutive relations are the result of the cognitive act and include the interaction between the categorial coordinates "result", on the one hand, and semantic nodes "subject", "object", "feature of object", "cognitive action", on the other:

- Heute nehme ich den Geruch nach verdorbenen Lebensmitteln deutlicher wahr als sonst, er ist, scheint mir, heute stärker als der, Geruch nach Lebendigem Frisch-geschlachtetem oder -gefischtem.
- Von Brombeeren bekommst du einen dunkelroten Mund, aber sie schmecken gut.

The perceptual situation has its spatial and temporal limits that are recognized and verbalized by a person, not only in a global sense of being, but also in relation to the subject of cognition.

Local relationships, fixed in the language fragment, contribute to expressing the object location towards any point, and connect the categorial nodes "place" and "ob-ject" or "place" and "subject":

- Über der gesamten Stadt liegt der leicht süßliche Geruch der Fischsoße Nuoc Mam, die in zahlreichen kleinen Betrieben entlang des Duong Dong Rivers produziert wird.
- Er sieht von Weitem Plantagen voller Kirsch- und Apfelbäume, die in diesem Jahr keine Früchte mehr tragen werden.

Forms of temporal relations reflected in the examples are as follows: the local-ization of the perception moment in time, on the one hand, and expression of the perceptual act duration, on the other. Thus, temporal relations connect the category of "time" with the categories of "subject" and "cognitive action".

- Schon am zweiten Morgen sah er – Salat – voller Ekel auf den harzigen Urinstein im Waschbecken, obwohl es doch Salat war, der seit mehreren Jahren aus Bequemlichkeit ins Waschbecken pinkelte, ohne es jemals mit geeigneten Mitteln zu reinigen.
- Malka hatte sie lange beobachtet, dann hatte sie sich unauffällig hinter eine Kundin geschoben, die gerade drei Eier kaufte.

5 Conclusion

Categorization as a mental phenomenon is the division of the cognizable objects to classes based on their similarities and differences, which determines the previous experience. Any act of categorization is the recognition of various objects by their essential or non-essential characteristics; it is the search of already existing groups (categories) in the conceptual system. The attributes of these groups can be partially or completely transferred to a new phenomenon.

Language reflects a person's view on the world; therefore, it shows the specificity of categorization as the basis of human cognition on the whole. The processing of new information takes place in accordance with the categories, which are the result of teaching and public practice and which are understandable to all members of the language society.

The unity of the categorial division of the world can be traced on the example of the objectified perceptual situation. The mental unit that fixes it in thinking is understood as a propositional information structure. Its main characteristic is interconnection of its parts named categorial nodes. From the point of view of objectification, the rules of thinking processes in the language, a sentence is a set of verbalized ideas that form a deep net of fragments in the mind. The linguistics position is characterized by the opinion that in each sentence there are significant categorial nodes realized at the speech level. The linguistic fragment reflects the specificity of getting perceptual knowledge: there are significant coordinates which have a verbalized form. Among them: "subject", "object", "cognitive action", "feature", "instrument", "result", "space", and "time". The set of categorial nodes verbalized in linguistic fragments varies, depending on linguistic and nonlinguistic factors.

The analysis of examples shows the presence of different intercategorial relations in the verbalized perceptual situation: subject-object, attributive, circumstantial, and pronominative.

Subject-object relations are the basis for a categorial division of the analyzed mental structure. From the point of view of linguistic objectification, this type of relationship is realized in one-, two- or three-component constructions which include only "object", "object" and "cognitive action", or all three components: "subject", "object" and "cognitive action". The choice of structure depends on motivation for expressing the subject: the optionality of the subject objectification is a typical feature of linguistic fixation of perceptual knowledge.

Attributive relations reflect the interaction between the "subject", "object", "instrument" and the categorial node "attribute" which consists in specializing the characteristics of these components.

Circumstantial relations involve the total-partitive, consecutive, local and temporal aspects. Total-partitive relations exist between the "subject" and the "instrument", local relations are realized between the categorial nodes "place" and "object"/"subject", temporal correlations are inherent to the components "time" and "subject"/"cognitive action". In consecutive connections it is difficult to define intercategorial relationships because they cannot be analyzed from the point of view of their realization duality: in the language fragments there are often presented interactions of more than one coordinates, among which "subject", "object", "cognitive action" etc. This can be explained by the fact that the result is described in the sentence as a rule, as a phenomenon, influenced by all the participants of the situation in total.

In the analyzed language fragments they can see the same coordinates; it points to the constant interaction of new information and accepted categorial net that reflects the laws and rules of the surrounding reality division by the human brain.

References

1. Boldyrev, N.N.: Kognitivnaya semantika [Cognitive semantics]. Tambovskij universitet, Tambov (2001). (in Russian)
2. Bondarko, A.W.: Funkcional'naya grammatika [Functional grammar]. In: Churilina, L.N. (ed.) Aktual'nye problemy sovremennoj lingvistiki [Actual problems of modern linguistics], pp. 63–86. Flinta-Nauka, Moscow (2009). (in Russian)
3. Bondarko, A.W.: K voprosu o perceptivnosti [On the question of perceptivity]. In: Apresyan, Yu.D. (ed.) Sokrovennye smysly: Slovo. Tekst. Kul'tura [The hidden meanings: Word. Text. Culture], pp. 276–282. Yazyki slavyanskoj kul'tury, Moscow (2004). (in Russian)
4. Bruner, J.: Beyond the Information Given: Studies in the Psychology of Knowing. W. W. Norton, Incorporated (eds.), New York (1973)
5. Fillmore, Ch.J.: The case for case. In: Zvegincev V.A. (ed.) New in Foreign Linguistics, pp. 369–495. Progress, Moscow (1981)
6. Furs, L.A.: Sintaksicheski reprezentiruemye koncepty [Syntactically representable concepts]. Extended abstract of a doctoral theses. Tambovskij gosudarstvennyj universitet im Derzhavina (2004). (in Russian)
7. Galich, G.G.: Kognitivnye strategii i yazykovye struktury [Cognitive strategies and language structures]. Omskij gosudarstvennyj universitet, Omsk (2011). (in Russian)
8. Galich, G.G.: O sistematizacii kategorij yazykovogo soderzhaniya s pozicij antropocentrizma [On the systematization of language content categories from the view of anthropocentrism]. Vestnik Omskogo universiteta 2(64), 372–375 (2012). (in Russian)
9. Ivin, A.A.: Filosofiya: Enciklopedicheskij slovar' [Philosophy: Encyclopedic Dictionary]. Gardariki, Moscow (2004). (in Russian)
10. Kasevich, V.B.: Semantika. Sintaksis. Morfologiya [Semantics. Syntax. Morphology]. In: Churilina, L.N.: Aktual'nye problemy sovremennoj lingvistiki [Actual problems of modern linguistics], pp. 344–368. Flinta-Nauka, Moscow (2009). (in Russian)

11. Khomyakova, E.G.: Kognitivno-informacionnaya situaciya kak instrument issledovaniya poznavatel'noj funkcii yazyka [Cognitive-informative situation as an instrument for studying the cognitive function of language]. In: Khomyakova E.G. (ed.) Percepciya. Refleksiya. Yazyk [Perception. Reflection. Language], pp. 236–250. Sankt-Peterburgskij gosudarstvennyj universitet, St. Petersburg (2010). (in Russian)

12. Klyoster, A.M., Shnyakina, N.Y.: K voprosu o kategorialnom chlenenii nauchnogo i obydennogo znaniya [Categorical division of scientific and everyday knowledge]. Problems of cognitive linguistics 1(54), 144–150 (2018). (in Russian)

13. Kruchinkina, N.D.: Kategorial'noe propozitivnoe semantiko-grammaticheskoe oformlenie konceptualizacii prototipov sobytij [Categorical propositional semantic-grammatical structure of conceptualization of prototypes of events]. In: Kubryakova, E.S. (ed.) Kognitivnye issledovaniya yazyka [Cognitive studies of the language], pp. 386–396. Institut yazykoznaniya RAN, Izdatel'skij dom TGU im. G.R. Derzhavina, Tambov (2010). (in Russian)

14. Kryazhevskikh, N.N.: Kategoriya v kognitivnoj lingvistike [The category in cognitive linguistics]. Vestnik Vyatskogo gosudarstvennogo universiteta 4(2), 12–15 (2010). (in Russian)

15. Kubryakova, E.S.: Glagoly dejstviya cherez ih kognitivnye harakteristiki [Verbs of action through their cognitive characteristics]. In: Arutyunova, N.D., Spiridonova, N.F. (eds.) Logicheskij analiz yazyka. Izbrannoe. 1988–1995, pp. 439 – 446. Indrik, Moscow (2003). (in Russian)

16. Kubryakova, E.S.: Kategorizaciya [Categorization]. In: Kubryakova, E.S. (ed.) Kratkij slovar' kognitivnyh terminov [A brief dictionary of cognitive terms], pp. 42–45. Filologicheskij fakul'tet MGU im. M. V. Lomonosova, Moscow (1997). (in Russian)

17. Kubryakova, E.S.: Yazyk i znanie: Na puti polucheniya znanij o yazyke: chasti rechi s kognitivnoj tochki zreniya. Rol' yazyka v poznanii mira [Language and knowledge. On the way of getting knowledge of the language: parts of speech from the cognitive point of view. The role of language in the world cognition]. Yazyki slavyanskoj kul'tury, Moscow (2004). (in Russian)

18. Lakoff, G.: Women, Fire and Dangerous Things: What Categories Reveal About the Mind. University of Chicago Press, Chicago (1987)

19. Levin, G.D.: Filosofskie kategorii v sovremennom diskurse [Philosophical categories in modern discourse]. Logos, Moscow (2007). (in Russian)

20. Luk'yanov, L.F.: Sushchnost' kategorii «svojstvo» [The essence of the category "property"]. Mysl', Moscow (1982). (in Russian)

21. Miletova, E.W.: Stanovlenie kategorii «kachestvo»v filosofii i yazyke [Formation of the category «quality» in philosophy and language]. Molodoj uchyonyj 1(2), 22–24 (2012). (in Russian)

22. Najsser, U.: Poznanie i real'nost' [Cognition and Reality]. Progress, Moscow (1981). (in Russian)

23. Norman, D.A., Bobrow, D.G.: On data limited and resource limited processes. Cognit. Psychol. 7(1), 44–64 (1975)

24. Panfilov, W.S.: Filosofskie problemy yazykoznaniya. Gnoseologicheskie aspekty. [Philosophical problems of linguistics. Epistemological aspects]. Nauka, Moscow (1977). (in Russian)

25. Pankrats, Ju.G.: Arhitektura kognicii [Architecture of cognition]. In: Kubryakova, E.S. (ed.) Kratkij slovar' kognitivnyh terminov [A brief dictionary of cognitive terms], pp. 12–13. Filologicheskij fakul'tet MGU im. M. V. Lomonosova, Moscow (1997). (in Russian)

26. Petrenko, V.F.: Osnovy psihosemantiki [Fundamentals of psychosemantics], 2nd edn. Piter, Saint Petersburg (2005). (in Russian)

27. Petrovskij, A.V., Yaroshevskij, M.G. (eds.): Kratkij psihologicheskij slovar' [Brief psychological dictionary]. http://psychology.academic.ru. Accessed 24 Mar 2015. (in Russian)
28. Rosch, E.: Principles of categorization. In: Rosch, E.H., Lloyd, B.B. (eds.) Cognition and categorization, pp. 27–48. Lawrence Erlbaum Associates, Publishers, Hillsdale, New Jersey (1978)
29. Shabes, V.Ya.: Sobytie i tekst [Event and text]. Vysshaya shkola, Moscow (1989). (in Russian)
30. Shafikov, S.G.: Kategorii i koncepty v lingvistike [Categories and concepts in linguistics]. Voprosy yazykoznaniya 2, 3–17 (2007). (in Russian)
31. Styopin, V.S.: O prognosticheskoj prirode filosofskogo znaniya [On the predictive nature of philosophical knowledge]. Voprosy filosofii 4, 39–53 (1986). (in Russian)
32. Trunova, O.V.: O kategorizacii, kategoriyah i kategorial'nom znachenii [About categorization, categories and categorical meaning]. In: Trunova, O.V. (ed.) Kommunikativno-paradigmaticheskie aspekty issledovaniya yazykovyh edinic [Communicative-paradigmatic aspects of the study of linguistic units], pp. 261–267. BGPU, Barnaul (2004). (in Russian)
33. Uyomov, A.I.: Veshchi, svojstva, otnosheniya [Things, properties, relations]. AN SSSR, Moscow (1963). (in Russian)
34. Velichkovskij, B.M.: Kognitivnaya nauka. Osnovy psihologii poznaniya [Cognitive science. Fundamentals of Psychology of Cognition]. Vol. 2. Academia, Moscow (2006). (in Russian)
35. Verhoturova, T.L.: Metakategoriya «nablyudatel'» v nauchnoj kartine mira [Metacategory "observer" in the scientific picture of the world]. Studia Linguistica Cognitiva. Vyp. I. Yazyk i poznanie: Metodologicheskie problemy i perspektivy [Language and cognition: Methodological problems and perspectives], 45–65 (2006). (in Russian)

Peculiarities of English Oil and Gas Terminology

Natalia V. Gorokhova$^{(\boxtimes)}$ (iD)

Gubkin Russian State University of Oil and Gas (National Research University),
119991 Moscow, Russian Federation
n.gorokhova@nxt.ru

Abstract. The terminological homonymy of the English technical terms of oil and gas terminology, formed on the semantic decay of a word, is the basis for lexical units with different denotations. Identification and analysis of homonymy in the terminology of special English vocabulary is one of the urgent tasks of the professional language. It plays an important role in many scientific contexts, for example, in power engineering, oil and gas fields development, production of fossil fuels, machinery engineering, etc. The lexical analysis of the terminological units of oil and gas sphere, based on the metaphorization and polysemy, shows that special terms may undergo semantic changes and transform into homonyms – a random coincidence of words identical in sound form and spelling but different in meaning. The homonyms, which name different phenomena, do not duplicate and do not contradict each other as one of the terms becomes more commonly used in the scientific environment, and subsequently the only one generally recognized by the specialists.

Keywords: Terminology · Homonymy · Polysemy ·
Semantic word transformation · General language

1 Introduction

The traditional notion of homonymy plays an important role in many scientific contexts, such as terminology, logical semantics and semiotics, it is a natural generalization of the corresponding linguistic concept. Homonymy is a graphic and (or) phonetic coincidence of words (signs, collocations and phrases) that have a different significance and (or) meaning [9].

Homonymy is the final case of polysemy – a phenomenon that is equally characteristic for a general literary language, and the terminological systems [4, 9, 25]. It is considered that homonymy is generated by random causes associated with the word formation in a language [15, 22]. Homonymy is very similar to polysemy in the same sound or graphic shape and correlates with several objects or phenomena of reality. However, polysemic connections between these realities are clearly recognized by the speakers, while homonymy has no links between the named realities in the modern language [11]. There is a continuous «interchange» of lexical units between the terminology and the common language: the words of general literary language lose some of their properties and turn into terms and, conversely, terms become units of a

© Springer Nature Switzerland AG 2019
Z. Anikina (Ed.): GGSSH 2019, AISC 907, pp. 367–372, 2019.
https://doi.org/10.1007/978-3-030-11473-2_38

common language [16]. The processes of terminologization and de-terminologization are the source of homonyms in various terminological systems [13]. Terminological homonymy differs from a similar phenomenon in a general literary language. In the language of science, homonymy occurs quite often, and it is the result of a semantic alteration of the word, when polysemy diverges so strongly that it becomes homonymy [15]. Homonymy, formed as a result of the coincidence of words different by origin, but identical in meaning, which takes place in general literary language, is not typical for the terminology of a speciality.

Terminological homonymy is fundamentally different from the similar phenomenon in general literary language by the following reasons: (1) terminology uses only one type of homonymy, namely the one that is the result of the semantic development of the word; (2) homonymy can be characterized only as an intersystem phenomenon: either these terms are units of different terminological systems or these terms are the consequence of the lexical-semantic formation, both have become homonyms from the generic words of general literary language [6, 19].

In the scientific and linguistic literature, the essence of homonymy is unambiguously understood. The view on terminological homonymy by different researchers is most often negative, since this phenomenon is a violation of the law of the sign [2]. Homonymy is called a «disease» of language, scientists believe that the relations of words that are identical in form and not related in meaning are irregular and exceptional, homonyms in all cases are an inadequate non-distinction of what should be different [16]. Often the negative attitude toward homonymy is due to the fact that it is difficult to draw a clear boundary between the phenomena of homonymy and polysemy: the former is the limiting factor of the latter [1]. Homonyms are often in apparent cognation. The clear common etymology of homonyms makes it necessary to qualify such cases as polysemy. In other words, homonymy is a «latent» polysemy [6, 11].

Thus, homonymy and polysemy are an indispensable attribute of natural language, enriching the general literary language with expressive means. Many researchers believe that in the scientific contexts homonymy is unacceptable, and in some cases it is even dangerous [18–20, 22]. Therefore, scientific and special needs are predominantly utilized by professionalisms (specially selected fragments of general literary language), less flexible than the language as a whole, but more adapted to the needs of the «served» area [9].

Some other researchers define homonymy as the sound and grammatical coincidence of linguistic units that are not semantically related to each other. The words-homonyms, in the opinion of these authors, are characterized primarily by the fact that they are compared with one or other phenomena of reality independently of each other, therefore there is no associative conceptual-semantic connection between them. When the lexical meaning of homonyms is clearly understood, they will not be confused [9, 21, 24, 25].

Homonyms is the result of a semantic split of one word, they lose the immutability of internal forms, representing lexical units with different meanings, belonging to different terminological fields [20]. The lack of internal autonomy of the term, its correlation with the terminological field, – these are the main criteria for distinguishing homonyms. Homonyms are admissible only when they refer to different areas [21]. The problem of polysemy and homonymy outside the special terminology loses its significance, because the terms identical in sound form and spelling do not practically occur in one branch of industry [12, 24]. Homonyms do not affect the communication process of specialists and do not influence on the verbal-semasiological level of their language [7, 13, 17, 23]. Homonyms occurring in one terminological system is rather a rare phenomenon.

2 Object and Methods

The object of the research is the terminological lexis of the vocabulary of the English language in oil and gas sphere.

To identify the homonymy of the English technical terms in the considered sphere of the language we selected the following methods:

- the method of continuous sampling from modern dictionaries and research materials,
- the method of contextual analysis using authentic text materials,
- the method of comparative analysis using an online search service,
- descriptive method which is necessary for identifying the peculiarities of the terminological units use and their transformations.

Such transformations contribute to the development of the languages and their continuous renewal. The processes that occur in the professional terminological field are actively manifested in English oil and gas terminology. The specialized language is actively using the appropriate terminology to a great extent simplifying and accelerating the process of communication.

3 Homonymy of the English Technical Terms of Oil and Gas Terminology

There are lexical units that have homonyms in oil and gas terminology, formed by metaphorical processes [5]:

horizon – (1) the line at which the earth's surface and the sky appear to meet (general language); (2) horizontal movement of the bit at any level (oil and gas terminology);

footprint – (1) the impression left by a foot or shoe on the ground or a surface (general language); (2) plan for the placement of pipeline equipment (oil and gas terminology);

formation – (1) the action of forming or process (general language); (2) reservoir layer during drilling operations (oil and gas terminology);

anchor – (1) a heavy object attached to a rope or chain and used to moor a vessel to the sea bottom (general language); (2) equipment for fixing lifting pipes to prevent movement of the downhole mechanism under the influence of the load (oil and gas terminology);

elevator – (1) a platform or compartment housed in a shaft for raising and lowering people or things to different floors or levels (general language); (2) the gripper for hanging the drill pipe during the lifting operations (oil and gas terminology);

candle – (1) a cylinder or block of wax or tallow with a central wick that is lit to produce light as it burns (general language); (2) ceramic filter in a cooling column (oil and gas terminology).

Such words of the common/general language, becoming terms, retain only the «phonetic envelope», the meaning of the same word is completely different; the basis of semantic derivation lies in the semantic transfer of the name through formal logical connections: similarity and contiguity. In other special terminological systems the words which previously were polysemantic lose their motivation for unity in the process of functioning [3, 10, 14]:

tine – (1) cultivator grip (agriculture); (2) fork-lift of a pipe-laying machine (oil and gas terminology);

cavern – (1) a cavity that arises in the body when the tissues caused by the disease are destroyed and deadened (medicine); (2) underground storage of liquid hydrocarbons (oil and gas terminology);

mat – (1) gym mat (sports); (2) bottom cushion (oil and gas terminology);

core – (1) the steel axis used in the mobile parts of the mechanisms of electrical appliances (instrument-making); (2) a sample of rock in the form of a cylinder, extracted from the well (oil and gas terminology);

shell – (1) a metal tube for a bullet into the opening of which a capsule is inserted (military); (2) blank for the pipes rolling (oil and gas terminology);

tire – (1) a rubber covering, typically inflated or surrounding an inflated inner tube, placed around a wheel to form a flexible contact with the road (car industry); (2) cable bandage (oil and gas terminology);

column – (1) part of the architectural structure in the form of a high pillar serving as a pedestal support, internal parts of the building (architecture); (2) rods (oil and gas terminology).

In some professional terminological systems, the semantics of the equally sounding terms is different:

blade – (1) bucket, blade (construction industry); (2) a guide knife for centering the edges of the strip in the forming cages during continuous welding (oil and gas terminology);

clinker – (1) boat with inlaid paneling (sports); (2) slag formation (oil and gas terminology);

conductor – (1) a person in charge of a train, streetcar, or other public conveyance (transport); (2) the first casing string (oil and gas terminology);

web – (1) the membrane of a duck, bat, etc. (zoology); (2) the crank pin of crankshaft (oil and gas terminology);

plug (1) large mass of rock or mineral of irregular cylindrical shape (geology); (2) an ejector for blowing out pipes (oil and gas terminology);

valve – (1) an old military rifle with screw holes in the trunk (military); (2) a short section with an external thread, used to connect pipes to each other or to connect them to tanks, vessels (oil and gas terminology).

4 Conclusion

Homonymy in oil and gas terminology is the result of the semantic formation of the sign, the fragmentation of different meanings of the polysemic word, the loss of the transitional parts of the whole. Terminological homonymy can be qualified as an intermediate phenomenon: the terms of the lexical semantic formation are called inter-system homonymy, the terms of various terminologies are considered as inter-scientific homonymy.

Homonymy in certain spheres of scientific and professional human activity is a phenomenon of little use, as a rule. Homonyms is the reality of different terminological fields, such units are unambiguous, and therefore do not impede the communication process due to the high professional competence of specialists.

References

1. Ahmanova, O.S.: Slovar' lingvisticheskih terminov [Dictionary of linguistic terms]. KomKniga, Moscow (2007). (in Russian)
2. Ahmanova, O.S.: Ocherki po obshchej i russkoj leksikologii [Essays on general and Russian lexicology]. Uchpedgiz, Moscow (1957). (in Russian)
3. Alefirenko, N.F.: Teoriya yazyka. Vvodnyj kurs. [Theory of language. Introductory course]. Akademiya, Moscow (2004). (in Russian)
4. Crystal, D.: A Dictionary of Linguistics and Phonetics, 6th edn. Blackwell Publishing, New York (2008)
5. Cuyckens, H., Dirven, R., Taylor, J.R.: Cognitive Approaches to Lexical Semantics. Walter de Gruyter, Berlin-New York (2003)
6. Danilenko, V.P.: Russkaya terminologiya. Opyt lingvisticheskogo opisaniya [Russian terminology. Experience of linguistic description]. Nauka, Moscow (1977). (in Russian)
7. Girutskij, A.A.: Vvedenie v yazykoznanie [Introduction to linguistics]. TetraSistems, Minsk (2001). (in Russian)
8. Gorohova, N.V.: Anglo-russkij slovar' terminov truboprovodnogo transporta [English-Russian Dictionary of Pipeline Transport Terms]. KAN, Omsk (2012). (in Russian)
9. Gorohova, N.V.: Omonimiya angloyazychnykh tekhnicheskikh terminov (na osnove terminopolya truboprovodnogo transporta) [The homonymy of English technical terms (based on the pipeline transport terminology)]. Vestnik LGU im. A.S. Pushkina 3(5), 95–99 (2015). (in Russian)
10. Korshak, A.A., Shammazov, A. M.: Osnovy neftegazovogo dela: ucheb. dlya vuzov [Basics of oil and gas business]. DizajnPoligrafServis, Ufa (2001). (in Russian)
11. Kreidler, ChW: Introducing English Semantics. Routledge, London (1998)
12. Lebedeva, N.B.: Mnogoslojnost' leksicheskoj semantiki i situatema kak polisituativnaya struktura [Multilayered lexical semantics and situational topic as a polysituative structure]. Vestnik CHelyabinskogo gosudarstvennogo universiteta. Filologiya. Iskusstvovedenie 5 (259), 92–97 (2012). (in Russian)
13. MacArthur, T.: Concise Oxford Companion to the English Language. Oxford University Press in Oxford, New York (2005)
14. Moiseev, M.V.: Leksikografiya kul'tury anglijskogo yazyka [Lexicography of English culture]. OmGU, Omsk (2006). (in Russian)

15. Reformatskij, A.A.: Vvedenie v yazykoznanie [Introduction to linguistics]. Prosveshchenie, Moscow (1967). (in Russian)
16. Reformatskij, A.A.: Vvedenie v yazykovedenie (in Russian) [Introduction to linguistics]. Aspekt Press, Moscow (1996)
17. Shajkevich, A.Ya.: Vvedenie v lingvistiku [Introduction to linguistics]. Akademiya, Moscow (2005). (in Russian)
18. Shields, C.: Order in Multiplicity: Homonymy in the Philosophy of Aristotle. Clarendon Press, Oxford (1999)
19. Shurygin, N.A.: Leksikologicheskaya terminologiya kak sistema [Lexical terminology as a system]. Izd-vo Nizhnevart. ped. instituta, Nizhnevartovsk (1997). (in Russian)
20. Slyusareva, N.A.: Terminologiya lingvistiki i metayazykovaya funkciya yazyka [Terminology of linguistics and language functions]. Voprosy yazykoznaniya 4, 69–76 (1979). (in Russian)
21. Sulejmanova, A.K.: Terminosistema neftyanogo dela i ee funkcionirovanie v professional'nom diskurse specialista: avtoref. dis. ... dokt. filol. nauk [The terminology of oil business and its functioning in the professional discourse of a specialist]. Bashkirskij gosudarstvennyj universitet, Ufa (2006). (in Russian)
22. Superanskaya, A.V., Podol'skaya, N.V., Vasil'eva, N.V.: Obshchaya terminologiya [General terminology]. Nauka, Moscow (1989). (in Russian)
23. Susov, I.P.: Vvedenie v yazykoznanie [Introduction to linguistics]. Vostok-Zapad, Moscow (2006). (in Russian)
24. Tatarinov, V.A.: Teoriya terminovedeniya [Theory of terminology]. Moskovskij licej, Moscow (1996). (in Russian)
25. Vinogradov, V.V.: Ob omonimii i smezhnyh s nej yavleniyah [On homonymy and related phenomena]. Voprosy yazykoznaniya 5, 3–17 (1960). (in Russian)

Coherence Features in Multimodal Electronic Literary Texts

Svetlana Kuchina$^{(\boxtimes)}$ ⓘ

Novosibirsk State Technical University, Novosibirsk 630120, Russia
s.kuchina@corp.nstu.ru

Abstract. The article deals with the coherence specificity in electronic literary texts. The author specifies the phenomenon of electronic literary text and covers several questions concerning its multimodal structure and the character of correlation between its elements. Multimodal structure of electronic literary texts represents specific textuality which is supported by the peculiar connections between its verbal and non-verbal elements. Multimodality involves components from various semiotic systems (text, sound, video, graphics, and animation), while participants of electronic discourse use various sensory organs to decode the information. Global cohesion in the electronic literary text is achieved by the direct and indirect denotative relevance of its verbal and non-verbal elements that conform (thematically or at the associative stage) to various categories of fictional worlds (characters, themes and motives). The research materials include several electronic literary texts (based on different platforms and technologies, such as Adobe Flash, HTML 5) that demonstrate the use of conceptually valid polycode elements in their structure.

Keywords: Electronic literature · Coherence · Multimodality

1 Electronic Literary Text: Definition and a Brief History of the Genre

Nowadays, in linguistics theory and practice, text studies go far beyond a closed text structure. Verbal code is the main source of text meaning, but it is not the dominant element of modern polycode communication. The rising semiotic complexity of modern communication is the most defining tendency today. Digital polycode space of the Internet changed significantly the esthetical benchmarks of modern times. The technology became significant in itself and these changes have implications for the text structure and its perception.

The history of electronic literary text is not very long. Emergence of electronic literary text resulted from a considerable technology development of 20th–21st centuries. The founder of the technological shift is Bush who suggested hypothetical proto-hypertext system Memex in 1945 [1]. Also, the theory and practice of hypertext in the literary project Xanadu by Nelson [2] played a significant role. The project was expected to result in a global system of various texts, the system would allow to compound new documents (texts) from pieces (transclusions) of other documents. These transclusions would help to activate all sources of the current document in real time.

© Springer Nature Switzerland AG 2019
Z. Anikina (Ed.): GGSSH 2019, AISC 907, pp. 373–381, 2019.
https://doi.org/10.1007/978-3-030-11473-2_39

Together with the technological development the concept of digital art and literature was rising. The phenomenon of electronic literary text became the realization of the postmodern text conception. The idea of "genitive" critics, which basic notions are avantext (authors draft copies), intertext (citations), metatext (second-order descriptions), and genotext (signifying infiniteness), was implemented in the electronic literary text structure.

Nelsons' ideas are also based on the deconstruction conception by Derrida, the phenomenon of indeterminacy, instability of meaning and constant semantic shifts in language [3]. A bit later Barthers transported Derrida's term "lexia" into the hypertext discourse, connected it with such notions as "blocks of signification" and "units of reading" [4]. The idea of rhizome became relevant in the hypertext discourse, too. Rhizome structure represents an open, nonlinear text structure [5].

First services and applications for hypertext appeared in the 80s of the 20th century. There were HyperCard Apple, Intermedia, HyperWave, WebThing, etc. First hypertext novels were published on the basis of the Storyspace project by Eastgate Systems, which was the first tool for complex and interlinked narratives. Storyspace started with works like Joyce's "Afternoon, a story" [6], Jackson's "Patchwork girl" [7] that became electronic hypertext classics nowadays.

In the 90s of the 20th century new media theorists made first attempts to interpret and systemize knowledge of electronic literary texts. Aarseth suggested the notion of cybertext, the nonlinear text structure which was seen as a machine producing signs that consist of the medium, the operator and the strings of signs that can be also divided into textones (strings of signs as they are in the text) and scriptones (strings of signs as they appear to the reader) [8]. Aarseth extrapolated cybertextuality not only to the electronic, but also to the codex narratives. The researcher thought that cybertextuality is an inner feature of the text mechanism which is aimed at permanent changes of its structure. Hayles pointed out that Aarseth did not take into the consideration the specificity of media. She suggested the MSA – Media Specific Analysis [9]. According to Hayles, the medium specificity influences the text content. Particular medium shapes the text that cannot be separated from its "words (and other semiotic components) and calls for the need to develop a theory that takes into consideration the medium as a crucial aspect of the content of a work [10, p. 89]. Hayles suggested the term technotext which she defined as "literary works that strengthen, foreground, and thematize the connections between themselves as material artifacts and the imaginative realm of verbal/semiotic signifiers they instantiate" [11, p. 25].

Thinking about electronic literary text specificity, Wardrip-Fruin defines the electronic literary text as a "work with important literary aspects that requires the use of digital computation" [12, p. 126]. However, it is worth noting that "the use of digital computation" is an important, but not determining aspect for electronic literary text. According to Zuern, the computation in electronic literature is essential not only to define the artifact specificity but also to emphasize the particular literary properties of the text [13]. So, electronic literary text is a born-digital text which verbal and non-verbal components unite into a coherent visual, structural, semantic and functional whole.

2   Coherence and Multimodality as Linguistic Categories

The basic characteristics of the text are the categories of integrity and connectivity. These text categories come from different terminological paradigms, and they are relevant subjects of extensive discussion when determining the criteria of textuality in the domestic and foreign linguistics (van Dijk, Galperin, Milevskaya, Paducheva, Troshina, Dressler, Bellert, etc.). In the foreign linguistic theory, the concepts of cohesion and coherence were introduced to refer to these terms, e.g. in the textuality model proposed by Dressler and Beaugrand [14], the basic features of the text are cohesion, coherence, intentionality, acceptability, informativity, situationality, and (typological) intertextuality. Here, cohesion is perceived as interconnection of surface (grammatical-syntactic, lexical, rhythmic and graphic) components of text structure, meanwhile coherence is a semantic and cognitive link in the cause-effect, temporal and referential aspects.

Correlation and coordination of verbal and non-verbal elements of the electronic literary text are manifested at the **semantic, syntactic** (semantic, syntactic, visual and graphical connectivity) and **pragmatic** (subject and conceptual integrity) levels of organization. The constant basis of all relations within the literary text is a semantic coordination of all its components and conformity to ideological and thematic content.

Typologically, all intertextual relations can be divided by:

1. The method for language representation (lexical, grammatical and syntactic levels);
2. The method for non-verbal representation (audio, video, animation and graphics);
3. The degree of expression (explicit and implicit);
4. Location/fixation in the structure of electronic literary text (anaphoric and cataphoric).

It is obvious that electronic communication is not limited by verbal means. A significant proportion is accounted for non-verbal elements (audio, video, graphics and animation). Analyzing the specifics of electronic discourse and its product – electronic text, researchers characterize the way information is presented in the network, as well as the nature of its transfer and the channel of perception through which the recipient receives and processes the transmittable data. Despite its extralinguistic nature, these factors have a significant influence on the structure and semantics of the electronic text. The type of information transfer in the electronic discourse, as well as the channel of perception through which the received information is processed, can be integrated by the concept of modality, which is a "sign system interpreted with the help of different senses" [15, 379–402 pp.]. Modalities are usually named from the point of view of the signal recipient and its sensory systems. The following types of modality include:

1. Visual (pictures, colour and visual images);
2. Audial – auditory (sounds, music, auditory images and intonations);
3. Kinesthetic – tactile (touch, muscle and skin feelings, inner sensations).

A key feature of electronic discourse is its multimodal character. Multimodality, in this case, involves different components from various semiotic systems in the composition of discursive elements (text, sound, video, graphics, and animation), while participants of electronic discourse use various sensory organs to decode information.

3   Electronic Literary Text Coherence Specificity: Pragmatic Level

The **pragmatic level** of correlation between verbal and non-verbal components of the electronic literary text (subject-conceptual integrity) ensures the coherence of its elements in an ideological and thematic manner that is directly related to the author's concept of a work of art.

This level of correlation between the electronic literary text elements is explained, on the one hand, by the coherence within the verbal component, based on the text-forming logical-semantic relations, the main technique of which is repetition of information in different parts of the text. And on the other hand, the integrity of the electronic literary text content is provided by micro and macro thematic correlations of non-verbal elements that are compositionally fixed after repeated text elements of the electronic piece (key words/phrases). This perception does not contradict the concept of van Dijk [16], according to which there are two main types of the text coherence: local and global, where local coherence is understood as the coherence of linear sequences (interphrasive unity) due to interphrasive syntactic relations, while the category of global coherence determines the unity of the text as a semantic whole, manifested through key words that thematically and conceptually unite the text and its fragments. The category of global connectivity of the text is actualized with the lexis related to one thematic circle (isotopy), the renomination of the title and elements of the subtitle summation in the subsequent text, and also with the repetition of syntactic structures [17, p. 26].

From the Babenko's point of view, "textual integrity is provided by its denotative space and the specific situation of its perception" [18, p. 41]. Integrity of the electronic literary text implies a direct and indirect denotative correlation of verbal and non-verbal components. With the direct denotative correlation, the verbal and non-verbal components of the narrative designate the same subjects/objects/situations that are realized in the literary work. With the mediated denotative correlation, verbal and non-verbal elements correspond with different subjects/objects/situations, realized in the literary work and connected thematically or associatively. However, it should be noted that, being a complex and multilevel sign, the text has a sufficiently flexible system of internal relations, in which the units of the lower levels of the system obey strict combination rules for its components, while at the higher levels of the linguistic system the conditions for connecting text parts are more free [18, p. 82].

Let's consider the interactive electronic literary text "In Absentia" by Carpenter [19] as an example. From the point of view of this literary text's structure, its electronic data are represented by a combination of verbal (small notes, made in prose and combined into a hypermedia text of six parts – 'Alouer', 'Avendre', 'Perdu', 'Trouve', 'Vide', and 'Home') and non-verbal (interactive parts from Google maps, as well as images, stylized photographs, and graffiti, reflecting the specificity of urbanized space) components.

Coherence of the verbal elements in "In Absentia" at the semantic level has an explicit character and is realized due to repetition of words from different lexical groups united by one theme. The main semantic component – the 'city', and everything that is related to the characteristics of the urban space and people's lives in it, is actualized in this piece with the help of vocabulary, which is thematically tied to a specific locus (Montreal).

At the lexical level it is expressed in the repetition of words that form a united paradigm. The repetitions include:

1. Paradigm of the nominants for the city/district: *Saint-Viateur Street, Mile End, Montreal, Clark Street Arcade street, Saint-Viateur Street, Clark Street, Saint-Urbain Street.*
2. A group of lexemes with names of urban realities: *Saint-Jean Baptiste party, coffee shop, copper-domed church.*
3. A group of "rental housing" themed lexemes: *new landlord, new apartment, roommate, owner, hostage, private property, rent, hydro bill,*
4. A group of "parts and objects of living space" themed lexemes: *ground floor, back alley, widow, door, wall, curtains, double room, building, holes in ceiling, cottage, new flooring, new furniture, duplex, condo, duplex.*
5. Groups of lexemes reflecting the sound characteristics of the city/district:

   – sounds of city streets: *Beer can rattled in the wind; the wind in our ears;*
   – sounds of music of different genres: *funk; country music; salsa; bad hip-hop;*
   – sounds neighbours make: *sneezes, grunts and telephone conversations; headboard bucks at the wall behind mine; my loud, cold, crumbling, old apartment; the dog barking his head off;*

6. Groups of lexemes reflecting the color esthetics of a city/district: *less colorful; snow park; pale-blue; giant pieces of ash; freshly painted front fence.*
7. Verbs that have the meaning of 'living in the city': shared an apartment, to flee, to pay rent, and live together.

The above thematic groups of words are directly related to the semantic explication of the conceptual content of the electronic artwork by Carpenter. In the text these words are distantly relative to each other, their placement correlates with the hypermedia structure of the artwork, in which verbal components are connected by hyperlinks attached to certain key words/images in the narrative.

If we consider the syntactics of verbal and non-verbal elements in "In Absentia", the correlation of "a certain part of the verbal component (chapter/fragment) – a non-verbal element" is realized: each verbal part/chapter of the work ('Alouer', 'Avendre', 'Perdu', 'Trouve', 'Vide', and 'Home') is attached to a certain segment of the Google maps and, at the same time, each part has integrated visual interactive components-links represented by visual images (a small static image of the front door of the house, an aircraft, a dog, a black and white image of the church, traffic signs, a visualized layout of rooms in an apartment, etc.), which lead to verbal components of this part of the work they are attached to. By activating (clicking) the interactive component, a recipient automatically opens the screen with the text; e.g., in the 'Perdu' part, a dog image links to a text fragment describing a large black dog walking in a park; an image of an orange tennis ball links to the story about long walks the character took with the dog, and games they played with the orange ball. In this case, we can also talk about the direct denotative correlation of the verbal and visual images in the artwork.

Semantic deployment becomes complete only at the moment when the components of a group of lexemes (keywords) used in one part or another of "In Absentia" unite in the consciousness of the recipient with the visual component. Google Maps are integrated into the narrative in such a way that the author inhabits the conditionally real areas of Montreal with the characters of the fictional world. The names of streets in "In Absentia" correlate with the names of real streets in Montreal, but they also correlate visually because components of the narrative in which the names of particular streets are mentioned are located in the corresponding segment of the electronic map of the city.

In the same way, visual representation of the urban landscape elements can be taken into account as photographs of graffiti or individual urban objects. In this case, we can also talk about direct denotative correlation, because verbal means of expressing ideological and thematic content directly correlate with the visual images of "In Absentia" (Fig. 1). Full realization of the literary concept of "In Absentia" (reconstruction of the complex intertwining of the characters' fates in the present and the past of the city) is possible only at the moment when a visual-graphic connection (Google maps, photographs and collages with images of Montreal streets) and verbal (stories of characters from the first person) of the literary image of Montreal. "In Absentia" by Carpenter implements an integrative type of relations between verbal and visual components, in which a visual component is an integral part of the textual whole, along with its lexical structure, providing coherence of components at the logical and semantic level.

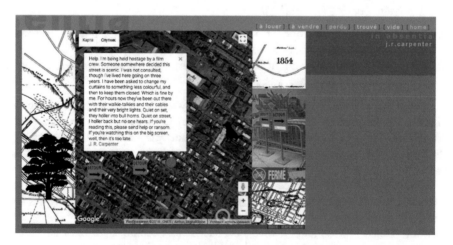

Fig. 1. In Absentia by J. Carpenter.

The indirect denotative correlation at the associative connectivity level can be considered in the electronic literary text "Redshift and Portalmetal" by Cardenas [19]. Non-verbal components of the main part in the artwork include a background in the form of a non-static landscape, executed in alarming shades of red, image.

In another part of "Redshift and Portalmetal" verbal component is typed by the background video of the main character, Roja, walking along the ocean shore; here the colour of choice changes dramatically from the alarming shades of red to the natural, but gloomy, gray colours of the coastal zone; the audio is represented by rhythmically repeating sounds resembling the growing thunder (Fig. 2).

Fig. 2. Redshift and Portalmetal by Cardenas.

Lexical, syntactic and graphic coherence of the verbal component from the electronic literary text is supported by the style of its polycode elements (video/audio series). The internal rhythmic structure of the text is sustained due to the author using syntactic anaphora and complete lexical repetition in almost every part of the work, e.g.:

And I couldn't breathe …
And the people couldn't talk to me…
And I felt I was losing my mind …
Or:
I had to leave,
And go into the stars,
And find a new planet,
In a new universe,
And let my planet die,
And live my own life. [20].

Emotional tension and drama of the character's experiences are strengthened and emphasized by gradation, e.g.:

I thought I could stay and save her, my planet
I thought I could just drop everything and stay and care for her and help her live a little longer [19].

The fading nature and the sounds indicating the approaching natural cataclysms, are equal (in relation to the verbal part) semantic components of "Redshift and Portalmetal" by Cardenas [20]. These non-verbal components are associated with verbal elements of the narrative at the associative and denotative levels, since the red background of the landscape and rhythmic sounds resembling thunder during a thunderstorm convey the feeling of an upcoming catastrophe not literally, but indirectly, by generating anxiety. The tension created by the author due to the polycode components acts as a kind of refrain to one of the main themes of the artwork, it focuses the attention on semantics of the main motive given in the verbal part – the motive for the death of the planet in contrast with the fragility and helplessness of the main character. Judging by the nature of fixing verbal and non-verbal components in "Redshift and Portalmetal" anaphoric type of the relations is also realized, background landscape and the video series appear on the screen a little earlier than the verbal component [20]. All elements (lexical, syntactic, graphic, audiovisual) making up the relations between verbal and non-verbal components in "Redshift and Portalmetal" by Cardenas are of very explicit nature.

4   Conclusion

In each electronic artwork we can talk about the special nature of textuality, provided by the connection of audio/video, graphics, animation with the content of a verbal component. In this case, we do not refer to the mechanical connection of audio/visual with the verbal code, but rather, as Chernyavskaya points out, to "the emergence of dynamic image relations with the 'cognitive module' of consciousness" [21, p. 64]. The coherence and integrity of the electronic literary text does not fully depend on the verbal code, which sometimes (given the increasing tendency of image visualization) is excessive. Integrity of the electronic text is born in the reader's consciousness, on the condition of particular cognitive operations – constructing an integral semantic value transmitted through verbal code and non-verbal elements of the electronic literary text.

References

1. Bush, V.: As We May Think. https://www.theatlantic.com/magazine/archive/1945/07/as-we-may-think/303881/. Accessed 10 June 2018
2. Nelson, T.: Project Xanadu. http://xanadu.com/. Accessed 10 June 2018
3. Derrida, J.: Deconstruction and the Possibility of Justice. Psychology Press, London (1992)
4. Barthers, R.: S/Z. Akademicheskiy Prospekt, Moscow (2009)
5. Deleuze, G.: Capitalism and schizophrenia. Thousand plateau. http://www.studfiles.ru/preview/5799260. Accessed 10 June 2018. (in Russian)
6. Joyce, M.: Afternoon, A Story. http://eastgate.com/catalog/Afternoon.html. Accessed 10 June 2018
7. Jackson, S.: Patchwork Girl. http://www.eastgate.com/catalog/PatchworkGirl.html. Accessed 10 June 2018
8. Aarseth, E.J.: Cybertext: Perspectives on Ergodic Literature. The Johns Hopkins University Press, Baltimore (1997)

9. Hayles, K.N.: Electronic Literature: What Is It? http://eliterature.org/pad/elp.html/. Accessed 10 June 2018
10. Rosario, D.G.: Electronic Poetry. Understanding Poetry in the Digital Environment. Jyväskylä University Printing House, Jyväskylä (2011)
11. Hayles, K.N.: Writing Machine. The MIT Press, Cambridge and London (2002)
12. Wardrip-Fruin, N.: Five Elements of Digital Literature. https://games.soe.ucsc.edu/sites/default/files/nwf-BC5-FiveElementsOfDigitalLiterature.pdf/. Accessed 10 June 2018
13. Zuern, J.: Figures in the interface: comparative methods in the study of digital literature. In: Simanowski, R., Schäfer, J., Gendolla, P. (eds.) Reading Moving Letters: Digital Literature in Research and Teaching (A Handbook), pp. 20–26. Trancript Verlag, Bielfeld (2010)
14. Dressler, W.: Text syntax. Novoye v zarubezhnoy lingvistike **8**, 111–137 (1978). (in Russian)
15. Forceville, C.: Non-verbal and multimodal metaphor in a cognitivist framework: agendas for research. In: Kristiansen, G., Achard, M., Dirven, R., Ruiz de Mendoza Ibàñez, F. (eds.) Cognitive Linguistics: Current Applications and Future Perspectives. Mouton de Gruyter, Berlin (2006)
16. van Dijk, T.A.: Strategies for understanding a coherent text. Novoye v zarubezhnoy lingvistike **23**, 153–207 (1998). (in Russian)
17. Valgina, N.C.: Text Theory. Logos, Moscow (2003)
18. Babenko, L.G., Kazarin, U.V.: Linguistic Analysis of Literary Text. Theory and Practice. Flinta, Moscow (2005). (in Russian)
19. Carpenter, J. R.: In Absentia. http://collection.eliterature.org/2/works/carpenter_inabsentia.html. Accessed 10 June 2018
20. Cardenas, M.: Redshift and Portalmetal. http://collection.eliterature.org/3/work.html?work=redshift-and-portalmetal/. Accessed 10 June 2018
21. Chernyavskaya, V.E.: Linguistics of the Text: Polycodularity, Intertextuality, Interdiscourse. LIBROCOM Book House, Moscow (2009)

Experience of Studying the Frame Verbalization in a Scientific Text (Based on the German Language Material)

Anna M. Klyoster[1]([⊠]) [iD] and Natalia Shnyakina[2] [iD]

[1] Omsk State Technical University, Mira Pr. 11, 644050 Omsk,
Russian Federation
annaklyoster@mail.ru
[2] Omsk State Pedagogical University, Naberezhnaja Tukhachevskogo 14,
644099 Omsk, Russian Federation
zeral@list.ru

Abstract. This article is devoted to verbalization peculiarities of the scientific knowledge in German. Language fragments, containing different definitions of the same concept, are used as research material. To distinguish between typical and variable knowledge about it, the terms protoframe and exoframe are used in the work, having a slots structure understood as a set of nodes and connections between them. Through the application of frame analysis, an attempt is made to identify the mandatory and optional frame nodes behind each definition. The correlation between the protoframe and the exoframe is considered in the aspect of dynamics of the speech-thinking processes, that is, within the framework of understanding the logic of the processes of generation and understanding of speech. The analysis presented in the work allows us to draw conclusions about the unity of the categorical and linguistic design of scientific concepts and the clarity of the knowledge structuring. In addition, the repetition in the sentence of the same conceptual structures indicates the presence, in the mindset of certain information, processing schemes, through which scientific knowledge acquires a unified form that facilitates the cognitive and communicative interaction of people within the same categorical coordinates.

Keywords: Protoframe · Exoframe · Dynamic frame · Static frame · Scientific knowledge · Verbal and cogitative processes

1 Introduction

The continuous development of various fields of knowledge, associated with the emergence of new achievements in scientific and technological progress, has led to considerable interest in the cognitive aspects of the generation and understanding of scientific text: studies concern thinking operations that provide transition from the conceptual level to the language level and vice versa. The stated problems are rather complicated: scientists try to describe internal algorithms of translation activity [7, 12, 16, 25] as well as the processes of speech production and understanding of the text [4, 8, 23, 24].

© Springer Nature Switzerland AG 2019
Z. Anikina (Ed.): GGSSH 2019, AISC 907, pp. 382–388, 2019.
https://doi.org/10.1007/978-3-030-11473-2_40

Scientific text, characterized by high degree of informativeness and orderliness of expressed concepts, performs verbalized knowledge and can be viewed as an objectified set of hierarchically and logically organized units of consciousness. This article carries on the tradition of studying the text from the positions of the frame structure [13, 16]. Through the frame analysis, an attempt is made to describe the scientific text as a situationally conditioned product of information generalization. Based on several definitions of the concept of "ergonomics", fixed in the scientific and technical literature, the analysis of the protoframe and exoframe interaction, reflecting the typical and variable representation of a person about the same scientific phenomenon, is conducted.

The following issues are consistently considered in the article: firstly, what is the frame as the format of knowledge and what are the traditions of frame analysis in cognitive linguistics; secondly, how the protoframe and the exoframe are related to each other as mental structures of the existence of scientific knowledge; thirdly, what patterns of scientific text structuring can be revealed as a result of analysis of the protoframe various language implementations.

2   A Frame as a Format of Knowledge Representation and Tradition of Frame Analysis in Cognitive Linguistics

A frame is one of the formats for storing a person's knowledge about the surrounding reality and a significant category of cognitive linguistics. As a way to schematize the experiment, the original frame was described by Minsky, who viewed it as a data structure for representing a stereotyped situation [15]. Minsky associated the ability of a person to process new information with the presence of certain rules, some stereotyped schemes that allow us to comprehend the experience in the mind: "a person, trying to learn a new situation for himself or to take a fresh look at already familiar things, selects from his memory some data structure (image) … in order to make it suitable for understanding a wider class of phenomena or processes by changing individual parts in it" [15, p. 212].

In cognitive linguistics, Minsky theory of frames has found wide distribution. Summarizing the various understandings, Kubryakova describes the term "frame" in various concepts. First, the frame is understood as a "system of choice of language tools - grammatical rules, lexical units, linguistic categories associated with the prototype scene," and secondly, frames are viewed as "props" by which we comprehend our own experience, "thirdly, as a "system of categories structured in accordance with a motivating context," then, as "a unit of knowledge organized around a certain concept … containing information about the essential, typical and possible for this concept" [11, p. 187]. It is easy to see that in the above definitions, the main meaning of the concept introduced by Minsky is the same: the frame is associated with a kind of standard structure imposed by the human consciousness on the information being learned of various types [15].

The scope of the frame analysis is quite diverse: it is used to analyze various types of text [2, 9, 14], separate lexical groups of the language system [1, 21], speech acts [17]. Also, this method is successfully used to study various terminology systems [10, 18, 20, 22]. The authors suggest that the scientific knowledge of a person about any subject area has a certain structure, the language objectification of which can vary depending on the extra-linguistic situation and the individual knowledge of the author of the text. Following the presented points of view in this paper, the understanding of the frame is based on the principle of hierarchical ordering of semantic information about an object or phenomenon within a certain terminology system, taking into account the relevance of the explicated knowledge in the text.

3   Protoframe and Exoframe as Mental Structures of Scientific Knowledge Existence

The use of a frame as a structure that regulates knowledge allows analyzing the information reflected in the language fragment as a collection of typical and new knowledge. In the process of recognizing or generating scientific text, a person deals with two types of frames that interact with each other. Different terms are used for their differentiation.

The interaction of known and new information is described in the concepts of translation. In the frame is considered as a mental construct, allowing to understand the dynamics of this process. The model of translation is based on the operation of knowledge frame structures. The prototypical frame structure stored in the interpreter's mind is understood by the authors as a static frame, while the situational realization of these representations in the process of reading the translated text is a dynamic frame. Interaction between them is understood by the authors of the theory as follows: "The dynamic frame is formed under the influence of verbal stimuli from the text by launching a special search system that isolates the correspondence between the information embedded in the text and the interpreter's cognitive knowledge that allows understanding the perceived information" [16, p. 92].

In the concepts of artificial intelligence modeling, the concepts "protoframe" and "exoframe" are applied [6, 19]. Protoframe is a stored in mind stereotype representation of a person about an object of reality; the structure, which includes the most significant information, through which the cognizable object is itself. Exoframe, in turn, is a newly created structure for displaying actual data, new information that is realized in accordance with the protoframe taking into account the situation.

To clarify the specifics of the existence of scientific knowledge, the terms protoframe and exoframe are used in the aspect of understanding and generating text. The essence of the process of speech generation is the transition from certain indefinite gestalt impressions to a dissociated proposition. This process originates at a deep level and involves the formation of a theme and a rema. Remus contains information about the phenomenon described and is understood as "a pre-semantic incompetitive complex in which both the situation and its participants are represented" [8, p. 347]. Awareness of this complex as a dismembered structure leads to the activation of a frame that sets the structure for the conceptualization of the cognizable phenomenon. "The speaker perceives the situation in terms of certain frames, and this means that the act of perception itself brings to life the frames with their terminals and, consequently, the potential semantic roles" [8, p. 347]. Understanding the text, on the contrary, begins with the activation of the frame. On the basis of language data, the candidate frame and the possible competition of frames are highlighted [3]. In the process of understanding the text, a person uses his generalized knowledge of the known object or phenomenon, that is, the protoframe. As Fillmore explains, free slots in the process of acquaintance with the text are gradually filled with information, more and more specifying the general picture [5]. In other words, the exoframe objectified in the text (dynamic frame) is essentially the result of the deployment of the protofram (static frame).

The formulated ideas reflect the specifics of speech-thinking processes and are used in the study of linguistic material through frame analysis.

4   Regularities in the Structuring of Scientific Text

A scientific text is a fragmentary objectification of a certain terminology system and can be viewed ambiguously: as a result of the process of thought verbalization and as a source material for its understanding. A consistent study of the logic of these processes provides information on the features of structuring the scientific text.

As the studied language fragments, the article uses texts on engineering psychology. Their frame organization is demonstrated by examining several definitions given to the concept of "ergonomics".

1. Ergonomie ist ein Fachgebiet, das sich in besonderer Weise mit der Interaktion von Mensch und Maschine befassen sollte.
2. Ergonomie optimiert Mensch -Maschine-Interaktion im Hinblick auf körperliche und geistige Aspekte.
3. Ergonomie ist die wissenschaftliche Disziplin, welche sich mit dem Verständnis der Wechselwirkungen zwischen Menschen und anderen Elementen eines Systems beschäftigt, sowie die berufliche Tätigkeit, welche die Theorie, Prinzipien, Daten und Methoden zur Gestaltung anwendet, um das menschliche Wohlbefinden und die gesamte Systemleistung zu verbessern.
4. Ergonomie nimmt sich insbesondere des interaktiven Aspekts an, wobei die Optimierung des Informationsaustausches zwischen Mensch und Maschine zentrale Bedeutung hat.

Frame analysis involves the identification of a slot structure objectified in the presented definitions of scientific knowledge and consists in isolating the mandatory and optional content. For this purpose, it is necessary to differentiate the upper and lower frame nodes described in the classic version of the frame theory by Minsky. "A frame can be imagined as a network consisting of nodes and connections between them. The "upper levels" of the frame are clearly defined, because they are formed by such concepts that are always true in relation to the intended situation. At lower levels, there are many special vertex terminals or "cells" that must be filled with characteristic examples or data" [15, p. 246]. The above quotation explains the principle of the functioning of a frame structure for the structuring of knowledge.

It seems that the upper nodes, reflecting the essence of the concept being realized, are the most significant and form the protoframe. In the examples considered, such nodes are "person", "machine" and "interaction between them". These nodes are present in each definition and characterize the concept of "ergonomics" itself.

The lower nodes with terminals filled with lexical units vary from definition to definition. In the first case, such nodes are the ideas "ergonomics is a region", "interaction is special." In the second text, additional meanings are realized: "the function of ergonomics - optimization of this interaction", "aspects of optimization-physical and spiritual". The third fragment contains information: "this is a scientific discipline," "it improves working conditions," "how scientific discipline has a theory, principles, data and methods". Finally, the additional meaning of the last definition is "optimization of information exchange". The superficial realization of scientific thought, reflected in the language fragment, is thus viewed as an exoframe, which includes the upper nodes of the protoframe and the lower nodes filled with specific information.

The analysis of examples has shown that to describe the essence of the phenomenon in the process of generating a scientific text, a person uses a stereotypical representation, that is, a "protoframe", imposing it on the situation that needs to be described. In this regard, each definition should be understood as a verbal implementation of the exoframe – a dynamic frame. The deployment of a static frame is carried out in the process of specifying knowledge, that is, clarifying the details and circumstances of the existence of the cognizable phenomenon. The structure of the analyzed definitions demonstrates the significance of individual concepts that are included in the structure of the exoframe. Understanding the utterance, on the contrary, is built in the direction of the dynamic, realized in the language fragment of the frame (exoframe) to the static (protoframe). When understanding the examples cited as an example, the person receiving information relies on his knowledge of the described scientific concept, that is the protoframe.

5   Conclusion

A frame is one of the formats for processing and storing knowledge. Its essence lies in the specific structuring of information: a lot of frames are stored in the mind, in accordance with which new information is recognized and objectified. This understanding of the frame caused the availability of various areas of application of frame analysis. This method is most productive in the field of studying the translation process, as well as the patterns of generation and understanding of the scientific text, which is a superficial verbalization of a part of a certain terminology system.

Fragment of the scientific text is understood as an objectified manifestation of the interaction of a typical and new knowledge - the protoframe and exoframe (static and dynamic frame) that provide the speech-processing processes. Protoframe is a generalized information structure that reflects specific knowledge in a particular subject area. In the process of generating an utterance, it forms a kind of skeleton consisting of the necessary upper nodes, through which the cognizable object is itself. When new information is imposed on the protoframe, conditioned by the situation and personal preferences of the speaker, an exoframe is formed in the consciousness, which includes both upper and lower nodes, filled with specific information. Understanding the text, on the contrary, is due to the search for a suitable protoframe.

As a result of the analysis of various language implementations of the protoframe, that is, definitions of the same concept, the features of the scientific text and patterns of its structuring can be revealed. First, the scientific text is a superficial implementation of the protoframe, fixed in the minds of all representatives of the scientific community and ensuring their successful communication within the same terminological coordinates. Secondly, meaningfully important components of knowledge about the described object are realized in each fragment, forming its conceptual structure; Additional information varies depending on the knowledge base of communicants and their individual language preferences. Third, the specificity of the deployment of the protoframe is to clarify individual concepts; Conjugation, on the contrary, is connected with the aspect of generalization, that is, understanding, while the generation of the text is based on the protoframe.

In conclusion, the proposed variant of frame analysis allows us to consider the text as a result of the speech-processing processes and to trace the specificity of its linguistic design in support of the features of its deterministic mental structures.

References

1. Babushkina, O.N.: Frejmovyj analiz ocenochnyh frazeologizmov, harakterizujushhih professional'nuju dejatel'nost' [The frame analysis of the estimated phraseological units that characterize professional activity]. Teoreticheskie i prikladnye aspekty izuchenija rechevoj dejatel'nosti [Theoretical and applied aspects of the study of speech activity] **6**, 40–45 (2011). (in Russian)
2. Butorin, S.V.: Frejmovyj podhod k analizu jazykovogo prostranstva nemeckogo romana-vospitanija [A frame approach to the analysis of the linguistic space of the German novel-education]. Izvestija Samarskogo nauchnogo centra Rossijskoj akademii nauk **12**(3–3), 761–765 (2010). (in Russian)
3. Charniak, E.: Context recognition in language comprehension. In: Lehnert, W.G., Ringle, M. H. (eds.) Strategies for Natural Language processing, pp. 435–454. Lawrence Erlbaum Associates, Hillsdale (1982)
4. Dijk van, T.A., Kintsch, W.: Strategii ponimanija svjaznogo teksta [Strategies of Discourse Comprehension]. Novoe v zarubezhnoj lingvistike: Kognitivnye aspekty jazyka [New in foreign linguistics: Cognitive aspects of language], vol. 23. Progress, Moscow (1988). (in Russian)
5. Fillmore, C.J.: An alternative to checklist theories of meaning. In: Proceedings of the First Annual Meeting of the Berkeley Linguistics Society, vol.1, pp. 123–131. Progress, Moscow (1975)
6. Gavrilova, T.A., Chervinskaya, K.R.: Izvlechenie i strukturirovanie znanij dlja jekspertnyh system [Extraction and structuring of knowledge for expert systems]. Radio and Communication, Moscow (1992). (in Russian)
7. Gusev, V.V.: Jempaticheskaja model' v formirovanii strategii perevoda [Empathetic model in the formation of the translation strategy]. Vestn. MSLU **480**, 26–41 (2003). (in Russian)
8. Kasevich, V.B.: Semantika. Sintaksis. Morfologiya [Semantics. Syntax. Morphology]. In: Churilina, L.N. (ed.) Aktual'nye problemy sovremennoj lingvistiki [Actual problems of modern linguistics], pp. 344–368. Flinta-Nauka, Moscow (2009). (in Russian)
9. Klyoster, A.M., Shnyakina, N.Y.: K voprosu o kategorial'nom chlenenii nauchnogo i obydennogo znaniya [Categorical division of scientific and everyday knowledge]. Voprosy kognitivnoj lingvistiki **1**(54), 144–150 (2018). (in Russian)

10. Klyoster, A.M.: Frejmovyj analiz nemeckoj terminosistemy inzhenernoj psihologii [Frame analysis of the German terminology of engineering psychology]. Vestnik Omskogo universiteta **3**, 249–255 (2011). (in Russian)
11. Kubryakova, E.S.: Frejm [Frame]. In: Kubryakova, E.S. (ed.) Kratkij slovar' kognitivnyh terminov [A brief dictionary of cognitive terms], pp. 187–189. Filologicheskij fakul'tet MGU im. M. V. Lomonosova, Moscow (1997). (in Russian)
12. Marchuk, JuN: Metody modelirovanija perevoda [Methods of translation modeling]. Nauka, Moscow (1985). (in Russian)
13. Maslennikova, E.M.: Frejmovoe predstavlenie semantiki teksta [Frame representation of text semantics]. Lingvisticheskij vestnik **2**, 114–124 (2000). (in Russian)
14. Merkulova, N.V.: Frejmovyj analiz teksta v aspekte onomastiki hu-dozhestvennogo diskursa [Frame analysis of the text in the aspect of onomastics of artistic discourse]. Nauchnyj vestnik Voronezhskogo gosudarstvennogo arhitekturno-stroitel'nogo universiteta: Sovre-mennye lingvisticheskie i metodiko-didakticheskie issledovanija [Modern linguistic and methodical-didactic studies] **10**, 22–28 (2008). (in Russian)
15. Minsky, M.L.: A framework for representing knowledge. In: Winston, P.H. (ed.) The Psychology of Computer Vision, pp. 211–277. McGraw-Hill, New York (1977)
16. Nefedova, L.A., Remhe, I.N.: Kognitivnye osobennosti perevoda nauchno-tehnicheskogo teksta [Cognitive features of the translation of the scientific and technical text]. Voprosy kognitivnoj lingvistiki **2**(100), 27–32 (2008). (in Russian)
17. Nikonova, Zh.V.: Osnovnye jetapy frejmovogo analiza rechevyh aktov (na materiale sovremennogo nemeckogo jazyka) [The main stages of the frame analysis of speech acts (on the basis of modern German language)]. Vestnik Nizhegorodskogo universiteta im. N.I. Lobachevskogo. Serija: Filologija. Iskusstvovedenie **6**, 224–228 (2008). (in Russian)
18. Novodranova, V.F.: Kognitivnye aspekty terminologii [Cognitive aspects of terminology]. In: Novodranova, V.F. (eds.) Kognitivnaya lingvistika: sovremennoe sostoyanie i perspek-tivy razvitiya [Cognitive linguistics: current state and prospects of development], pp. 13–16. Izd-vo Tambov. gos. un-ta., Tambov (1998). (in Russian)
19. Pospelov, D.A.: Prikladnaja semiotika i iskusstvennyj intellekt [Applied Semiotics and Artificial Intelligence]. Programmnye produkty i sistemy [Software products and systems] (1996). http://swsys.ru/index.php?page=article&id=1079. Accessed 11 May 2018. (in Russian)
20. Shirokolobova, A.G.: Frejmovyj analiz terminosistemy gidrotehniki russkogo i anglijskogo jazykov [Frame analysis of the terminology of hydraulic engineering of Russian and English languages]. Voprosy kognitivnoj lingvistiki **1**, 52–56 (2011). (in Russian)
21. Starostina, E.V.: Frejmovyj analiz russkih glagolov povedenija [Frame analysis of Russian verbs of behavior]: Extended abstract of a doctoral theses. Saratov State University (2004). (in Russian)
22. Sytnikova, T.A.: Frejmovyj analiz terminologii predmetnoj oblasti (na primere anglojazy-chnoj komp'juternoj tehnicheskoj terminosistemy) [Frame analysis of domain terminology (using the English computer technical terminology system as an example)] (2008). https://cyberleninka.ru/article/v/freymovyy-analiz-terminologii-predmetnoy-oblasti-na-primere-angloyazychnoy-kompyuterno-tehnicheskoy-terminosistemy. Accessed 14 Aug 2018. (in Russian)
23. Ventsov, A.V., Kasevich, V.B.: Problemy vosprijatija rechi [Problems of speech perception]. URSS Editorial, Moscow (2003). (in Russian)
24. Zalevskaya, A.A.: Vvedenie v psiholingvistiku [Introduction to Psycholinguistics]. Rossijsk. gos. gumanit. un-t, Moscow (2007). (in Russian)
25. Zwilling, M.Ya.: Kognitivnye modeli i perevod [Cognitive models and translation]. Vestn. MGLU **480**, 21–26 (2003). (in Russian)

Nomen Structures in English Pipeline Terminology

Natalia V. Gorokhova[1](✉) and Irina N. Kubyshko[2]

[1] Gubkin Russian State University of Oil and Gas
(National Research University), Moscow 119991, Russian Federation
n.gorokhova@nxt.ru
[2] Razumovsky Moscow State University of Technologies and Management
(the First Cossack University), Moscow 109004, Russian Federation
irina-kub@yandex.ru

Abstract. Nomen structure in special terminology is one of the main issues in modern linguistics. It is one of the numerous language groups which attracts interest of philologists and can be the basis for studying in the context of various professional formats. The criterion for distinguishing between a term and a nomen is the consideration of their roles in the field of special communication. For differentiation between terms and nomens, the following functions are used: identification, labeling, subjectivity, motivation, nomination, etc. Nomen structure is considered as a typical form of professional knowledge which possesses all these functions and contains grapheme and digital components, in contrast to the term. Through the application of the terminological analysis of numerous nomen structures in English pipeline terminology, an attempt is made to identify the main characteristics of nomens in professional sphere. In addition, the analysis presented in the work allows conclusions about the linguistic design of nomens and clarification of the near-and-far periphery of terminological nomenclature.

Keywords: Term · Terminology · Nomen · Lexis · Pipeline transport

1 Introduction

The idea that a special language structure is represented by different components (terms, nomens, and other constituents) is discussed in a number of Russian [2, 4, 5, 9, 11, 14, 27] and foreign scientific papers. Knowledge about the nature of nomens, their types and structures can be complemented by the data obtained from modern linguistic papers [1, 9, 13, 23, 28]. This category is much newer than the terminology. As a special lexical class it was formed not earlier than in the XVIII century [7, 15, 20, 22], initially only for natural sciences where terminology was a tool that fixed the nomenclature. Currently, professional terminology is computable since it reflects the system of its concepts. A terminological field or a terminological context is important for the term. The nomenclature is not associated with concepts, it is more nominative, relying sometimes on purely external similarities (for example: S-pipe, V-tube, J-pipeline, X-system). Nomens are freely used out of context, since the properties of

© Springer Nature Switzerland AG 2019
Z. Anikina (Ed.): GGSSH 2019, AISC 907, pp. 389–397, 2019.
https://doi.org/10.1007/978-3-030-11473-2_41

things do not change in scientific or everyday communication. There can be nomenclature of natural sciences, technology, production, trade, economy, etc. Special nomens are developed in biology to denote numerous species of plants and animals, in chemistry – to determine millions of chemical compounds, in geography – to define places on the globe. Technical nomenclatures are real and substantive. They are designed to refer to numerous parts of machines and devices [13]. Sometimes it is difficult to separate the nomen and the term; the main difference between them is the fact that the term expresses essential features of a concept while the nomenclature name is an object, a single concept.

A nomen is a lexical unit by means of which we refer to a visible object and perceive this object without realizing its exact place in the language system [29]. In the system ordering of special terminology, a linguist is faced with the problem of differentiation of terms and related units, and the process and results of terminological work depend on the determination of the distinctive features of terms in relation to other categories of special vocabulary [12, 24, 25, 29].

2   Subjects and Methods

The object of the research is the terminological base of the special English language vocabulary of pipeline transport.

Methods relevant for such a study are stipulated by the aim of identifying nomen structures that are manifested in the considered language system, and can be presented in the form of the method of continuous sampling from modern dictionaries and research materials, method of contextual analysis using authentic texts, method of comparative analysis using an online search service, as well as the descriptive method required for identification of the peculiarities of the use of terminological units.

Modern society which is developing in the digital world of mass communications is more capable of making an impact on the development of professional language. The terminological transformation processes are actively manifested in English pipeline transport terminology, which, undoubtedly, affects the system of the vocabulary and grammatical structure of the special language in general.

3   Term and Nomen Concept in Special Terminology

The English terminology of pipeline transport is formed by different terminological units. Depending on the degree of codification and information content of pipeline terminology, these terminological units can be represented by the traditional classification in three registers [17]:

1. Register of general *scientific* terms
2. Register of general *technical* terms
3. Register of *special* terms

Such a division is presented in the works of many outstanding scholars [3, 10, 16, 28]. These registers compose terminology, i.e. a set of language elements and their relationships. However, this stratification is considered as the initial analysis of scientific systematization of terminology at a word level and does not reflect the whole variety of word forms of the terminological field, with the register of nomens being one of its important components [9].

The modern interpretation of nomen brings it closer to the term by its main characteristics. It should be noted that the mentioned names are often positioned, coexisting in the terminological space. Opposition of the term to the nomen can be represented in Table 1.

Table 1. Functional difference between the term and the nomen.

Term	Nomen
Important in theoretical aspect	Important in applied aspect
Performs the function of the concept reflection, pointing an idealized entity	Performs the labeling function by naming the type of real objects
Strives for the objectivity of information reflection in researcher's mind	Reflects the subjective perception of the world by a researcher

The parameter of the subjectivity/objectivity of information reflection seems to be somewhat controversial, therefore it is probably more correct to determine lexical units by the type of motivation, i.e. units that correspond to the reflection of the subjective perception of the world by a researcher [30] or semantically unmotivated units [14]. Despite the opposition to the above parameters, the term and the nomen have a common point: both are associated with the need to reflect a concept.

A set of terms, names or nomenclature marks is called a nomen, but the definition of this type of nominations is not completely clear. It defines the nomenclature as a set of special terms used in the scientific field; names of typical objects of this science [3]. It is also considered as a set of names used in any branch of science, production, etc. (in contrast to terminology, which also contains the designation of abstract concepts and categories) [19].

This interpretation unites nomens and terms on the basis of the specificity/abstractness of a designated concept. At the same time, nomens appear as a variety of terms (a set of special terms). In fact, it eliminates the distinction between the two types of terminological field elements. It seems that in comparison with the units of other types, nomens are definitely closer to terms. It can be assumed that nomens are included in the near-nuclear part of the terminological field.

The criterion of differentiation between a term and a nomen is their role in the sphere of special communication. If terms allow isolating or idealizing abstraction of high level, nomens are necessary in the process of identification, labelling of a real object, and generalization is minimal. While the term is characterized by the function of the reflection concepts (terms «service» theory), the nomenclature mark is created in order to name, label individual items (nomens are associated with the implementation of theoretical knowledge in terms of application). The nominative function is the main one for a nomenclature unit, while the main function of a term is the significative function [11].

Functional specificity of nomens determines the originality of their forms. If terms at the stage of creation seek to reflect the objectivity of information and, therefore, tend to be motivated, nomens tend to be artificially classified, they reflect the subjective perception of the human world [14]. The level of the subjective choice of names or titles while creating a nomen is so high that the degree of desemantization approaches the onomastic vocabulary [13]. Nomen, in this case, is the name of specific products of economy, reproduced on the same sample a specified number of times, as well as the names of single concepts [11]. Names of mass production products are required to characterize a separate product but when a certain model or type of a product is considered.

Apparently, despite the obvious proximity of nomens to onims, it is still necessary to distinguish the first from the second ones on the basis of their differentiation by the level of generalization: a nomen denotes a certain type, a variety of objects (although in the sphere of operation it can be used for the nomination of a single representative of the type); onim denotes a single, specific object (although potentially it can be used as designation of a certain type of objects that will replicate copies of this single object). Nomens joining in nomenclature form the system of abstract and conventional symbols [4].

Many units of nomenclature are complex words consisting of two components: (1) word–nuclear, performing the function of the genus index, the index of belonging of an object to a group of similar objects, the determinant of the place of this nomen in the system; (2) digital or graphemic, which indicates the presence of the object features that distinguish it from a set of similar objects, and acts as a differentiating indicator. However, nomen is considered as an element of the terminological system with corresponding semantic functions [9].

The variety of types of objects, the possibility of a constant increase in their number (especially the products of human activity, for example, the production of various devices, mechanisms, units, etc.) lead to the structural mobility of terminology and maintain the openness of the nomen system in a terminological field. The nomen system is not closed and is not as strictly organized as the system of terms. It is possible to add or exclude the whole blocks from a terminological field without changing the system as a whole [27]. The appearance of nomens or their disappearance from a circle of special nominations occurs at different rates – from several months to many years. The terminological system may be accompanied by several changes of nomenclature [30].

In different terminological fields the difference between a term and a nomen can be greater or lesser. This differentiation is not always evident. The term – nomen oppositions revealed in synchronies are obviously supported by the existence of different purposes of functioning of these nominations. The only purpose of the nomenclature is to provide the most convenient form to designate objects, things, without direct relation to the needs of theoretical thought [10]. Nomens are freely used outside of the context, since properties of the named things do not change the form regardless of their sphere of communication or usage [22].

4 Nomenclature in Pipeline Transport Terminology

Pipeline transport as a special technical field of the power engineering industry develops and exists due to the material base that possesses various equipment, mechanisms, consumables, etc. Involvement of a large number of enterprises in the development of new pipeline equipment has led not only to the appearance of new systems, technologies, programs, setups, but also to the growth of terminological systems. Moreover, different companies give their own names to their inventions. As a result, machines designed to move, aim, lower, and lay pipes, including pipelayers, excavators, loaders, tractors, graders, bulldozers, etc., and characterized by approximately the same functions, have different names [9]. In addition, rapid development of technologies requires constant modifications of equipment, which results in a large number of versions of one developed series.

The structural components of a nomen comprise graphemic and digital parts where the graphemic unit is the compression of the generic term for this nomen, and the digital unit is representation of the main technical characteristics of an object [1, 12]. The system of nomenclature units is relative, but their classification on the basis of thematic selection is quite possible:

1. Names of machines and systems [21]:

    - *PL61, PL72, PL83, PL87* – pipelayer (Caterpillar)
    - *PL3005D, PL4809D* – pipelayer (Volvo)
    - *PL52, PL52 3, PL22* – pipelayer (Liebherr)
    - *PL95H* – pipelayer (Delta)
    - *D355 C(28), D355 C(09), D355 C(34)* – pipelayer (Kamatsu).

2. Process and technology names [18]:

    - *OSR* – Global Industry Oil Spill Response
    - *HiPo* – highly potential incident
    - *EMSolutions* – PipeLine Machinery International (PLM) introduces Cat®
      Equipment Management Solutions for the pipeline construction industry
    - *Cat Tier 4 Interim* – technologies to reduce carbon footprint left by the engines
      and equipment for North America.

3. Name of a product:

    - *SRB* – sulfate-reducing bacteria.

4. Name of equipment and related materials [8]:

    - *Modular S – centrifugal pump*
    - *NSL – centrifugal pump*
    - *ESL– centrifugal pump*
    - *SA – centrifugal pump*
    - *Rotan CD, Rotan CC, Rotan GP, Rotan HD, Rotan ED – gear pump*
    - *6G-7270, 6S-4728, 7 W-1947, 9 W-2883 – tubes and pipes.*

5. Registered mark:

   - *AccuGrade* ® – Grade Control System
   - *RealAduio*® – Streaming technology
   - *Distributed Component Object Model*® – Distributed object technology
   - *COM*® – Component technology.

6. Types of standards [26]:

   - *ASME/ANSI B16.5* – Pipe flanges and flange connections
   - *ASME/ANSI B31.3* – Pipelines of chemical plants
   - *ASME/ANSI B31.4* – Pipeline transportation systems.

The graphemic part of the above nomens is the standard adopted by the American Society of Mechanical Engineers (ASME) and the American National Standards Institute (ANSI). Thus, the graphemic part's motivation is obvious. Typically, it is the compression of the corresponding (generic for a given nomen) term expressed by a complex word or phrase, and, therefore, it performs a dual role: establishes the place of a nomen in a series of homogeneous units and indicates the thematic area to which it belongs, and the main design features of the called object.

There are European standards set by the German Institute for Standardization (DIN – Deutsches Institut für Normung) and European Committee for Standardization (EN – European Committee for Standardization), for example, DIN/EN 1092-1 – Flanges. The digital part of the nomen shows the main technical characteristics of the object and distinguishes it from a set of similar objects. There is a reason to believe that this structure is typical of most technical terms: ISO 50001 – Energy management, ISO 31000 – Management of risks, where ISO (International Organization for Standardization) is a developer and publisher of international standards.

7. Names of documents/acts:

   - *ISGOTT* – International Safety Guide for Oil Tankers and Terminals
   - *SOLAS* – International Convention for the Safety of Life at Sea
   - *MARPOL 73/78* – International Convention for the Prevention of Pollution from Ships.

8. Names of single concepts:

Asian stream	*EastWay*	*Orient stream*
Blue diamond	*Eurasian wave*	*Pacific stream*
Blue stream	*Far East Stream*	*Smile*
Diamond	*Great East Stream*	*SISTERHOOD*
Drugnavostok	*Helium*	*SUPER Line stream*
East Stream	*Milky Way*	*TRANSGASFORMATION*
Eastern connector	*Ocean stream*	*Urajio*

5 Conclusion

Pipeline transport nomens are specific, and, typically, it is possible to determine by their grapheme part what kind of object they name, and, knowing the terminology, it is easy to determine the nomenclature system to which they relate, where the numbers mostly indicate modification. The most part of a nomen is associated with the term and the concept it names [12]. This fact has made it possible to perform a thematic selection of groups identified in the pipeline transport area, and to predict the further appearance of nomens with the preserved grapheme part and the modified digital part. A mixed method is both semantic and syntactic; abbreviations are also used in the formation of the pipeline transport nomenclature.

It should be recognized that systematization of nomenclature is very problematic due to the classification place. The same name can occupy different places in different codification grids. For example, in the system of industry classifications there is a tendency to divide the same objects by different parameters. If we proceed from the existing system of normative documents, only insulation materials can be classified according to eleven parameters [13]. In this case, classification depends on the practical needs of people [12], and the nomenclature, showing a particular layer of a specific language, hundredfold higher than the number of terms. Without well-developed and systemically ordered terminology and nomenclature, the progress of science is impossible.

The modern pipeline transport terminology comprises nomens, thereby adding to its lexical composition, but these nomens reveal a certain complexity of decoding. The terminological system forms the core of a terminological field where you can enter the rest part of the special category; the latter, in turn, makes the near-and-far periphery and reflects a number of oppositions that arise on different grounds.

Thus, the analysis of a number of criteria used for the differentiation of terms and nomenclature marks demonstrates that, due to the active development of terminology, the boundaries between nomens and terms are quite unstable. In most cases, they are complex words consisting of the graphemic part (compression of the generic term for a given unit) and the digital part (showing the main characteristics and distinctive features of an object). The fixed functions of each part allow the development of regularities of the formation of technical nomens. The nature of the connection of nomenclature with terms suggests that nomens are necessary for an easy generation of a large number of names of specific products and single concepts.

References

1. Aleksandrova, G.N.: Sootnoshenie znakovoj struktury i funkcij terminologicheskih i kvaziterminologicheskih yazykovyh edinic: avtoref. dis. ...cand. filol. nauk [Correlation of sign structure and functions of terminological and quasi-terminological language units]. Samara State Pedagogical University, Samara (2006). (in Russian)
2. Alekseeva, L.M.: What is a term. Russian terminology science. Termnet Pubisher, Vienna (2004)

3. Ahmanova, O.S.: Slovar' lingvisticheskih terminov [Dictionary of linguistic terms]. KomKniga, Moscow (2007). (in Russian)
4. Belova, A.Y.: O razgranichenii terminov i nomenov [On the delineation of terms and nomenclature]. In: Aktual'nye problemy stilelogii i terminovedeniya: tezisy mezhdunar. konf., posv. 80-letiyu prof. B.N. Golovina, pp. 15–16. Nizhegorodsky gos. Universitet, N. Novgorod (1996). (in Russian)
5. Budagov, R.A.: Novye slova i znacheniya [New words and meanings]. Chelovek i ego yazyk **2**, 147–153 (1976). (in Russian)
6. Bulatov, A.I.: Anglo-russkij slovar' po nefti i gazu [English-Russian Dictionary of Oil and Gas]. RUSSO, Moscow (2002). (in Russian)
7. Chupryna, O.G., Zhukov, I.A.: Motivy nominacii astronomicheskih terminov s koloronimom v sovremennom anglijskom yazyke [Motives for the nomination of astronomical terms with colorimony in modern English]. Vestnik MGPU. Ser. « Filologiya. Teoriya yazyka. Yazykovoe obrazovanie » **3**(19), 50–54 (2015). (in Russian)
8. Gorohova, N.V.: Anglo-russkij slovar' terminov truboprovodnogo transporta [English-Russian dictionary of pipeline transport terms]. KAN, Omsk (2012). (in Russian)
9. Gorohova, N.V.: Angloyazychnaya terminologiya truboprovodnogo transporta v sociolingvisticheskom osveshchenii: avtoref. dis.… kand. filol. nauk [English-language terminology of pipeline transport in sociolinguistic coverage]. OmGTU, Omsk (2012). (in Russian)
10. Glushko, M.M.: Yazyk anglijskoj nauchnoj prozy: avtoref. dis.… dokt. filol. nauk [Language of English scientific prose]. MGU, Moscow (1980). (in Russian)
11. Grinyov, S.V.: Osnovy leksikograficheskogo opisaniya terminosistem: dis. …dokt. filol. nauk [Basics of lexicographical description of terminology]. MGU, Moscow (1990). (in Russian)
12. Grinyov, S.V.: Terminology in the Era of Globalisation Russian Terminology Science. Termnet Pubisher, Vienna (2004)
13. Grinyov-Grinevich, S.V.: Terminovedenie: ucheb. posobie dlya stud. vyssh. ucheb. zavedenij [Terminology]. Akademiya, Moscow (2008). (in Russian)
14. Hodakova, A.G.: Terminy i nomeny [Terms and nomens]. Vestnik Nizhegorodskogo universiteta im. N.I. Lobachevskogo **4**(1), 411–416 (2012). (in Russian)
15. Kozhina, M.N.: O specifike hudozhestvennoj i nauchnoj rechi v aspekte funkcional'noj stilistiki [On the specificity of artistic and scientific speech in the aspect of functional stylistics]. Gos. Universitet, Perm' (1966). (in Russian)
16. Kulikova, I.S.: Obuchayushchij slovar' lingvisticheskih terminov [Teaching dictionary of linguistic terms]. Nauka, St.Petersburg, Moscow (2004). (in Russian)
17. Lavrova, A.N.: O pod'yazyke organicheskoj himii: monografiya dokt. filol. nauk [About the sublanguage of organic chemistry]. Nizhegorodsky gos. Universitet, N. Novgorod (1994). (in Russian)
18. Lysyanyj, K.K., Ianov, V.A.: Tolkovyj slovar' neftegazovyh ob"ektov truboprovodnogo transporta [Explanatory dictionary of oil and gas facilities of pipeline transport]. Tyumen', Cessiya (2005). (in Russian)
19. Maruzo, Z.H.: Slovar' lingvisticheskih terminov (in Russian) [Dictionary of linguistic terms]. Izd-vo inostr. lit, Moscow (1960)
20. Mitrofanova, O.D.: Yazyk nauchno-tekhnicheskoj literatury [Language of scientific and technical literature]. MGU, Moscow (1973). (in Russian)
21. Pipeline equipment. http://www.plmcat.com/pipeline-equipment/pipelayers. Accessed 03 Feb 2016
22. Razinkina, N.M.: Razvitie yazyka anglijskoj nauchnoj literatury [Development of the language of English scientific literature]. Nauka, Moscow (1978). (in Russian)

23. Rozhnova, I.A.: Neologizmy v anglijskoj terminologii poligraficheskogo proizvodstva: avtoref. dis. ... kand. filol. nauk [Neologisms in the English terminology of printing production]. OmGTU, Omsk (2005). (in Russian)

24. Serebrennikov, B.A.: O materialisticheskom podhode k yavleniyam yazyka [On the materialistic approach to the phenomena of language]. Nauka, Moscow (1983). (in Russian)

25. Shelov, S.D.: Ob opredelenii lingvisticheskih terminov (opyt tipologii i interpretacii) [On the definition of linguistic terms (experience of typology and interpretation)]. Voprosy yazykoznaniy 3, 21–31 (1990). (in Russian)

26. Skrynnik, Yu.N.: Neft'. Gaz. Oborudovanie: terminologicheskij slovar' [Oil. Gas. Equipment: terminological dictionary]. Nedra, Moscow (2004). (in Russian)

27. Superanskaya, A.V., Podol'skaya, N.V., Vasil'eva, N.V.: Obshchaya terminologiya [General terminology]. Voprosy teorii. Nauka, Moscow (1989). (in Russian)

28. Tatarinov, V.A.: Obshchee terminovedenie: ehnciklopedicheskij slovar'. Rossijskoe terminologicheskoe obshchestvo RossTerm [General Terminology: Encyclopedic Dictionary]. Moskovskij Licej, Moscow (2006). (in Russian)

29. Tihonova, I.B.: Kognitivnoe modelirovanie professional'noj terminosistemy (na materiale anglijskoj terminologii neftepererabotki): avtoref. dis. ...kand. filol. nauk [Cognitive modeling of professional terminology (on the basis of the English terminology of oil refining)]. OmGTU, Omsk (2010). (in Russian)

30. Vinokur, G.O.: O nekotoryh yavleniyah slovoobrazovaniya v russkoj tekhnicheskoj terminologii [On the phenomena of word formation in Russian technical terminology]. Trudy MIFLI V, 3–54 (1939). (in Russian)

Farewell to the Body: The Interpretation Field of the Concept 'Modern Body' in the Scientific Discourse of David Le Breton's *L'adieu Au Corps*

Olga Melnichuk◉ and Tatiana Melnichuk(✉)◉

North-Eastern Federal University, Belinsky Street 58, 677000 Yakutsk, Sakha, Russian Federation
madrid03@mail.ru, ta.melnichuk@s-vfu.ru

Abstract. Due to the progress made in medicine, chemistry, pharmacology, cosmetology and technological sciences in the 20th century, the attitude towards the human body and its perception have changed, which has led to new attributes of the 'body' concept in the modern worldview. Semantic, semantic-cognitive and cognitive methods are used to identify the interpretive (evaluative and pragmatic) attributes of the 'modern body' concept and describe the interpretation field of this concept as part of the Western scientific corporeal discourse encapsulated in David Le Breton's monograph 'L'adieu au corps'. The analysis of the markers representing the interpretive attributes reveals the tendency towards the disappearance of the body, contraposing the body against the 'I', as well as generally negative attitudes towards the body. The body and the 'I' are presented as being separate, rather than united. New body practices and body techniques serve the purpose of changing the body in order to increase personal confidence and reinforce the individual's identity.

Keywords: Concept · Modern body · Interpretation field · Evaluative zone · Pragmatic zone · Bodilessness

1 Introduction

The monograph of the famous French sociologist and anthropologist David Le Breton *L'adieu au corps* [1] was chosen for this study since it provides a broad panorama of modern research in the fields of medicine, genetics, biology, cybernetics, robotics, artificial intelligence, philosophy, and sociology. All these research areas relate in different ways to the problem of the body, therefore the monograph provides a synthesis of texts of modern scientific corporeal discourse.

The problem of the body and corporeality attracted the interest of philosophers, anthropologists, culturologists, historians, sociologists, and art historians in the 20th century. In the 21st century, researchers observe fundamental changes in the individual's attitude to the body, which indicates changes in consciousness and the mind. Over the past 10 years, the modern body has been an object of art theory and cultural, philosophical, and sociological studies, where the attention is focused on the concepts

© Springer Nature Switzerland AG 2019
Z. Anikina (Ed.): GGSSH 2019, AISC 907, pp. 398–410, 2019.
https://doi.org/10.1007/978-3-030-11473-2_42

of the postmodern body, cultural body and corporeality as sociocultural phenomena. Scholars have turned their interest to 'the body-centrism of modern culture' and semiotics of the modern body in art [see 1–11, etc.]. Although the present research does not in any way claim to be an exhaustive study of the modern body concept from a linguistic viewpoint, it should be noted that so far we have encountered very few contemporary linguistic studies of the concept of modern body. One example is a paragraph in Tataru's dissertation on the representation of the 'human body' concept in the modernist texts of J. Joyce and V. Woolf, where the author identifies two types of the body – the physical social body represented as a puppet structure and the imaginary asocial body represented as a way of escaping reality [12].

If we look into the evolution of the relationship between an individual and their body, it becomes clear that in the Middle Ages an individual and the body were perceived as an integrated whole. The body was considered an untouchable manifestation of a person. In the 10th–11th centuries, the word *corps* ('body') functioned as a kind of personal pronoun ('my body' meant 'I', 'his body' meant 'he', etc.), i.e. the word 'body' was used to denote a person [13–15]. In the Middle Ages, the body and soul were inseparable. According to the medieval notion, the human was made of the material and mortal body and the immaterial and immortal soul. The radical separation of the soul from the body, according to the historian Jacques Le Goff, began in the era of classical rationalism of the 17th century. The opposition of the body and mind began with Descartes. The Cartesian division of matter and spirit and the understanding of the body as a machine had for a long time formed the basis of the Western culture and Western idea of personality [16, 17]. By the end of the 20th century, French and American sociologists had started looking into the relationship between the modern individual and the body, and came to the conclusion that in the modern version of the dualism, the body was contrasted with the individual as a whole, rather than with the individual's mind or soul as had been the case in the past [1, 2, 18]. What is, then, the modern attitude of an individual to the body?

We propose a linguo-cognitive approach to this problem. The purpose of this article is to reveal how the interpretation field of the concept 'modern body' is represented and described in modern scientific corporeal discourse, based on Le Breton's *L'adieu au corps*.

We identify the concept 'modern body' (along with other concepts, for example, 'female body', 'the other's body', etc.) as part of the macro-concept 'body', since it has its own field structure. In our research, we adhere to the theory of the concept and its research methods developed by Sternin, since it currently appears to be the most convincing and comprehensive theory that provides a clear interpretation of the concept's essence and structure and describes the stages of its research. The concept is understood as a linguistically modeled unit of the 'national cognitive consciousness', a unit that shapes and at the same time describes the national sphere of concepts [19].

The keyword of the concept (also known as the 'name' of the concept) in our case is the word *corps* ('body'), and the substitute words are the pronouns *il* ('it', subjective case), *le* ('it', objective case), *celui-ci* ('it', subjective case), the adjective *ce dernier* ('the latter'), the synonym *chair* ('flesh') and its substitute pronoun *elle* ('she'), as well as the somatic words *visage* ('face'), *peau* ('skin'), and others. A total of 820 markers were found containing the keyword and substitute words. All the attribute forming markers can be divided into three groups:

(1) lexemes with a common seme (synonyms);
(2) thematic vocabulary;
(3) contextual synonyms.

All the definitions given in the semantic and semantic-cognitive analyses are taken from the electronic dictionary *Le Grand Robert de la langue française* [20].

2   Interpretation Field of the 'Modern Body' Concept

The interpretation field of a concept was defined by I. A. Sternin as the totality of the cognitive attributes of a concept. As human consciousness comprehends the cognitive attributes, they serve to interpret the image of a concept [19].

In our research, we consider two zones of the interpretation field of the 'modern body' concept – the evaluative zone and the pragmatic zone.

The evaluative zone combines the interpretive attributes (IA) which express various levels of evaluation: general, aesthetic, moral, emotional, etc. Evaluation is understood here as recognition by an individual of the significance of the object and categorization of its properties and attributes as positive or negative by identifying its advantages and disadvantages. They are evaluated on the basis of individual needs, aspirations and goals, which, in turn, originate from sensory experience, emotions or existing standards [19, 21, 22].

The pragmatic zone includes the interpretive attributes expressing the pragmatic attitude of an individual to a conceptualized phenomenon, and the knowledge related to the specifics of how this phenomenon can be employed [19].

Social sciences refer to the ability to use the body as 'body techniques' and 'body practices'. Some researchers use the two terms interchangeably [23], while others ascribe different meanings to each of them. Mauss, who was the first to propose the body techniques classification, understood them as the traditional ways by which people in different societies use their bodies [24]. *Dictionnaire du corps en sciences humaines et sociales* ('Dictionary of the body in the humanities and social sciences') differentiates between body practices, body techniques, social practices and sports practices [25]. From a philosophical point of view, there are socio-cultural practices that represent a whole range of methods and mechanisms for organizing and (or) self-organizing people's activities to meet individual and collective needs, and there are body techniques that constitute individual operational components of the actual implementation of practices, so that these techniques are a kind of algorithmic constructs of both actions on the body and actions of the body [26].

Among the body techniques are ways of movement, rest, sleep, body care, nutrition, etc. Body practices usually include bodybuilding, use of drugs, use of alcohol, piercing, suicide, tattoos, body therapy, plastic surgery, diet, exercise, etc. [18, 25].

Body practices and body techniques not only satisfy the vital needs of an individual, but, as Baranov points out, they contribute to the bodily change, spiritual change, and formation of individuality, allowing the individual to learn his/her capabilities and purpose in society; they are an instrument enabling an individual to adapt to everyday life and the world around [23].

3 Evaluative Zone of the 'Modern Body' Concept

We have identified 275 markers that form the following five interpretive attributes (IA) of the evaluative zone: 'Bodilessness', 'Uselessness of the Body', 'Negative Attitude to the Body', 'Positive Attitude to the Body', 'Separation of the Body from the Individual'.

The analysis indicates that the modern body largely appears to be superfluous and useless in scientific corporeal discourse. In the metaphorical sense, the body is disappearing and is bound to be destroyed. The body is hated and despised; it is imperfect, short-lived, ugly, and slow. On the one hand, the body is separated from the individual, i.e. it becomes an autonomous subject, the 'individualization and psychologization' of the body take place [18]. Yet, the body constitutes an integrated whole with an individual; it is glorified, loved, and is of a great value for an individual.

3.1 IA Bodilessness (108)

This interpretive attribute is formed by 108 markers, among which the most represented are the ones that use linguistic means with common synonymous semes: *suppression* ('removal'), *exclusion* ('exclusion'), *disparition* ('disappearance'). The most numerous markers within this group are the ones with the prepositions *sans* ('without') and *hors* ('outside').

The preposition *sans* is defined as '*préposition qui exprime l'absence, le manque, la privation ou l'exclusion*' [20] ('preposition expressing absence or lack of something, deprivation or exclusion'). The lexeme *exclusion* ('exclusion') contains the seme *suppression* ('removal').

Seventeen markers contain the preposition *sans* ('without'), e.g., *un monde sans corps* (3) – 'disembodied world', *un monde sans chair* (1) – 'a world without flesh', *sexualité sans corps* (3) – 'bodiless sexuality', *l'amour sans corps* (1) – 'bodiless love', *l'érotisme sans corps* (1) – 'bodiless eroticism', *la communication sans vis-age, sans chair* (1) – 'faceless, fleshless communication'. This group also includes the marker expressed by the verb *se passer de* ('do without') containing the seme *vivre sans* ('live without'): *se passer du corps* (2) – 'do without a body'.

The second most frequent seme is the preposition *hors* ('outside') with the meaning *sens local d'extériorité, d'exclusion* ('local sense of externality, exclusion'), which is found in 15 markers: *enfants... conçus hors corps* (2) – 'children conceived outside the body', *l'embryon hors corps* (2) – 'the embryo outside the body', *un érotisme hors corps* (1) – 'eroticism outside the body', *une humanité hors corps* (1) – 'mankind outside the body', *la communication hors corps* (1) – 'communication outside the body', etc.

The lexeme *effacement* ('deletion') is defined as *destruction* ('destruction'), *suppression* ('removal'). → *Affaiblissement* ('weakening'), *disparition* ('disappearance'), *évanouissement* ('disappearance') [20]. Five markers with the word *effacement* ('disappearance') have been identified, e.g., *effacement du corps* – 'disappearance of the body', and one marker with the verb *s'effacer* ('fade away'): *le corps s'efface* – 'the body fades away'.

A separate group consists of the markers (7) with the noun *disparition* ('disappearance') and the verb *disparaître* ('disappear'): *la disparition probable du corps humain* (1) –'probable disappearance of the human body', *une disparition de la chair* (1) – 'disappearance of the flesh', *le corps va disparaître bientôt* (1) – 'the body will soon disappear', *il est amené à disparaître* (1) – 'it is bound to disappear', etc.

In addition, this interpretive attribute is objectified by:

– the noun *fin* ('end'), which is used as a synonym for the noun *disparition* ('disappearance'), e.g., *la fin du corps* (5) – 'the end of the body';
– the lexemes *supression* ('removal') and *supprimer* ('remove'), which are present in six markers, e.g., *la volonté de supprimer le corps* – 'the desire to get rid of the body', *la suppression du corps* – 'elimination of the body';
– the synonymous verbs *éliminer* ('eliminate') (6), *liquider* ('liquidate') (2), *abolir* ('abolish') (2) with the semes *faire disparaître* ('make disappear, eradicate'), *supprimer* ('remove') in six markers, e.g., *éliminer le corps* – 'eliminate the body', *liquider ou de transformer le corps* – 'liquidate or transform the body', *abolir le corps* – 'abolish the body'.

We also include in this group the means of expressing the absence of something and getting rid of something, which in this case function as contextual synonyms, e.g., *soustraction du corps* – 'getting rid of the body', *l'homme dépourvu de corps* – 'a person devoid of a body', *se débarasser de la chair superflue et encombrante* – 'get rid of superfluous and cumbersome flesh'.

3.2   IA Negative Attitude to the Body (76)

The most frequent in this group are 12 markers with the lexemes containing the common seme 'contempt', e.g., the noun *mépris* (6) ('contempt'), the adjective *méprisable* (1) ('despicable') and the synonymous noun *dénigrement* ('denigration') (4): *mépris du corps* – 'contempt for the body', *le corps humain est méprisable* – 'the human body is despicable', *le dénigrement du corps* – 'denigration of the body', etc.

The second most recurrent emotion is *haine* ('hatred') (8): *la haine du corps* (4) – 'hatred of the body', *la haine avouée du corps* – 'open hatred of the body', *la haine farouche du corps* – 'fierce hatred of the body', etc.

Negativity is also presented through 20 markers containing the lexemes that express a negative attitude towards the body, e.g., *la méfiance à l'égard du corps* – 'mistrust of the body', *souillure du corps* – 'defilement of the body', *le statut déprécié du corps* – 'devalued status of the body', *combattre le corps* – 'fight the body', *le corps est le péché originel, la tâche sur une humanité* – 'the body is the original sin, a spot on humanity'.

We also include in this group 36 markers with the lexemes that denote the characteristics of the body perceived as negative from a human viewpoint:

– imperfection (6), e.g., *l'imperfection du corps* – 'imperfection of the body', *un corps inachevé et imparfait* – 'an incomplete and imperfect body', *insuffisances du corps* – 'deficiencies of the body', *les defauts du corps* – 'the defects of the body';
– transience (4), e.g., *un corps imparfait et voué á la mort* – 'an imperfect and doomed to death body', *le corps humain est... voué au pourissement* – 'the human body is doomed to decay';
– slowness (3), e.g., *il est lent* – 'it is slow', *la réponse lente des muscles du corps* – 'slow reaction of the body muscles';
– ugliness (2), e.g., *leurs corps leur paraissent laids et vieux* – 'their bodies seem to them ugly and old';
– fragility (2), e.g., *la fragilité du corps de l'Autre* – 'fragility of the body of the Other';
– object and source of suffering (2), e.g., *ce corps... souffrant* – 'this suffering body', *le corps est la source... de toutes les souffrances* – 'the body is the source of all the suffering';
– burdensomeness, cumbersomeness (2), e.g., *un corps encombrant* – 'a bulky body';
– other markers, not included in any of the groups above, e.g., *le corps est... une perte de temps, un gâchis* – 'the body is a waste of time, a mess', *ce corps banal, quotidien* – 'this banal, everyday body', *le corps humain est sans grâce, dérisoire dans sa forme* – 'the human body is without grace, ridiculous in its form'.

3.3   IA Uselessness of the Body (32)

The uselessness of the body is represented by 32 markers.

Seven markers contain the synonymous adjectives *surnuméraire* ('superfluous') defined as *qui est en surnombre, en trop* [20] ('something that is in excess, too much'), and *en trop* ('too much'), e.g., *le corps est surnuméraire, le corps en trop* – 'the body is supernumerary, too much of the body'.

The adjective *superflu* ('superfluous, unnecessary'), used twice in the text, is defined as *qui n'est pas indispensable* ➜ *Inutile, oiseux, vain* [20] ('something that is not necessary, useless, idle, vain'). It is synonymous with the adjective *inutile* ('useless'), e.g., *le corps devient superflu* – 'the body becomes unnecessary'.

Five markers contain the adjective *inutile* ('useless') and lexemes with the seme *utile* ('useful'), e.g., *un corps encombrant et inutile* – 'a useless and burdensome body', *le corps a perdu toute l'utilité* – 'the body has lost all purpose', *peu utiliser leurs corps* – 'barely use their bodies'.

Other 18 markers use the synonyms that express the contextual meaning of uselessness, e.g., *les corps ne sont rien* – 'the bodies are nothing', *le corps ne sert plus à rien* – 'the body is no longer useful', *le corps n'est plus nécessaire* – 'the body is no longer needed', *oublier leurs corps* – 'forget their bodies'.

3.4   IA Positive Attitude to the Body (27)

Positive attitude to the body is represented by:

- markers (7) with the lexemes that express the contextual unity of the body and the individual, e.g., *nous sommes toujours de chair, nous restons de chair* – 'we are always of flesh, we remain flesh', *l'egalité à son corps* – 'equality with your body', *quitter l'epaisseur de son corps serait quitter la chair du monde* – 'leaving the thickness of your body would mean leaving the flesh of the world';
- markers (5) containing the lexemes with the common seme *célébrer* ('celebrate'), such as *hymne* ('hymn, praise'), *éloge* ('praise'), *glorieux* ('glorious'), e.g., *corps glorieux* (3) – 'glorious body', *un hymne au corps* – 'a hymn to the body', *cet ouvrage est bien un éloge sans réserve du corps* – 'this work is the unconditional praise of the body';
- markers (4) with the lexemes containing the common seme *adoration* ('worship, adoration'): the verb *aimer* ('love'), defined as 'avoir un sentiment d'adoration' [20] ('to worship'), and the noun *culte* ('cult'), defined as 'admiration mêlée de vénération, parfois d'adoration' [20] (admiration mixed with reverence, sometimes with worship). For example, *aimer le corps de son partenaire* – 'loving the body of one's partner', *j'aime... mon corps* – 'I love my body', *le culte du corps* – 'the cult of the body';
- markers (2) with the noun *passion* ('passion'): e.g., *la passion du corps* – 'the passion for the body';
- markers (4) with the lexemes that express the common meaning 'value of the body', e.g., *le corps acquiert de l'importance* – 'the body becomes important', *le corps est essentiel* – 'the body is essential', *le corps est associé à une valeur incontestable* – 'the body is associated with an undeniable value';
- markers (3) with the lexemes denoting the merits of the body, e.g., *la perfection du corps est le seul salut* – 'perfection of the body is the only salvation', *l'ouverture du corps* – 'the openness of the body', *l'individualité du corps* – 'the individuality of the body'.

3.5   IA Separation of the Body from the Individual (20)

This attribute, represented by 20 markers, includes lexical means that do not share a common seme, but within the context of *L'adieu au corps* they acquire the meaning of 'separation of the body from the individual', thus becoming contextual synonyms:

- the lexemes *dissocier* ('dissociate') and *dissociation* ('dissociation') used in six markers, e.g., *le corps est dissocié de l'homme* (1) – 'the body is separated from the person';
- the lexemes *autonomisation* ('acquisition of independence') and *s'autonomiser* ('become independent') (2), e.g., *Il s'est autonomisé du sujet* (1) – 'it separated from the subject';
- the nouns *perte* ('loss'), *rupture* ('breaking'), the participle *isolé* ('isolated'), the verb *opposer* ('oppose'), and others, e.g., *la perte de la chair* (1) – 'loss of flesh', *la rupture concréte entre l'homme et son corps* (1) – 'the concrete gap between an individual and a body'.

4 Pragmatic Zone of the 'Modern Body' Concept

We have identified 90 markers that form the following pragmatic attributes (PA) of the pragmatic zone: 'Correcting the Body', 'Experimenting on the Body', 'Cutting the Body', 'Dismantling the Body', 'Performing Actions on the Virtual Body', 'Caring about the Body', 'Building the Body', 'Completing the Body', 'Other Actions'.

The analysis of the pragmatic attributes of the 'modern body' concept in modern scientific corporeal discourse indicates that a progress in science and medicine (particularly, in the fields of plastic surgery, chemistry, pharmacology and cosmetic industry) has led to the appearance of new body techniques and body practices such as 'changing the body', 'correcting the body', 'completing the body', 'dismantling the body' (body organs replacement, prosthetics, etc.). The body is cut, sawed (note that in this context the verbs 'cut' and 'saw' do not refer to amputations or surgical operations based on medical indications), genetically recoded, experimented on (consider, for example, the world-famous performance artists Fakir Musafar, Stelarc or Orlan who see the body as a mere anachronism, a shell, an accessory).

4.1 PA Correcting the Body (25)

This attribute is formed by the markers that contain verbs with the common seme *changer* ('change, modify'), such as *changer* (8) ('change'), mainly as a gerund, *modifier* (5) ('modify'), *réformer* (2) ('change for the better'), e.g., *en changeant son corps* (5) – 'changing one's body', *en changeant la forme de son corps* – 'changing the shape of one's body', *changer son corps* – 'change one's body', *en changeant les traits de leur visage ou l'aspect de leur corps* – 'changing the facial features or the appearance of their body', *modifier leurs corps* – 'modify their bodies', *le fragment de corps à modifier* – 'the body fragment to be changed', *réformer le corps humain* –'reform the human body'.

Additionally, this group includes the markers with the synonymous verbs *corriger* (3) and *rectifier* (3) ('correct, repair, put right'), having the common semes *reformer* ('reform') and *modifier* ('modify'), e.g., *corriger le corps humain* – 'correct the human body', *corriger le corps* – 'correct the body', *rectifier un corps mal ajusté* – 'correct an ill-fitting body', *rectifié par la science notre corps sera parfait* – 'rectified by science, our body will become perfect'; as well as markers with contextual synonyms, e.g., *donner à son corps (son visage) la forme qui lui convient* – 'to give one's body (one's face) a suitable form'.

We also include in this group the markers with the contextual synonyms expressing actions that result in changes in the body, e.g., *celui-ci est éventré,... scié* – 'the body is disemboweled,... sawed', *le corps est brûlé, mutilé, percé,... tatoué, entravé dans les vêtements inappropriés* – 'the body is burned, mutilated, pierced,... tattooed, restrained in inappropriate clothing', *Michael Jackson... s'est fait remodeler le visage, décrêper les cheveux, éclaircir la peau* – 'Michael Jackson... had his face remodelled, his hair straightened, his skin lightened', *le corps est... recodé génétiquement* – 'the body... is genetically recoded'.

4.2   PA Experimenting on the Body (8)

This attribute is formed by the markers that express actions performed on the body with the aim of shocking and experimenting on its capabilities, the case study of Stelarck, Fakir Musafar and Orlan: *il… s'enfonce des aiguilles dans le corps* – 'he… sticks a needle into his body', *il s'attache à un mur, une lourde chaîne autour du corps* – 'he clings to a wall, a heavy chain around the body', *il recouvre l'intégralité de son coprs d'une peinture dorée* – 'he covers his whole body with gold paint', *une série de longues pointes de métal pénétrant son corps* – 'a set of long metal spikes penetrating his body', *il se suspend à des crochets fichés sur tout son corps* – 'he hangs on hooks pierced through his body', *le corps se choisit dans son contenu et surout dans sa forme* – 'the body is chosen according to its content and, especially, to its form', *le corps embroché et rattaché à des câbles* – 'the body is securely fixed and attached to the cables', *il pousse le corps à ses limites physiques et psychologiques* – 'he pushes the body to its physical and psychological limits'.

4.3   PA Cutting the Body (5)

This attribute comprises the markers that contain the lexemes with the common seme *couper* ('cut'), namely, past participles and infinitives of the verbs *entailler* (1) ('slash, cut'), *découper* (3) ('carve, cut up'), *taillader* (1) ('slash, chop'): *celui-ci est… entaillé, découpé* – 'the body is notched, cut', *le corps est… tailladé* – 'the body is slashed', *elle fait découper son corps* – 'she (Orlan) carves her body', etc.

4.4   PA Dismantling the Body (5)

The linguistic means representing this attribute are: the noun *démantèlement* ('dismantling, taking apart'), e.g., *le démantèlement du corps* – 'disassembling the body'; the past participle of the verb *démanteler* ('dismantle'), e.g., *celui-ci est… démantelé* – 'it is dismantled'; and the contextual synonyms, e.g., *le corps est… décomposé* – 'the body… is disassembled', *le morcellement du corps* – 'dismembering the body'.

4.5   PA Performing Actions on the Virtual Body (4)

The common denominator for the markers that form this attribute is the vocabulary and the context related to the topic of cyberspace: the verb *virtualiser* ('virtualize'), e.g., *Stelarc virtualise son corps* – 'Stelarc virtualizes his body'; the adjective *électronique* ('electronic'), e.g., *un remaniement électronique du corps* – 'electronic alteration of the body'; the verbs *mettre en mémoire* ('put in memory') and *garder* ('keep'), e.g., *remodeler son corps à sa guise, le metre en mémoire et le garder à disposition ou en remanier la forme et les apparences selon son humeur* – 'reshape his body (the body of an imaginary partner) as he pleases, put it into memory and keep it available or change its form and appearance according to his mood'; *un usage inédit du corps* – 'the novel use of the body (in cyberspace)'.

4.6   PA Caring About the Body (4)

This attribute includes the markers (4) with the verb *se soucier* ('care'), e.g., *se soucier d'un corps* – 'worry about the body', *se soucier davantage de son corps* – 'care more about one's body', *on se soucie du corps* – 'we care about the body'.

4.7   PA Building the Body (4)

This group includes the markers with the lexemes containing the common seme *construire* ('construct, build'), namely, the verbs *reconstruire* ('rebuild'), e.g., *reconstruire le nouveau corps à partir de son ADN ou d'un autre corps* – 'rebuild the new body from his DNA or from another body'; *bâtir* ('build'), e.g., *d'avoir bâti son corps* – 'having built his body'; and the noun *bâtisseur* ('builder'), e.g., *le body builder, le bâtisseur de corps* – 'the body builder, the builder of the body'.

4.8   PA Completing the Body (4)

In this group, the markers contain the verb *compléter* ('complete'), e.g., *compléter un corps insuffisant en lui-même à incarner l'identité personnelle* – 'to complete a body insufficient in itself in order to embody the personal identity', *l'ambiance intellectuelle d'un corps inachevé et imparfait dont l'individu doit compléter la forme* – 'the intellectual atmosphere of an unfinished and imperfect body whose form the individual must complete', *le compléter ou le rendre conforme à l'idée que l'on s'en fait* – 'complete it or make it conform to the idea that we have', *placer les implants souscutanés au niveau du front* – 'place the implants under the forehead skin'.

4.9   PA Other Actions (31)

This category comprises other markers with the lexemes that do not have a common seme or common meaning and express various actions performed on the body, e.g.:

– *le corps est scannérisé, purifié, géré, remanié, renaturé, artificialisé… ou éliminé, stigmatisé au nom de l' 'esprit' ou du 'mauvais' gène* – 'the body is scanned, purified, managed, redesigned, annealed, artificialized… or eliminated, stigmatized in the name of the 'spirit' or 'bad' gene';
– *certains percés aiment l'idée d'avoir du métal dans le corps* – 'those with piercing like the idea of having metal in their bodies';
– *maintenir le corps en vie* – 'to keep the body alive';
– *colonisation du corps par la technique* – 'colonization of the body by technology';
– *le corps est soumis à un design* – 'the body is subject to design';
– *des corps photographiés ou filmés* – 'bodies are photographed or filmed';
– *gérer son propre corps* – 'manage your own body';
– *commercialiser les produits de son corps* – 'commercialize the products of his body';
– *sursignifier le corps* – 'oversignify the body';
– *se communique au corps entier* – 'communicates with the whole body';
– *penser le corps* – 'conceive the body'.

5   Conclusion

It can be noted that the majority of the interpretive attributes of the 'modern body' concept in modern scientific corporeal discourse reflect an individual's negative attitude towards the body. The prepositions *sans* and *hors*, nouns and verbs with the semes 'destruction' and 'deletion' reveal a tendency to eliminate the body (in the metaphorical sense) in scientific discourse, especially in technological sciences, genetics, cybernetics and robotics. Contextual lexical and grammatical means (the adjective *probable* ('probable'), noun *volonté* ('wish'), future tenses, etc.) indicate that disappearance of the body is what awaits us in the future – perhaps, the near future. At the same time, the body is already disappearing: in vitro fertilization has made the female body unnecessary; the virtual body of the Other provides Internet users with sex and eroticism without the physical body of a partner and, by making the physical body useless, destroys it.

People experience contempt (*mépris, dénigrement*) and hatred (*haine*) for the body. There are several reasons that could be employed as an explanation: firstly, imperfection of the body (*imperfection, imparfait*); secondly, the fear of death, because the body is transient, short-lived (*voué á la mort, voué au pourissement*); thirdly, the body is cumbersome (*encombrant*) and slow (*lent*), which greatly irritates the internauts who have learned the freedom of a virtual body. The interpretive attribute 'Body is Useless' is mainly realized in the discourse related to artificial intelligence and robotics.

All this leads to the separation of the body and the individual. In his monograph, Le Breton uses the lexemes *homme* ('man, person')/*sujet* ('subject'), although it seems that the point here is the separation of the body and the 'I'. The category of 'I' is a complex and controversial issue; the concept of 'I' is generally understood as a structural unit which represents the final product of the self-comprehension process [27]. It is possible that the case here is not so much in separation but opposition of the body and the 'I' (in spite of the fact that the vocabulary of separation is used in the monograph).

The pragmatic attributes of the 'modern body' concept indicate that the new body practices and body techniques are aimed at changing the body in order to strengthen the personal identity, to give confidence to the 'I' of an individual. The body and the 'I' no longer constitute an integral whole; the body becomes an instrument for self-expression and self-affirmation of the 'I'. As we have found out in our other research on the cognitive image of the 'modern body' concept, the body is being reduced to the level of an object, a thing, as evidenced by the fact that the core of the 'modern body' concept contains 80 markers of the cognitive attribute 'The Body is an Object/Thing'.

In fact, this perception is, apparently, intrinsic not only to modern scientific discourse. The advertising discourse (in particular, plastic surgery advertising) also forms similar pragmatic attributes. Patients of plastic surgeons have their bones shaved (changing/correcting the body) and silicone implants inserted (completing the body). In this regard, it is interesting to identify the content of the 'modern body' concept in naive consciousness, as well as the content of the concepts 'female body' and 'male body' both in naive consciousness and in different types of discourse in comparison with the previous eras; how much the content of these concepts has changed in the context of new sports techniques (e.g., bodybuilding) and increasing popularity of plastic surgery for both men and women.

Only 27 markers employ the linguistic means that express a positive attitude to the body in the Western scientific corporeal discourse. An individual loves his/her body, praises it, and attributes a special value to it. The most important characteristic of this group, however, is the unity of the body and individual, because an individual is made of flesh through which he/she is able to perceive the world. It should be noted that the majority of the markers expressing a positive attitude are identified in the comments written by David Le Breton himself.

References

1. Le Breton, D.: L'adieu au corps. Editions Métailé, Paris (2013)
2. Le Breton, D.: La transgression comme une voie de salut: le corps mis à mal. In: Que reste-il de nos tabou? 15e forum Le Monde Le Mans 24 au 26 Octobre 2003, pp. 43–51. Presses Universitaires de Rennes, Rennes (2004)
3. Timoshenko, M.A.: Chelovecheskoe telo kak kul'turnaya forma [Human body as a cultural form]. Ph.D. thesis. Novgorod State Pedagogical University (2009). (in Russian)
4. Stepanov, M.A.: Opyt myshleniya tela: k ehpistemologii Ditmara Kampera [Experience of the body mentality: on Dietmar Kamper's epistemology]. Ph.D. thesis. Saint-Petersburg State University (2011). (in Russian)
5. Kiseeva, E.V.: Telo kak oblast' ehksperimenta v sovremennom plasticheskom teatre [Body as a field for experiment in the modern plastic theater]. Istoricheskie, filosofskie, politicheskie i yuridicheskie nauki, kul'turologiya i iskusstvovedenie. Voprosy teorii i praktiki 2(28), 83–86 (2013). (in Russian)
6. Shmeleva, N.V.: Usilenie ehffekta sovershenstva tela v kontekste sovremennoj kul'-tury [Intensification of body perfection effect in modern culture]. Al'manah sovremennoj nauki i obrazovaniya 5–6(84), 149–151 (2014). (in Russian)
7. Kovaleva, E., Spirina, M.O.: Reprezentaciya zhenskogo tela v sovremennoj joge: teo-reticheskij i metodologicheskij podhody [Female body representation in modern yoga: theoretical and methodological approaches]. http://jour.isras.ru/index.php/inter/article/view/4358. Accessed 09 Sep 2016. (in Russian)
8. Chekmareva, M.A.: Aktual'naya antichnost'. Kul't tela v zerkale sovremennogo iskus-stva [Relevant Antiquity. Body cult reflected in the modern art]. Trudy istoricheskogo fakul'teta Sankt-Peterburgskogo universiteta. 22, 189–198 (2015). (in Russian)
9. Kovaleva, L.S.: The problem of the body and corporality of the modern man in cinematography. J. Sib. Fed. Univ. Hum. Soc. Sci. 6(9), 1466–1473 (2016). (in Russian)
10. Makarov, A.I., Toropova, A.A.: Otchuzhdennye tela: traktovka koncepta telesnosti v postmodernizme [Body alienation: interpretation of the concept of corporeality in post-modernism]. Vestnik Volgogr. gos. un-ta. 4(34), 16–26 (2016). (in Russian)
11. Shchekleina, A.S.: Vzaimodejstvie tela i duha v sovremennom tancteatre [Interaction of body and spirit in modern dance theater]. In: Sovremennyj tanec: diskurs i praktiki (sbornik statej), pp. 136–151. Gumanitarnij Universitet, Ekaterinburg (2017). (in Russian)
12. Tataru, L.V.: Tochka zreniya i ritm kompozicii narrativnogo teksta (na materiale proizvedenij Dzh. Dzhojsa i V. Vulf) [Point of view and compositional rhythm of the narrative text: case study of J. Joyce and V. Woolf]. Doctoral thesis. Saratov State University (2009). (in Russian)

13. Le Breton, D.: Corps et sociétés. Essai de sociologie et d'anthropologie du corps. Librairie des méridiens, Paris (1985)
14. Greimas, A.J., Keane, T.M.: Dictionnaire du moyen français. La Renaissance. Larousse, Paris (1992)
15. Rey, A.: Dictionnaires historique de la langue française. Dictionnaire Le Robert, Paris (2000)
16. Le Goff, J., Truong, N.: Une histoire du corps au Moyen Age. Flammarion, Paris (2008)
17. Carvallo-Plus, S.: La construction du concept de corps par la tradition philosophique antique et classique. Préparartion de l'ENS-LSH à l'agrégation de sciences sociales. http://socio.ens-lyon.fr/agregation/corps/corps_conf_cavallo.php. Accessed 26 Sept 2014
18. Kitabgi, S., Hanifi, I.: Introduction: La sociologie et le corps: Géologie d'un champs d'analyse. In: Ciosi-Houcke, L., Pierre, M. (eds.) Le corps sens dessus dessous. Regards des sciences sociales sur le corps. L'Harmattan, Paris (2003)
19. Sternin, I.A.: Struktura koncepta [Stucture of the concept]. In: Izbrannye raboty. Teoreticheskie i prikladnye problemy yazykoznaniya, pp. 172–184. Istoki, Voronezh (2008). (in Russian)
20. Dictionnaires Le Robert – Le Grand Robert de la langue française, https://www.lerobert.com . Accessed 16 May 2018
21. Prihod'ko, A.I.: Poliparadigmal'nyj harakter kategorii ocenki [Multiparadigmatic character of the category of evaluation]. In: Yazykovaya lichnost' i ehffektivnaya kommunikaciya v sovremennom polikul'turnom mire. Sbornik statej po itogam III mezhd.nauchno-prakt. konferencii, pp. 61–69. Izd-vo BGU, Minsk (2018). (in Russian)
22. Dormidontova, O.A.: Kategoriya ocenki i ocenochnaya kategorizaciya s pozicij sovremennoj lingvistiki [Category of evaluation and evaluative categorization in modern linguistics]. Al'manah sovremennoj nauki i obrazovaniya 2(21), 47–49 (2009). (in Russian)
23. Baranov, V.A.: Telesnye praktiki kak sposob lichnostnogo konstruirovaniya [Body practices as a method of constructing personality]. Tavricheskij nauchnyj obozrevatel' 12(7), 7–12 (2016). (in Russian)
24. Mauss, M.: Obshchestva. Obmen. Lichnost': Trudy po social'noj antropologii [Societies. Exchange. Personality: Works on social anthropology]. Nauka, Moscow (1996). (in Russian)
25. Andrieu, B.: Dictionnaire du corps en sciences humaines et sociales. CNRS EDITIONS, Paris (2006)
26. Pron'kina, A.V.: Praktiki tela i telesnye praktiki: opyt opredeleniya fenomenologicheskih granic [Practices of the body and body practices: on defining the phenomenological borders]. Obshchestvo: filosofiya, istoriya, kul'tura 11, 47–49 (2016). (in Russian)
27. Zhdanov, A.A.: Sootnoshenie ponyatij « Ya-koncepciya » i « Samosoznanie » [Correlation of concepts 'I-conception' and 'self-consciousness']. Provincial'nye nauchnye zapiski 1, 80–84 (2016). (in Russian)

Chulym Turkic is a Uralian Kipchak Language, According to the Leipzig–Jakarta List

Innokentiy N. Novgorodov[1]([✉]) [iD],
Nurmagomed E. Gadzhiakhmedov[2] [iD], Mussa B. Ketenchiev[3] [iD],
Natalya V. Kropotova[4] [iD], and Valeriya M. Lemskaya[5,6] [iD]

[1] North-Eastern Federal University, Belinsky Str. 58, 677000 Yakutsk,
Russian Federation
i.n.novgorodov@mail.ru

[2] Dagestan State University, M. Gadzhieva Str. 37, 367000 Makhachkala,
Russian Federation
nurl@yandex.ru

[3] Kabardino-Balkarian State University, Tchernyshevskaya Str. 173,
360004 Nalchik, Russian Federation
ketenchiev@mail.ru

[4] Crimean Engineering and Pedagogical University, Per. Uchebniy 8,
295015 Simferopol, Russian Federation
hilal@mail.ru

[5] Tomsk State Pedagogical University, Kievskaya 60, 634061 Tomsk,
Russian Federation
lemskaya@gmail.com

[6] Tomsk State University, Lenina 36, 634050 Tomsk, Russian Federation

Abstract. Background. This article is about the relationship of the Chulym Turkic language to the Kipchak Turkic languages. The article sets questions to which of the modern Kipchak languages Chulym Turkic is related or whether it is a separate Kipchak language group. The relationship of the Chulym Turkic language to Cuman (Ponto-Caspian) is also studied. The Chulym Turks are the people of the South East of the West-Siberian Plain. The number of the Chulym Turks is around 365 people in Russia.

Materials and Methods. Research materials are words of the Leipzig–Jakarta list, phonetic and morphology data of the Turkic languages. The Leipzig–Jakarta list is a 100-word list to test the degree of the relationship between languages. The most resistant words were taken from dictionaries, publications and native speakers. In this survey the comparative method is used as the main method. The quantitative method is applied to count the similarities and discrepancies in the Leipzig-Jakarta list of the Turkic languages.

Discussions. Previously we understood Chulym Turkic to be of the Kipchak Turkic language origin, according to the Leipzig-Jakarta list.

Conclusions. Authors conclude that the Chulym Turkic language differs from the Cuman (Ponto-Caspian) Turkic languages in 2 items. Homogeneity of the

© Springer Nature Switzerland AG 2019
Z. Anikina (Ed.): GGSSH 2019, AISC 907, pp. 411–419, 2019.
https://doi.org/10.1007/978-3-030-11473-2_43

Kipchak languages as comparable objects, according to the Leipzig-Jakarta list, is revealed. Detailed analysis of the list of vocabulary and grammar data reveals that the Chulym Turkic language belongs to the Uralian Kipchak languages.

Keywords: Languages · Leipzig-Jakarta list · Grammar data · Turkic · Kipchak · Cuman · Uralian · Chulym turkic

Abbreviations

alt.	Altai
az.	Azerbaijani
bash.	Bashkir
chin.	Chinese
chul.	Chulym
i.e.	Indo-European
kar.	Karaim
kaz.	Kazakh
Kir.	Kazakh dialect
kkal.	Karakalpak
kbal.	Karachay-Balkar
kum.	Kumyk
Küär.	Kueyric dialect
kyrg.	Kyrgyz
mo.	Written Mongolian
nog.	Nogai
skr.	Sanskrit
stat.	SiberianTatar
tat.	Tatar
tel.	Teleut
Tel.	Teleut dialect
tu.	Turkic
turk.	Turkmenian
wyug.	Western Yugur
yak.	Yakut
[Dulson]	Chulym field data of A. P. Dulson

1   Introduction

Altaic Studies are of the current interest within the modern Euro-Asian research held by the leading countries worldwide in the context of their geopolitics. To study processes of the Altaic language community convergence [2, 4, 27, 28], it is reasonable to explore the Leipzig–Jakarta list [2, 10] and after that to conduct a comprehensive study of the Turkic, Mongolian, Tungus-Manchu, Korean and Japonic peoples, taking into consideration the achievements of linguistics, history and genetics.

The Leipzig-Jakarta list on several Turkic languages has already been published [9, 10]. Previously we arrived at the conclusion that the Turkic languages are divided into two main groups according to the Leipzig–Jakarta list [9]. The first one is the Yakut and Kipchak languages, and the second one – the Chuvash and Oghuz languages.

As for Chulym Turkic, we identified it to be of the Kipchak Turkic language origin according to the Leipzig–Jakarta list [9]. Herein we raise a question: which of the modern Kipchak languages Chulym Turkic belongs to, or is it a separate Kipchak language group?

As it is known, the modern Kipchak languages are classified into different groups: peripheral (Kyrgyz and Altai), Cuman or Ponto-Caspian (Karachay-Balkar, Kumyk, Karaim), Tatar-Bashkir or Uralian (Tatar, Bashkir) and Kangly or Aralo-Caspian (Kazakh, Karakalpak, Nogai) [24, p. 35].

In our publications we formed a judgment that the Chulym Turkic language was more similar to the peripheral Kipchak languages than the Uralian ones and that the Teleut and Siberian Tatar idioms are separate Kipchak Turkic languages, whereas the former is peripheral and the latter is Uralian [11, p. 292]. Also, we determined that the Chulym Turkic language differs from the Kangly (Kazakh, Karakalpak and Nogai) Kipchak languages considerably [12, p. 118].

As we have not studied the relationship of Chulym Turkic to the Cuman (Ponto-Caspian) Kipchak languages according to the Leipzig-Jakarta list yet, in this article we will briefly address this question, so a few lines about their speakers should be written.

The Chulym Turks are the people in the south-east of the West-Siberian Plain that inhabit the lower and middle flow of the Chulym River. The majority of the community resides in the Russian Federation's Teguldet Region of the Tomsk Oblast and Tjuxtet Region of the Krasnoyarsky Krai, mainly in the villages of Pasechnoye and Chindat.

The number of Chulym Turks is around 365 people, according to official statistics [29]. The Chulym Turkic language included the Middle and Lower Chulym dialects; the Middle dialect's sub-groups are Tutal and Melet with the differentiation going back to the historical existence of indigenous provinces, i.e. "volosts". At present the Lower Chulym dialect is considered to be totally extinct.

The Karachay-Balkar language is one of the Cuman (Ponto-Caspian) Kipchak Turkic languages spoken by the Karachay-Balkars the majority of whom live in the North Caucasus (the number of the Karachays is 218403 and that of the Balkars is 112924, according to the latest All-Russian census of 2010 [29]). The Karachay-Balkar language is the state language of the Karachay-Cherkess Republic, Russia. Karachay-Balkar has 2 dialects: *člj* and *clz*.

The Kumyk language is one of the Cuman (Ponto-Caspian) Kipchak Turkic languages spoken by the Kumyks (503060 people, according to the latest All-Russian census of 2010 [29]), the majority of whom live in the North Caucasus. Kumyk is the state language of the Republic of Dagestan, Russia. It has 5 dialects: Khasavyurt, Buinaksk, Khaitag, Podgorniy and Terek.

The Karaim language is one of the Cuman (Ponto-Caspian) Kipchak Turkic languages spoken by the Karaims who live in Russia (Crimea), Poland, Lithuania and Ukraine. The total number of the Karaim population is about 2000 people [18, p. 27]. According to the 2014 census, 535 Karaims live in Crimea [3, p. 108]. The 2010 All-Russian census identified 205 Karaims [29]. The UNESCO data indicate that Karaim in

Crimea became extinct, whereas in Lithuania there are 50 speakers and in Ukraine 6; the database is available online at [30]. Karaim has 3 dialects: Crimean, Trakai and Halich.

Thus, the Cuman or Ponto-Caspian Kipchak languages spread in the European territories of Russia (Caucasus, Crimea), Ukraine and Lithuania.

Historically the Kipchaks dwelled in feudal states that formed after the collapse of the Yenisei Kyrgyz Khaganate, Golden Horde and Nogai Horde.

2 Materials and Methods

Published materials are used in the present study of the Chulym Turkic language [9]. The materials of the Karachay-Balkar, Kumyk and Karaim languages are taken from different publications and recorded from native speakers. The Leipzig-Jakarta list is a 100-word list to test the degree of language relationship by comparing words that are resistant to borrowing [9, 25]. The indicated 100 most resistant words are used here to establish the relationship of the Chulym Turkic language to the Cuman and Kipchak Turkic languages.

Here, the Leipzig-Jakarta list is the data for the quantitative method adopted. It is used to count the similarities and discrepancies in the Leipzig-Jakarta list of the Turkic languages in order to reveal the degree of homogeneity of comparable objects which is important in studying the relationship of different idioms and their genetic carriers.

For the convenience of analysis the Leipzig-Jakarta list was taken from open electronic sources rather than [25] due to material arrangement: e.g., the alphabetical order of vocabulary and a single lexeme for identifying each vocabulary sample.

Before presenting materials of the Leipzig-Jakarta list, it should be noted that '50' is the number of the Leipzig-Jakarta list item; 'leaf' – its meaning; (8.56) – the index number of the World loanword database available online at [26].

3 Discussions

As analysis shows, the Leipzig-Jakarta list of Chulym Turkic differs from that of the Cuman (Ponto-Caspian) (Karachay-Balkar, Kumyk and Karaim) languages in 2 items (50, 99), e.g.:

50 'leaf' (8.56) chul. *pür* (<tu. [17, p. 92a, 20, p. 296]), *kaak, kak* (<tu. [5, p. 161]); kbal. *čapraq* (<tu. [22, p. 130]); kum. *yapïraq* (<tu.); kar. *yapraq* (<tu.) etc.

As for discrepancies in the Leipzig-Jakarta list of the Cuman (Ponto-Caspian) Kipchak languages, we cannot find them between the Karachay-Balkar and Kumyk idioms. Differences could be found in synonymy in one language and absence of it in the other, also in synonyms and phonology of the matching words, e.g.:

1 'ant' (3. 817) kbal. *qumursxa* (<tu. [6, p.140]); kbal. *gumulžuk* (<kbal. *gumul-* 'to run cowering' + derivational affix *-žuk*); kum. *xomursya* (<tu.);

71 'sand' (1.215) kbal. *qum* 'песок' (<tu. [6, p. 133]; kbal. *üzmes* (<kbal. *üz-* 'to tear up'(cf. wyug. *yüz-* 'to break off' (<tu. [19, p. 621]) + derivational affixes *-ma* + *-z*); kum. *xum* (<tu.), *qayïr* (<tu. [5, p. 217]) etc.

The difference between Karachay-Balkar, Kumyk and Karaim is found in 1 item (47), e.g.:

47 'knee' (4.36) kbal. *tobuq* (<tu. **top* 'orb' [23, p. 197]); kum. *tobuq* (<tu.); kar. *tiz* (<tu. [21, p. 336]).

We concluded previously that the Chulym Turkic language was more similar to the peripheral Kipchak languages than the Uralian ones [11, p. 292]. Also we recognized that the Chulym Turkic language differed from the Kangly or Aralo-Caspian (Kazakh, Karakalpak and Nogai) Kipchak languages considerably [12, p. 118]. Now taking into consideration Radloff's dictionary "Versuch eines Wörterbuch der Türk-Dialekte" [13–16] and the data of "Etymological dictionary of the Turkic languages" [5–7, 19–22] we identify the difference of Chulym Turkic language from the peripheral (Kyrgyz, Altai and Teleut) Kipchak languages in 1 item (16) and from the Kangly (Kazakh, Karakalpak and Nogai) Kipchak languages in 2 items (16, 19), e.g.:

16 'child (kin term)' (2.43) chul. *oyïlan, uylan* (<tu. [19, p. 411]), *käč, käš'* (<tu. [5, p. 75]); kyrg. *bala* (<tu. [20, p. 47] <i.e.: skr. *bāla* 'child', 'boy' [8, p. 376]); alt. *bala* (<tu. <i.-e.); tel. *pala* (<tu. <i.e.);

19 'to cry/to weep' (16.37) chul. *sïxtala-, sïkta-* (<tu. [7, p. 390]); [Dulson] *ïk-* (<tu. **ïk* 'sob', 'to cry/to weep' [19, p. 650]); kaz. *žïlau* (<tu. [19, p. 79]), *eŋïriu* (<tu. [19, p. 366]), *egïlu* (<tu. [19, p. 79]); kkal. *žïlaǫ* (<tu. [19, p. 79]), *eŋreǫ* (<tu. [19, p. 366]), *öksiǫ* (<tu. [19, p. 503]), *sïŋsiǫ*, cf. yak. *sïŋsïy-* 'to sob' (Turkic borrowing in Yakut); nog. *yïlav* (*yïla-*) (<tu. [19, p. 79]), *o'ksu'v* (*o'ksi-*) (<tu. [19, p. 503], etc.

The Kangly and Cuman Leipzig-Jakarta lists have full accordance. The same could be said about the Kangly and peripheral data. The Cuman and peripheral rosters have a discrepancy in 1 item only (99), e.g.:

99 'yesterday' (14.49) kyrg. *kečee* (<tu. [21, p. 40]); alt. *keče* (<tu.); tel. *keče* (<tu.); kbal. *tünene* (<tu. [21, p. 315]); kum. *tünegün* (<tu.); kar. *tünegin* (<tu.).

Formerly we inferred that the Kazakh Leipzig-Jakarta list differs from that of Karakalpak in 1 item (11) [12, p. 118] and Teleut differentiates from that of Altai also in 1 item (1) [11, p. 289]. But now considering Radloff's data published in his dictionary in 1893 in the first volume on pages 871, 1825 [13], and in 1905 in the third volume on page 2103 [15], we have come to the conclusion that the Kazakh & Karakalpak, and Teleut & Altai Leipzig-Jakarta lists have full accordance. It should be noted that the abbreviation Kir. in Radloff's dictionary means the Kyrgiz dialect [13, p. XVII] because before 1917 the name Kyrgiz represented the language of the present-day Kazakhs, and the abbreviation Tel. signified the Teleut dialect [13, p. XVIII], which nowadays is known as the Teleut language, e.g.:

11 'to blow (intransitive)' (10.38) kaz. *soɣu* (<tu. [7, p. 288]), Kir. *es-* (<tu. [19, p. 553]) [13, p. 871], Kir. *ür-* (<tu. [19, p. 635]) [13, p. 1825]; kkal. *üpleǫ* (onomatopoeia of wind blowing *üp* + derivational affix *-la*), cf. turk. *üflemek*, az. *üflämäk*, kkal. *ürle* (<tu.), *esiǫ* (<tu.), *hüǫleǫ* (onomatopoeia of wind blowing *hüǫ* + derivational affix *-la*); nog. *u'ru'v* (*u'r-*) (<tu.), nog. *esu'v* (*es-*) id. (<tu.);

1 'ant' (3. 817) alt. *čïmalï* (<mo.: *čubali, čumali* 'ant'); tel. *qüzürüm* (<tel. **qusurum* <**qumurs* <**qïmïrs* (cf. turk. *qïmïrsa-* (<**qïmïrs* + a-) 'to creep, to swarm about insects' (<tu. [6, p. 141]) > yak. **qïmïrsakač* (<*qïmïrsa-* + -kač) > yak. *qïmïrdacas*), Tel. *čïmalï* (<mo.) [15, p. 2103].

As for discrepancies in the Leipzig-Jakarta list between the Chulym Turkic and the Uralian (Tatar, Bashkir and Siberian Tatar) data, they were found in 6 items (7, 14, 16, 27, 50, 95) [11, p. 292]. We can hardly say though that we have found them because careful analysis of Radloff's dictionary [13–16] and the "Etymological dictionary of the Turkic languages" [5–7, 19–22] data show that there are no real differences in the Leipzig-Jakarta list in the Chulym Turkic and the Uralian languages, e.g.:

95 'wide' (12.61) chul. *jalbak* (<tu. [22, p. 100]); tat. *kiŋ* (<tu. [21, p. 46]); stat. *kiŋ* (<tu.); bash. *kiŋ* (<tu.), *iŋle* (<tu. [19, p. 352]), *yaðï* (<tu. [22, p. 155]).

In Radloff's dictionary published in 1899 in the second volume on page 1067 [14] one can find the word *käŋ* 'wide' with the abbreviation Küär. According to W. Radloff's abbreviation system, Küär. means the Kueyric (Küärik) dialect [13, p. XVII] of the Turkic community. Radloff himself translated the Kueyric dialect from German into Russian as the Chulym dialect [13, p. XVII]. Today, the latter is identified as the Chulym Turkic language. Radloff recorded his Kueyric data during his trip to the River Chulym in 1863 [1, p. 446]. Based on the Küär. *käŋ* 'wide', our conclusion implies that historically the Chulym Turkic language matched the Uralian Kipchak in these words: tat. *kiŋ*, stat. *kiŋ*, bash. *kiŋ* 'wide'.

Here we should also mention that the Uralian Kipchak (Tatar, Bashkir and Siberian Tatar) language data from the Leipzig-Jakarta list [11, p. 290] have full accordance with the Uralian and Kangly Leipzig-Jakarta lists. The same could be said about the Uralian and peripheral data. The Cuman and Uralian rosters have a discrepancy in 1 item only (99), e.g.:

99 'yesterday' (14.49) tat. *kičä* (<tu. [21, p. 40]); bash. *kisä* (<tu.); stat. *kicä*; kbal. *tünene* (<tu. [21, p. 315]); kum. *tünegün* (<tu.); kar. *tünegin* (<tu.).

Thus, the Chulym Turkic language and the Uralian ones (Tatar, Bashkir and Siberian Tatar) reveal their unity, according to the Leipzig-Jakarta list.

In general, homogeneity of the Kipchak languages as comparable objects, according to the Leipzig-Jakarta list, is revealed. Only one discrepancy (item 99) could be found between the Uralian & Cuman, and peripheral & Cuman groups of the Kipchak languages. Also, only one difference could be noticed among the Kangly languages (item 99) [12, p. 118] and the Cuman ones (item 47).

To study the question of the unity of Chulym Turkic and the Uralian languages, one should address other data for analysis.

According to Shcherbak's classification, the Uralian Kipchak and Chulym Turkic languages have differences in phonetics and morphology [24, pp. 34, 35, 39, 40].

Shcherbak's synchronous classification of the Turkic languages was established with consideration of phonetics, morphology and vocabulary signs: (1) the phonetic appearance of the words *ajak* 'leg', *taɣ* 'mountain', *aɣïz* 'mouth', *oɣul* 'son', 'boy', 'guy', *jaŋak* 'cheek'; (2) the forms of the past perfect, or unobvious tense (the first person singular), and the forms expressing the possibility and the impossibility of performing an action (the first person singular, the past categorical tense); (3) words with the meanings of 'to rest', 'to love, to consider appropriate', 'myself' (the first person, singular) [24, p. 24].

The phonetic differences in appearance of the words *ajak* 'leg', *taɣ* 'mountain', *aɣïz* 'mouth', *oɣul* 'son', 'boy', 'guy', *jaŋak* 'cheek' between the Chulym Turkic language and the Uralian ones may be explained in some cases by the obsolete state of the latter which is reflected in Chulym, e.g.: chul. voiceless *č-* in *čaak* 'cheek' and voiced *j-* in tat. *jaŋak* id. In other cases discrepancies could be interpreted by the innovative processes, e.g.: the falling of the consonant *ŋ* between vowels in chul. *čaak* 'cheek', cf. tat. *jaŋak* id.; different reflection of the Proto Turkic consonant *-δ- in *aδak* 'leg' in the Uralian languages: bash. *aδak* (<*azak), cf. chul. *azak*. Also, inequalities could be interpreted by the influence of the Uralian languages on Chulym, e.g.: voiced *j-* instead of the voiceless *č-* in chul. *jaak* 'cheek'. In morphology one can find in lieu of dissimilarities the same and similar forms that are present in the Chulym Turkic language and the Uralian ones, e.g.: the forms of the past perfect (the first person singular) chul., bash., stat. *alγam, alγanïm* 'I had taken'; the forms expressing the possibility and the impossibility of performing an action (the first person singular, the past categorical tense) chul. * қälip aldïm* 'I could come', bash. *қilä aldïm*, stat. *қilä altïm* id., chul. * қälip albadïm* 'I could not come', bash. *қilä almadïm*, stat. *қilä almatïm* id. Here we have similar constructions of adverbial participles that are represented by the affixes *-ïp* and *-ə*: chul. * қälip* (*қəl-* 'come' (<tu. **kel-*) + *-ip*) and bash. *қilä* (*қil-* 'come' (<tu. **kel-*) + *-ä*). In vocabulary, likeness instead of dissimilarities also could be found, e.g.: chul. *tïn-* 'to rest' and stat. *tïnïk-* id.; chul. *кïn-* 'to love, to consider appropriate' originating from the Proto Turkic *кïn-* 'to do something with passion' [6, p. 216] have a match in tat. *кïn-* 'to do', cf. yak. *gïn-* id.; chul. words *pozum, poyïm, özüm* (<tu. *ōs) 'myself' (the first person singular) have correspondence with the tat. *üzïm* (<tu. *ōs), bash. *üδïm* (<tu.), stat. *üsäm* (<tu.) id.

Analysis of the Leipzig-Jakarta list and signs of Shcherbak's synchronous classification shows the unity of the Chulym Turkic language and the Uralian ones.

4   Conclusion

The Leipzig-Jakarta list of Chulym Turkic differs from that of the Cuman (Ponto-Caspian) (Karachay-Balkar, Kumyk and Karaim) languages in 2 items (50 and 99).

Homogeneity of the Kipchak languages as comparable objects, according to the Leipzig-Jakarta list, is revealed in this article. Only one discrepancy (item 99) could be found between the Uralian & Cuman, and peripheral & Cuman groups of the Kipchak languages. Also, only one difference could be noticed among the Kangly languages (item 99) and the Cuman ones (item 47).

The Leipzig-Jakarta list and Shcherbak's synchronous classification data indicate the existence of a unity between Chulym Turkic and the Uralian Kipchak languages.

Acknowledgments. We sincerely thank the Russian Federation Government for the grant RScF 18-18-00501 Digital Dialectologic Atlas of Turkic languages in Russia; this publication was prepared within the framework of this research project. We also sincerely thank Shimon Yuhnevich and Lidiya Mashkevich who checked and added items to the Karaim materials. We also sincerely thank Linara Ishkildina who checked and added the Bashkir materials.

References

1. Dulson, A.: Chulymsko-tyurkskij yazyk [The Chulym language]. In: Baskakov, N. (ed.) Yazyki narodov SSSR [The languages of the USSR] II, pp. 446–466 (1966). (in Russian)
2. Erdal, M.: The Turkic-Mongolic relationship in view of the Leipzig-Jakarta list. In: Novgorodov, I. (ed.) Unpublished Proceedings of 2016 International Symposium "The Leipzig-Jakarta List of the Turkic Languages as a Source of the Interdisciplinary Comprehensive Studies" (2016)
3. Itogi perepisi naseleniya v krymskom federalnom okruge. Federalnaya sluzhba gosudarstvennoj statistiki [Results of the population census in the Crimean Federal District. Federal service of state statistics]. Statistika Rossii. Moscow (2015). (in Russian)
4. Janhunen, J.: The altaic hypothesis: what is it about? In: Novgorodov, I. (ed.) Unpublished Proceedings of 2016 International Symposium "The Leipzig-Jakarta List of the Turkic Languages as a Source of the Interdisciplinary Comprehensive Studies" (2016)
5. Levitskaya, L., Dybo, A., Rassadin, V.: Etimologicheskiy slovar tyurkskikh yazykov. Obshchetyurkskie i mezhtyurkskie leksicheskie osnovy na bukvy "K", "Q" [Etymological dictionary of the Turkic languages. All-Turkic and cross-Turkic lexical stems starting with letters "K" and "Q"]. Yazyki russkoy kultury. Moscow (1997). (in Russian)
6. Levitskaya, L., Dybo, A., Rassadin, V.: Etimologicheskiy slovar tyurkskikh yazykov. Obshchetyurkskie i mezhtyurkskie leksicheskie osnovy na bukvy "K" [Etymological dictionary of the Turkic languages. All-Turkic and cross-Turkic lexical stems starting with letter "K"]. Yazyki russkoy kultury. Moscow (2000). (in Russian)
7. Levitskaya, L., Blagova, G., Dybo, A., Nasilov, D.: Etimologicheskiy slovar tyurkskikh yazykov. Obshchetyurkskie i mezhtyurkskie leksicheskie osnovy na bukvy "L", "M", "N", "P", "S" [Etymological dictionary of the Turkic languages. All-Turkic and cross-Turkic lexical stems starting with letter "L", "M", "N", "P", and "S"]. Vostochnaya literatura RAN. Moscow (2003). (in Russian)
8. Munkasci, B.: Beiträge zu den alten Lehnwörtern im Türkischen. Keleti-Szemple 2(3), 376–379 (1905)
9. Novgorodov, I., Lemskaya, V., Gainutdinova, A., Ishkildina, L.: The Chulym Turkic language is of the Kipchak Turkic language origin according to the Leipzig-Jakarta list. Türkbilig 29, 1–18 (2015)
10. Novgorodov, I.: Ustojchivyi slovarnyi fond Tyurkskikh yazykov [The Leipzig-Jakarta list of the Turkic languages]. SMIK, Yakutsk (2016). (in Russian)
11. Novgorodov, I., Efremov, N., Ivanov, S., Lemskaya, V.: A relationship of Chulym Turkic to the peripheral and Uralian Kipchak languages according to the Leipzig–Jakarta list. In: Filchenko, A., Anikina, Z. (eds.) International Conference on Linguistic and Cultural Studies LKTI 2017: Linguistic and Cultural Studies: Traditions and Innovations. Advances in Intelligent Systems and Computing (677), pp. 286–295. Springer Nature, Cham (2018)
12. Novgorodov, I., Efremov, N., Gainutdinova, A., Ishkildina, L., Kukaeva, S., Tazhibayeva, S., Erezhepova, D., Lemskaya, V.: A relationship of Chulym Turkic to the Kangly (Aralo-Caspian) Kipchak languages according to the Leipzig–Jakarta list. In: Zhang, H. (ed.) Proceedings of 2018 4th Education and Education Research International Conference on Social Sciences and Interdisciplinary Studies. Advances in Education Sciences, vol. 20, pp. 115–120. Singapore Management and Sports Science Institute, Singapore (2018)
13. Radloff, W.: Opyt slovarya tyurkskih narechij [Experience of the Dictionary of the Turkic dialects], vol. I. Imperatorskaya akademiya nauk, Saint Petersburg (1893). (in Russian)
14. Radloff, W.: Opyt slovarya tyurkskih narechij [Experience of the Dictionary of the Turkic dialects], vol. II. Imperatorskaya akademiya nauk, Saint Petersburg (1899). (in Russian)

15. Radloff, W.: Opyt slovarya tyurkskih narechij [Experience of the Dictionary of the Turkic dialects], vol. III. Imperatorskaya akademiya nauk, Saint Petersburg (1905). (in Russian)
16. Radloff, W.: Opyt slovarya tyurkskih narechij [Experience of the Dictionary of the Turkic dialects], vol. IV. Imperatorskaya akademiya nauk, Saint Petersburg (1911). (in Russian)
17. Räsäsnen, M.: Versuch einen etymologischen Wörterbuchs der Türksprachen. Suomalais-Ugrilainen Seura, Helsinki (1969)
18. Sarach M., Kazas, M.: Antropologiya i demografiya karaimov tyurkov. Karaimskaya narodnaya ehnciklopediya [Anthropology and demography of the Karaims. The Karaim folk encyclopedia], vol. VI. Tortuga-Klub, Moscow (2000). (in Russian)
19. Sevortyan, E.: Etimologicheskiy slovar' tyurkskikh yazykov. Obshchetyurkskie i mezhtyurkskie osnovy na glasnye [Etymological dictionary of the Turkic languages. All-Turkic and cross-Turkic stems starting with vowels]. Nauka, Moscow (1974). (in Russian)
20. Sevortyan, E.: Etimologicheskiy slovar tyurkskikh yazykov. Obshchetyurkskie i mezhtyurkskie osnovy na bukvu "B" [Etymological dictionary of the Turkic languages. All-Turkic and cross-Turkic stems starting with letter "B"]. Nauka, Moscow (1978). (in Russian)
21. Sevortyan, E.: Etimologicheskiy slovar tyurkskikh yazykov. Obshchetyurkskie i mezhtyurkskie osnovy na bukvy "V", "G", "D" [Etymological dictionary of the Turkic languages. All-Turkic and cross-Turkic stems starting with letters "V", "G", and "D"]. Nauka, Moscow (1980). (in Russian)
22. Sevortyan, E., Levitskaya, L.: Etimologicheskiy slovar tyurkskikh yazykov. Obshchetyurkskie i mezhtyurkskie osnovy na bukvy "J", "ZH", "Y" [Etymological dictionary of the Turkic languages. All-Turkic and cross-Turkic stems starting with letters "J", "ZH", and "Y"]. Nauka, Moscow (1989). (in Russian)
23. Shcherbak, A.: Sravnitelnaya fonetika tyurkskih yazykov [Comparative phonetics of the Turkic languages]. Nauka, Moscow (1970). (in Russian)
24. Shcherbak, A.: Vvedeniye v sravnitelnoye izucheniye tyurkskikh yazykov [Introduction to the Comparative Study of the Turkic Languages]. Nauka, St. Petersburg (1994). (in Russian)
25. Tadmor, U.: The Leipzig-Jakarta list of basic vocabulary. In: Haspelmath, M., Tadmor, U. (eds.) Loanwords in the World's Languages: A Comparative Handbook, pp. 68–75. Mouton de Gruyter, Berlin (2009)
26. The World Loanword Database (WOLD). http://wold.clld.org/meaning. Last accessed 09 Nov 2018
27. Vovin, A.: Koreo-Japonica: A Re-evaluation of a Common Genetic Origin. Hawai'i Studies on Korea. University of Hawai'i Press, Honolulu (2010)
28. Vovin, A.: Lexical and paradigmatic morphological criteria for the establishing language genetic relationships. In: Novgorodov, I. (ed.) Unpublished Proceedings of 2016 International Symposium "The Leipzig-Jakarta List of the Turkic Languages as a Source of the Interdisciplinary Comprehensive Studies" (2016)
29. Vserossiyskaya perepis naseleniya [All-Russian Census]. http://www.gks.ru/free_doc/new_site/perepis2010/croc/perepis_itogi1612.htm. Last accessed 09 Nov 2018. (in Russian)
30. UNESCO Atlas of the World's Languages in Danger. http://www.unesco.org/languages-atlas/index.php. Last accessed 09 Nov 2018

Structural and Functional Characteristics of the Tabloid Misinforming Headlines (Based on "the Sun" Articles Analysis)

Natalya Saburova[✉] [ID] and Claudia Fedorova [ID]

M.K. Ammosov North-Eastern Federal University, 677000 Yakutsk,
Russian Federation
natalya_saburova@inbox.ru, fkill0252@gmail.com

Abstract. The article focuses on misinformation used in a tabloid headline in structural and functional aspects. Tabloid has always been one of the most popular forms of media, both printed and online, where misinforming headlines serve as an effective means of drawing the audience's attention. Theoretical basis of the problem dealt with in the article is presented in the classification of misinforming headlines, elaborated by A.S. Podchasov, as well as in works on media linguistics and media language ofby T.G. Dobrosklonskaya and K.Yu. Ovcharenko. Material for the study was obtained from the online version of "The Sun", one of the most popular British tabloids, which circulates in both printed and online formats. The text selected for the article is analyzed structurally (on the level of index page headline, main headline, lede, main text and illustrations) and semantically (whence key verbal and non-verbal elements will be interpreted from the prospective of their meaning in the context of the headline and the text). In the course of analysis we compare the meaning formed by the headline to the meaning formed at the textual level. Thus, presence and type of misinformation found in the headline of the given text is identified.

Keywords: Media · Tabloid · Text · Headline · Misinformation · Structure · Function

1 Introduction

Recently a significant rise of interest in analyzing modern media language and specifics of its functioning from researchers in various fields of study could be noticed in the field of linguistics. Growth and fast advance of newer media forms, as well as rise of new genres and the impact it has on modern culture and society, makes further research in this field of particular importance.

Tabloid, being one of the most popular and in many cases influential types of media, has successfully moved from traditional print form to online format. This change brought along a number of both structural and semantic changes which are actualized practically on all levels of the text. Along with this, one aspect of tabloid periodicals, both printed and online, remains unchanged, this being tabloids' focus on drawing potential readers' attention by all possible means. Headlines have always been

© Springer Nature Switzerland AG 2019
Z. Anikina (Ed.): GGSSH 2019, AISC 907, pp. 420–425, 2019.
https://doi.org/10.1007/978-3-030-11473-2_44

one of the most powerful means to guarantee the audience interest; and newer online media characteristics and features serve efficiently to enforce this interest.

Focus of this article is misinformation incorporated in the tabloid headline and actualized on all structural levels of online media tabloid text, and used in order to draw the attention of potential readers. Analysis of headline, text, and illustrative material in one article demonstrates some of the important functional characteristics of misinformation as a means of drawing attention to the text.

2 Literature Review

Dobrosklonskaya [1] and Solganik [2], Russian researchers of the contemporary media language identify the following media functions:

1. Informational
2. Ideological
3. Regulative
4. Entertaining

Entertainment has been one of its key functions from the beginning of media. From this aspect, tabloids have always been actualizing at least two of the abovementioned functions - those of informing and entertaining. The principle of infotainment – combination of "serious", informative and analytical content with "lighter" material (show business news, quizzes, popular memes, etc.) has since spread and could be found in such popular online media sites as "Buzzfeed", "Huffington Post", "The Daily Beast" and many others.

The growing ubiquitous entertaining content easily makes tabloids one of the most popular forms of modern media. When reviewing the definitions of the term "tabloid", one may clearly see two trends: the format description and the evaluative description of content (or evaluative trend). Format trend describes the size, font, color scheme and illustrative features of tabloids, while evaluative trend could be attributed to the nature of content. For instance, Prytkov defines tabloids as periodicals "with shorter texts, large headlines, marked by the use of bright colors and contrasting color schemes, large illustrations, use of exaggerative means, low familiar tone, and pronouncedly colloquial vocabulary and grammar. Thematically tabloids tend to focus on sensational topics (news on celebrities, scandals, catastrophes, tragedies, and the like). Advertisement plays key role in supporting tabloid circulation" [3, p. 213].

Such definition leads one to conclude that functionally tabloids dwell on three aspects:

1. Informative (definitive of a media outlet)
2. Entertaining (definitive of the type of media)
3. Drawing readers'/viewers' attention (focus on commercial aspect).

The third function (as the one that actualizes the evaluative aspect of tabloids' content) could in its turn be defined by the notion of sensationalism. Sensationalism in the form of a sensational news (often provided by famous political and public figures), often serves as a primary means of drawing the audience attention, thus guaranteeing the targeted amount of reads, views, or "clicks" (in online versions of periodicals) and providing advertising revenues.

Ovcharenko identifies eight ways of creating a sensation, usually being actualized on every major level of a text:

1. Personification: a news story associated with a famous person;
2. Tabooed themes: stories focused on sex scandals, death, gore, or extreme violence;
3. Hyperbolizing and excessive detailing;
4. Deliberate dramatization, excessive focus on negative elements of a story;
5. Extreme subjectivity in presenting a story and making judgments;
6. Deliberate familiarity of tone;
7. Use of misinforming headings;
8. Presence of key or exclusivity markers [4].

Since headline is the focus of the article, let us consider some of the key characteristics of headline as an important structural and semantic element of any media text. Headline is the element with which a text "introduces" itself to a potential reader. It has a rather complex task of both attracting the audience and in a brief and contracted form introducing the contents of the text (ideally in true-to-text form and meaning). Regardless of genre, type of periodical, or theme, each headline has three aspects: semantic, structural, and stylistic. In their turn, the prominence and character of functioning of each aspect does largely depend on genre and thematic features of a given text. Overall, informative texts (news) tend to have more semantically transparent and structurally/stylistically simple headlines, while analytical texts and features are more complex in all three aspects. From this perspective, tabloid texts present an interesting example, since generically they are linked to news (traditionally a more "serious" genre), and thematically they predominantly focus on light entertaining and sensational topics.

Functionally, Smirnova singles out three functions of media headline: informative (it informs readers of a subject), advertising (it is aimed at drawing the audience), and graphic (actualized via non-verbal means) [5]. Graphic function could also be viewed as a means of actualizing the advertising function through size, types and coloring of font. In case of tabloid headlines, both the informative and advertising functions are equally prominent with graphic one expanding beyond font in case of online version of periodicals (large colorful photos on the index pages, and inserted videos).

According to Podchasov, misinformation in headings is frequently used as a way of creating sensation, ultimately aimed at drawing audiences' attention. He defines misinformation in media as erroneous, inaccurate, or false information used to deliberately mislead the audience for a certain effect (depending on a number of factors and objectives, from ideological drives to commercial interests). The classification compiled by the researcher is based on the level of connection between headline and main body and includes three types of misinforming headlines used in media:

1. "Incomplete" headlines, reflecting single aspect of the main body of text and semantically "failing" to be representative of the whole text;
2. Headlines containing deliberate distortions of facts, presented in the main body;
3. "Defective" headlines, apparently not linked to any elements of the main body [6].

3   Methodology

The text selected to be analyzed for the article was published in the online version of the popular British tabloid "The Sun" on May 15th 2018. The subject of the text is (then) forthcoming wedding of Prince Harry and American actress Meghan Markle. This event largely dominated the British media over the course of the last two years. It is also worth noting that nearly all the events involving members of the royal family quickly become center of media attention in Great Britain (as well as nearly all over the world).

The headline on article page of the site states: *"WEDDING CRASHERS Meghan Markle's nephews and sister-in-law arrive in London ahead of royal wedding – despite NOT being invited"* [7].

The first element of the main heading is known as an **exclusivity marker**. It is one of the most characteristic features of "The Sun" format. It usually contains a pun and/or a reference to another popular culture event or phenomenon. The marker is usually emphasized with a large, bold and brightly colored font and is immediately followed by the main heading. In our case the phrase *"Wedding crashers"* directly alludes to a popular 2006 Hollywood comedy blockbuster centered on wedding events. The phrase itself is a modified word "gate-crasher": a person who arrives uninvited at an event in hopes of getting through. Since the topic of the text is royal wedding (as is indicated by the accompanying illustrations showing Prince Harry and Meghan Markle), the meaning of the marker clearly has sensationalist implications and serve to draw the readers' attention.

The **main heading** provides necessary details and at the same time raises the sensationalist tone by focusing on Meghan Markle's family: *"Meghan Markle's nephews and sister-in-law…"*. The negative element is printed in capital letters (*"NOT"*) which directly links it to the marker on both graphic and semantic level. Thus, the main heading of the text contains two scandalous implications: (1) Meghan Markle's family members dare to arrive at the royal wedding without invitation; and (2) members of Meghan Markle's family were not even invited. Both implications focus reader's attention on the potentially negative aspects of the event, and in particular, on Meghan Markle's family.

Subheading provides further details on the identity of the "nephews and sister-in-law": "Tracy Dooley and her two sons Tyler and Thomas are believed to be 'special correspondents' for Good Morning Britain and will broadcast from Windsor on Saturday" [7]. At the same time the second part of the structure somewhat contradicts both the meaning and the tone of the marker and the heading: reader learns that the aim of the family's trip to Great Britain is indeed not attending Harry and Meghan's wedding. Misinformation on this structural level of the text is actualized via consequential **detailing** from the exclusivity marker to the subheading.

The **main body** of the text largely consists of illustrations; large-scale photos depicting the eponymous family members at the airport, and official photographs of the royal couple. The details found in the opening paragraphs of the text mostly repeat the information provided by the main heading and subheading.

Let us consider the examples: *"MEGHAN Markle's **nephews and sister-in-law have touched down** in the UK ahead of the Royal wedding - despite **not being invited**"; "Tracy Dooley and her two sons Tyler and Thomas were spotted **arriving at Heathrow with 13 suitcases** yesterday"; "The trio has previously revealed **they weren't asked** to come to Meghan's wedding to Prince Harry this Saturday but could be hoping for a last-minute invite"* [7]. Repetitions ("nephews and sister-in-law", "not being invited") and paraphrasing ("have touched down", "arriving at Heathrow", "they weren't asked to come...") of the headline's elements make an effect of semantic excessiveness characteristic of tabloids. Readers also learn some new details of the family's stay: *"They are staying at the H10 hotel in Waterloo, where rooms are available for **as little as £70 a night**..."* [7]. The seemingly excessive details concerning hotel fees along with information on the size of the family's luggage ("13 suitcases") could be implying the family's limited financial capability combined with high hopes of being invited.

Significant portion of the text centers on a figure not directly mentioned in the headline, Thomas Markle, who at that moment was known not to attend the wedding due to health problem. Thomas Markle is introduced through comments made by his grandson. The remarks cited in the text are rather dramatic in tone and focus on Thomas Markle's poor health and dire emotional state: *"My grandfather is **in pieces** – he is **broken**. He is **not doing well**"; "...he has been **thrusted into all this** with the royals"; "It is **overwhelming**. My grandfather just **wants to be left alone** – he just **wants the piece and the quiet back**"* [7]. The given segment of the text clearly lacks any significant informative value since this piece of information is nothing new to the readers; it is aimed at making an emotional appeal to the audience through dramatization built on the use of emotionally charged vocabulary. One may suggest that unexpected mentioning of Meghan Markle's father is aimed at enhancing readers' interest following the excessive repetitions and insignificant details in the opening part of the text.

The **closing segment** of the text is also presented as a citation of the comment made by Meghan Markle's sister-in-law and addresses the titular issue of not being invited. Although formally it references the headline, the meaning of the comment contradicts the former's meaning: *"I don't think we're going to get an invitation, and **that's fine**"; "**We're OK** with that and we're **supporting her**. We're so **proud of her**"* [7]. The tone of the comment also contradicts the implied negativity of the headline.

4   Conclusion

The analyzed text contains an example of misinformation used in a tabloid headline. Analysis of the headline and main body of the article allows concluding that, according to Podchasov's classification, the given example presents **a combination of headline type 1 and 2**. The **headline** claims that Meghan Markle's family members arrived despite not being invited, while the **main body** adds details which shift this focus: they were not invited; they did arrive to the UK; but are not intending and never intended to "gate-crash" the royal wedding. The semantic aspect which dominates the headline (through the marker and the meaning of the main heading) could not be considered as representative of the situation described in the text (type 1). This "partial" representation of the text's contents in the headline also dwells on a certain distortion of the fact provided in the text. The arrival of the family members to Great Britain with an aim of attending an event different from the royal wedding (although associated with it) is presented through the marker and the main headline as something different by means of exaggeration and implications (type 2). Functionally the misinformation used in the text is aimed at building a sensation, guaranteeing readers' attention. Of eight sensation-building means presented in Ovcharenko's classification, five are used in the text: *personification, hyperbolizing/exaggeration, dramatization, misinformation,* and *exclusivity markers*. Misinformation, being used in the headline, becomes one of the key sensation-building means along with personification, while exclusivity marker becomes an effective element of building a misinforming headline.

References

1. Dobrosklonskaya, T.G.: Medialingvistika: sistemnyi podkhod k izucheniyu yazyka SMI. [Medialinguistics: systemic approach to the study of mass-media language]. Flinta, Nauka, Moscow (2008). (in Russian)
2. Solganik, G.Y.: Leksika gazety: funkcional'ny aspekt [Vocabulary of a Newspaper: Functional Aspect]. Vysshaya shkola, Moscow (1981). (in Russian)
3. Prytkov, A.V.: Kachestvennaya i bul'varnaya pressa v sisteme SMI [Quality press and tabloids in the system of mass-media]. Sci. J. Voronezh State Univ. Series "Philology. Journalism" **2**, 211–215 (2011). (in Russian)
4. Ovcharenko, K.Yu.: Sensatsiya kak osobyi tip novosti [Sensation as a special type of news]. Sci. J. Donetsk Inst. Soc. Stud. Series "Philology" **5**, 1–5 (2009). (in Russian)
5. Smirnova, E.A.: Funktsii gazetnogo zagolovka (na primere anglo-americanskoi pressy) [Functions of newspaper headline (on the example of English-American press)]. In: Proceedings of International Conference "Person–Word–Society", pp. 32–34. Mogilev State University, Mogilev (2009). (in Russian)
6. Podchasov, A.S.: Dezorientiruiuschiye zagolovki v sovremennykh gazetakh [Misinforming headlines in modern newspapers]. Russkaya rech' [Russian Language Journal] **3**, 52–54. Moscow (2000). (in Russian)
7. Christodoulou, H.: WEDDING CRASHERS Meghan Markle's nephews and sister-in-law arrive in London ahead of royal wedding – despite NOT being invited, https://www.thesun.co.uk/news/6289626/meghan-markle-nephews-sister-in-law-london-royal-wedding/ (2018). Last accessed 18 Aug 2018

Application of Speech Activity Theories to the Process of Teaching Humanities

Margarita V. Tsyguleva[1]($\boxtimes$) , Helena V. Tsoupikova[1] ,
Maria A. Fedorova[2] , and Irina N. Efimenko[1]

[1] Omsk State Automobile and Highway University,
Prospekt Mira 5, 644080 Omsk, Russian Federation
m.v.tsyguleva@gmail.com, sidorova_ma79@mail.ru,
efimenko_1951@bk.ru
[2] Omsk State Technical University, Prospekt Mira 11,
644050 Omsk, Russian Federation
chisel43@yandex.ru

Abstract. The article presents significant achievements of the Russian science in speech activity (Harris and Chomsky's transformation and generative model, Vygotsky's dynamic speech activity model, Chomsky and Miller's stochastic model of speech activity, Osgood's theory of speech production levels, Zhinkin's universal subject code model of speech production, etc.). The aim of the article is to propose methodical application of the described mechanisms for the process of speech production. The paper reveals advantages of scientific research and represents methodical solutions that provide for successful development of students' speech and communicative, linguistic and language competences as well as their self-reflective skills. The authors conclude that it can be done if students choose associates to the words, for example, reaction tasks; build sentences with associative word pairs; build texts on the basis of an associative word chain; combine any two specific words with the help of an associative bond and describe a mechanism of associative transfers; organize logical and semantic text content in the form of supporting schemes being analogues of universal subject codes (USC); decompresssupporting schemes into a great number of text variants; produce and control their own intellectual operations; make a prognosis of communication and compare this prognosis with particular communicative conditions, etc.

Keywords: Speech activity · Inner speech · Universal subject code · Competence · Speech variation

1 Introduction

Communicative training of students reveals significant problems associated with underdeveloped thinking and speech. We suppose that the main problems are as follows:

1. Content and structure of education do not meet the requirements of the society for modern specialists who should be able to adequately and effectively establish inter- and crosscultural communication
2. Linguistic courses are fragmented and isolated to some extent; integrating and systematizing subjects are unavailable to students since these subjects are not included in the university curriculum.
3. Traditional presentation of learning materials on language semantics is far from being justifiable in a higher school. The reasons are:
   3.1. Language education is aimed at developing language competence and partially speech competence, with the goals and objectives of communication being not real;
   3.2. Operation of verbal and cogitative mechanisms in the process of understanding (interpretation) and generating texts (text production) is not taken into account;
   3.3. Students are not acquainted with a number of systemic semasiological categories while studying humanities; it leads to communicative degradation and underdevelopment of text production abilities.

Formation of a new, cognitive and communicative paradigm which is based on the anthropocentric approach presupposes learning a language by using it, with a human factor and social conditions of communication taken into account [4]. The communication structure includes a speaker, listener, object and text where text is a universal mechanism of restructuring knowledge into information (on the part of the speaker) and comprehending and transforming the information into cognitive structures (on the part of the listener). Therefore, aspects of communication are the language which is studied by systemic linguistics and speech in its relation to the reality (external speech) and thinking (inner speech) which is studied by communicative and cognitive linguistics.

2 Theory

The necessity for developing students' communicative abilities and verbal skills is fixed in Federal Educational Standards and practically realized through Speech Study subjects, optional and elective courses such as "Technologies of Russian language business communication". This course is aimed at broadening and strengthening students' language knowledge, developing their abilities to fluently use the language in different spheres of communication and, most importantly, in professional business communication. Some specific goals of studying this subject can include:

- cognitive goals (formation of linguistic and communicative competences);
- practical goals (formation of language and speech competence);
- goals common for all the subjects: pedagogical (aesthetic development of students by means of a language) and educational (developing logical thinking and ability to autonomously work with language and speech material).

The goals determine the principles of selecting the course content:

- knowledge of language theory (terms and notions, language facts, grammar and vocabulary);

- knowledge of speech activity theory (functional style, communication ethics and speaking etiquette, text stylistics);
- knowledge of speech activity strategies in various communication spheres;
- speech skills and abilities (usage of functional equivalents, including those at the text level) [15].

All the goals listed above are connected with a current and socially significant problem of teaching effective communication to students, i.e. developing an ability to compose a text with due regard to such factors as a speech situation (communication conditions), purposes and motives of a communication act, means of providing adequacy of expressing/comprehending the message meaning, and an appropriate selection of the message's stylistic and genre form.

3   Methodology

The problem can be solved if the following methods and techniquesare employed: imitation exercises; messages and texts construction; creative and constructive exercises; multiple-choice questions associated with communication in different conditions; tasks on editing the style or genre of the text, on completeness of information and degree of its comprehensibility for various interlocutors.

Furthermore, as we can see, monologue is also a communicative act representing interaction of the communicating subjects: while creating a text, the speaker mentally produces/predicts the interpretative efforts of the listener and devises an appropriate tactics of his/her speech activity for the interlocutor to better comprehend the speech.

4   Modeling

While producing monologues or dialogues, the communication agents actively perform the following functions:

- a communicative function which provides interaction between the agents during their communication;
- an informative function, the agents' self-fulfillment while organizing information into a text form;
- a cognitive function, when interlocutors take into account the peculiarities of each other and obtain knowledge about each other;
- an emotive function adding emotions and expression to the text information;
- a creative function which allows the speaker to transform information so that the resulting text could affect the interlocutor.

When preparing an oral text, the speaker produces the so-called "mental speech act" [5] which forms the strategy and tactics of an external utterance in the inner speech:

- evaluating the prospects and the retrospective of a communicative situation;
- selecting the linguistic means to formalize the thought;

- choosing the communication topic and style depending on the situation and participants of communication;
- forecasting the results: interlocutor's evaluation of the message, his response to the information and the way of its expression [12].

Yet in 1995, Bakhtin [1] recognized and approved of the fact that the processes of understanding do not work without inner speech. Moreover, the processes of speech production also require an active work of inner speech and thinking. The majority of models for speech processes (scheme descriptions of speech activity which contain its main characteristics) describe its working mechanism. Each model has both positive and negative features, although the positive ones allow us to optimize the process of communicative competences formation and improve the teaching process aimed at creating and anticipating texts.

According to Harris and Chomsky's transformation and generative model [2], the sentence structure consists of its inner structure forming the meaning of the sentence and a surface structure, a form, which conveys this meaning. Transformation of simple core syntactic structures leads to the creation of new meanings. Therefore, the core structure "I usually do it" can be transformed with changing the status of the sentence from the affirmative to the interrogative or negative one ("Do I usually do it?", "I do not usually do it.").

It is also possible to transform the sentence by changing the voice ("It is usually done by me.", "Is it usually done by me?", "It is not usually done by me.", "Is it not usually done by me?").

While comparing the variants created, the speaker chooses the most appropriate external form to express the idea. Besides, such transformation helps students to actively strengthen their grammar knowledge (voice, tense formation, word order). Methodical application of this model can presuppose transformation ofan utterance to specify its meaning or to find the meaning equivalents that can clarify the speaker's idea andinterpret it adequately (If I do it right, you mean?).

Another model that is of great interest to methodologists is Vygotsky's dynamic speech activity model [16]. According to this model, the idea is recoded into the meaning (thought to word) passing through an inner speech stage. The scientist distinguishes three stages of verbal and cogitative activity: thinking (thought), inner speech, and external speech (word). The text production process includes consequent stages of the motive, thought and inner speech (where the thought is being formed, being prepared for verbalization) and external speech (verbal expression of the idea).

The dynamic speech activity model is the basis used in methodology for the construction of the theoretical framework for communication, formation and development of learner's communicative competence.

According to Chomsky and Miller's stochastic (scholastic) model of speech activity [10], speech is a sequence of elements. Appearance of a new element in a speech chain is dependent on the elements available. As a result, a language is considered as a countable set of states which can be fixed in the appropriate dictionaries by means of associative experiments.

This model can be applied inlanguage teaching methodology in the following way:

- choosing associates to the words, for example, reaction tasks;
- constructing sentences with associative word pairs;
- building texts on the basis of an associative word chain;
- combining any two specific words with the help of an associative bond and describing the mechanism of associative transfers [14].

This work is necessary for teaching students to understand objectively predetermined and individual mechanisms of associating and enrich their vocabulary as well as refresh grammar structures for producing texts of different types.

Another model which significanceformethodologists cannot be overestimated is a theory of speech production levels developed by Osgood [11]. The researcher distinguishes several levels of speech responses, i.e. systems of person's reactions to speech and other stimuli that are directly and indirectly expressed in the external speech:

- receptive (zero) level;
- integrative (first) level;
- representative (second) level;
- self-stimulating (third) level;
- level of motor coding (fourth).

The levels reflect the stages of speech production (motivational, semantic, stage of sequences, and integrative stages), with the first four levels (from a zero level to the third one), as we can guess, being realized in the internal speech, and the last one (the fourth)—in the external speech. It is highlighted in the theory that speech is produced on several levels simultaneously. and it can take from a second to several years to realize this process since it depends on the complexity of the task to be solved.

Stimuli are recoded into nerve impulses at the level of reception and cause the recipient's response. During their integration, these nerve impulses form a gestalt, a probable result of apprehension. This probability depends on the listener's apprehension experience [3].

Then, the obtained gestalt is associated with the nonverbal stimuli at the representative level where the meaning arises. The self-stimulating level turns the process "outwards" but the level of motor coding transforms motor schemata into behavioral patterns [7].

With regard to language teaching methodology, the theory mentioned above is applicable for work based on matching the stages of text construction with speech reaction levels, which will be helpful for developing students' self-reflection skills, promoting clear understanding of their verbal and cogitative activity.

Vygotsky's theory was expounded by Luria [9] who developed an inner schema of an utterance. Speech production was presented by him in the following way: an original inner schema (inner meaning), the so-called primary "semantic record", whose form is similar to that of the reduced utterance, is transformed into inner speech, into the "system of organized speech meanings", and then is transferred into an external utterance. This schema allows us to draw a methodical conclusion: when teaching speech production, it is necessary to appeal to word combinations and sentences but not only to keywords, since speech is a system of syntagmata according to Luria [9].

However, different communicators willpossess various "depths of text reading" or degrees of adequate interpretation.

Linguists, psycholinguists, researchers of the theory of speech activity, text theory, language (native and foreign) teaching methodology are greatly interested in Zhinkin's universal subject code model of speech production [18]. The meaning (text idea) is emerged in a universal subject code which presentation can be formalized in the unlimited amounts of messages. "Natural language usage is possible only through the phase of the inner speech" [18, p. 118].

Inner speech systematizes the ways of thinking in the form of spatial schemes, visualization, intonation aftersounds and single words. Inner speech is very subjective; it is a mediator between a communicator's individual thinking and a common language. The major component of speech activity and inner speech is universal subject codes that determine the selection of the speech content, the motive and the topic. Universal subject codes are a natural characteristic of the human brain in general, irrespective of the language a person speaks. Moreover, as the scientist notes, the communicators' speech is represented not as single sentences but as whole texts characterized by all the distinctive features [17].

Methodical implementation of Zhinkin's model presupposes:

- organization of logical and semantic text content in the form of supporting schemes being analogues of universal subject codes (USC) during comprehension and interpretation of the text and providing the adequacy of the text content comprehension;
- decompression of supporting schemes into a great number of text variants develops the variability of students' speech, makes it rich and eloquent. While defining the appropriateness of the text variant depending on the communication conditions, the student learns to effectively construct his speech [13].

The process of speech production is described in Leontiev's activity model [6] in the following way:

The first stage is the inner programming of the text. The programming units are universal subject codes. This stage includes differentiation of the speech situation (communication conditions) and the topic of the text (what speech situation is presented in the text). Besides, a distinction is also made between the elements of the text topic: an image of the subject (topic) and semantic realization of the subject with the elements of the personal meaning (rheme). A communicator performs various intellectual operations such as inclusion, enumeration, etc. at the stage of inner programming.

Leontiev calls the second stage 'grammatical-syntactic realization'. This stage is subdivided into a text-grammatical stage (when personal meaning is recoded into an objective code); phenogrammatical substage (where semantic features are distributed and code elements are linearly organized into messages); syntactic programming stage (grammatical characteristics are distributed according to the code elements); syntactic control stage (comparison of the available prognosis with a particular communicative situation).

The third stage, according to Leontiev, is motor programming and the fourth one is manifestation of the meaning in external speech.

Methodical conclusions that should be done on the basis of this model are as follows:

- students should be taught to produce and control their own intellectual operations;
- learners should know how and in what variants it is possible to make a prognosis of communication, how and why they must compare this prognosis with particular communicative conditions.

The Levelt's model "from intentions to articulation" [8] describes the mental processes of conceptualization, formulation, articulation, control and also the systems which actualize these processes such as conceptualizer, formulator, articulator and monitoring.

According to Levelt, the process of text production passes through the stages of intention, information selection, structuring, and connection of new information with the earlier stated one. The process results in a speech utterance. It is the conceptualizer that is responsible for the occurrence of all these stages.

The formulator aimed at recoding the conceptual information into the language one produces stages of lexis selection, construction of a morphological and syntactic text framework.

The articulator extracts the sequential inner speech blocks from an articulation buffer and constructs an articulatory plan of the text. Hence, the external speech is produced.

The monitoring is responsible for self-control, self-attitude and self-correction both in the external and internal planes of verbal and cogitative activity.

The model described above can be used in methodology for the development of such learning technologies that train students to perform a sequence of intellectual operations, information processes, to follow the connection between new and old information for more exact characterization of the information features, to formulate an utterance in different ways and develop their self-reflection.

5   Conclusion

Modern trends in the development of our society, rapid exchange of information, particularly, at the international level set new requirements for experts in the sphere of communication. These requirements go beyond the expert's professional competency and cover personal traits of the specialist who is expected to be a highly qualified, harmoniously developed and versatile person being ready for successful self-actualization in the society.

In harmony with these requirements, higher education system establishes a goal of training professionals with a set of profound and socially important knowledge, skills, and abilities. It motivates future professionals to consciously obtain knowledge and skills, develop intellectually, and become the linguistic personae of the new society. To reach the set goal, it is necessary for educators and researchers to search for the optimal educational impact on learners, with the focus on self-education.

References

1. Bakhtin, M.M.: Chelovek v mire slova [A man in the world of words]. Russian Open University, Moscow (1995). (in Russian)
2. Chomsky, N.: New Horizons in the Study of Language and Mind. The Press Syndicate of the University of Cambridge, Cambridge (2001)
3. Galich, G.G., Klyoster, A.M.: Cognitive and pragmatic interpretation of terminological fragments in the professional discourse. Adv. Intell. Syst. Comput. **677**, 242–249 (2018)
4. Klyoster, A.M., Shnyakina, N.J.: The event concept categorial network. Adv. Intell. Syst. Comput. **677**, 260–265 (2018)
5. Kurtseva, Z.I.: Golossovestiimental'nyjrechevojpostupok [Voice of conscience andmental-speechact]. School of Professor Ladyzhenskaya, Moscow (2005). (in Russian)
6. Leontiev, A.A.: Deyatelnyi um (Deyatelnost, Znak, Lichnost [The active mind (Activity, Sign, Personality)]. Smysl, Moscow (2001). (in Russian)
7. Leontiev, A.A.: Osnovy psikholingvistiki [Basics of psycholinguistics]. Smysl, Moscow (2003). (in Russian)
8. Levelt, W.J.M.: Speaking: from Intention to Articulation. MIT Press, Cambridge (1989)
9. Luriya, A.R.: Yazyk i soznaniye [Language and consciousness]. Moscow State University Publishing, Moscow (1998). (in Russian)
10. Miller, G.A., Chomsky, N.: Finitary Models of Language User, Handbook of Mathematical Psychology, vol. 2. Wiley, New York (1963)
11. Osgood, ChE: Lectures on Language Performance. Springer and GmbH & Co. KG, Berlin and Heidelberg (1980)
12. Tsoupikova, H.V.: Organizatsiya raboty nad tekstom s tselyu vyyavleniya ego postsuppozitsii v khode osvoyeniya spetskursa «Semasiologiya» [Organization of text work to find text postsupposition in the Semasiology course]. Omsk Sci. Bull. **5**(91), 208–212 (2010). (in Russian)
13. Tsoupikova, H.V.: Formirovaniye i sovershenstvovaniye kognitivnykh i kommunikativnykh umenij studentov v tekhnologii optimizatsii uchebnogo protsessa vuza v usloviyakh sovremennogo informatsionnogo vzaimodeistviya [Developing and mastering students' cognitive and communicative skills in optimization technology of university educational process in the context of current information exchange]. SibADI, Omsk (2016) (in Russian)
14. Tsoupikova, H.V.: Formirovaniye kommunikativnoj kompetentsii studenta sredstvami semasiologii kak uchebnoy distsipliny (organizatsionnyye strategii obucheniya) [Developing student's communicative competence by means of semasiology as an academic subject (organizational learning strategies)]. SibADI, Omsk (2010) (in Russian)
15. Tsoupikova, H.V.: Kompetentnostnyj podkhod k obucheniyu yazyku I rechi uchaschikhsya vysshej shkoly [Competence-based approach to language and speech teaching of higher school students]. Collection of research papers of young scientists, postgraduate students and student **10**, 273–280 (2013). (in Russian)
16. Vygotsky, L.S.: Myshlenie i rech [Thinking and speech]. Pedagogika, Moscow (1982). (in Russian)
17. Zhinkin, N.I.: Rech kak provodnik informatsii [Speech as an information conduit]. Nauka, Moscow (1982). (in Russian)
18. Zhinkin, N.I.: Yazyk – rech – tvorchestvo [Language – speech – creation]. Labirint, Moscow (1998). (in Russian)

Part-of-Speech Affiliation of Pre-christian Personal Names

Anna Vasileva$^{(\boxtimes)}$ (iD)

North-Eastern Federal University, Yakutsk 677000, Russian Federation
silong84@mail.ru

Abstract. All personal names have certain characteristics distinctive of nouns. However in some cases (particularly well reflected in non-calendar names) other parts of speech serve as anthropo-stems. Functioning of verbs and adjectives as personal names is a characteristic feature of Turkic onomastikons; particularly, anthroponyms (personal names, surnames, and nicknames). The article focuses on 1210 anthroponymic units (namely, their anthropo-stems) analyzed from the perspective of their part-of-speech affiliation in the system of Yakut language. The compiled corpus is based on the name-list of pre-Christian Yakut personal names by the historian F.G. Safronov. The analysis resulted in the following data on the Yakut personal names' part-of-speech affiliation: 433 verbal anthropo-stems, 254 adjectival anthropo-stems, 467 noun anthropo-stems, and 32 units with other parts of speech (adverbial, numeral, onomatopoeic). The remaining 24 units are represented by complex names, the analysis of which lies outside of the scope of this study. Therefore, although the number of noun anthropo-stems in pre-Christian anthroponymicon prevails, the number of names formed from other parts of speech (verbs and adjective) is still significant enough, which in its turn reflects the old-Turkic tradition of naming.

Keywords: Anthroponyms · Yakut language · Naming · Parts of speech

1 Introduction

Personal names of any ethnic group or nationality represent the most valuable source of linguo-cultural data and information. Anthroponymy, a field of research focusing on the analysis of people's names (names, surnames, patronymics, nicknames and other elements) studies anthroponymic units from a number of perspectives, interdisciplinary principle being of particular importance. Within the linguistic aspect, personal names are traditionally considered in accordance with the "structure (morphology) – semantics – cultural information" scheme. The article focuses on one of the structural aspects of Yakut personal names; particularly, that of anthropo-stem and their affiliation with particular parts of speech. The main objective of the research is to describe the part-of-speech affiliation of the Yakut Pre-Christian name list.

Z. Anikina (Ed.): GGSSH 2019, AISC 907, pp. 434–439, 2019.
https://doi.org/10.1007/978-3-030-11473-2_46

2   Literature Review

Study of anthroponomy in the Republic Sakha (Yakutia) has been conducted in a few historical and philological research works. Mikhail Spiridonovich Ivanov – Bagdaryyn Syulbe [1] wrote 13 books on toponymy, 2 books on anthroponymy and a big number of articles focusing on contemporary state and history of the local toponymic and anthroponymyc research.

Creation of electronic reference system of the Yakut personal names and surnames compiled by Monastyrev, Ivanov, Yefremov in 2014 [8] became an important milestone in anthroponymy. Material for the compilation was based on the Ivanov catalogue, as well as the materials compiled by the authors. The compilation's target audience is staff of registry offices who could use it as a spelling reference; it includes 3225 Yakut personal names and surnames.

As for historical anthroponymy in Yakut linguistics, Skryabina presented a Candidate of Science degree in Philology thesis in Tomsk in 1983 titled "Formation of local anthroponymic system (on the material of Russian official documents in Yakutia)". In the thesis the author states that "a big part in formation of the Yakut anthroponymic system was played by the Russian language … By early XIX century Yakut names acquire the role of second or accompanying names. Specific feature of the XVII–XVIII centuries' Yakut name compilation is the tradition of using two or more names to refer to one individual in official documents of the period where first name is Christian, while second is a Yakut name by which a bearer is publicly known" [7, p. 12].

Samsonov's several books focus on the onomastic studies, namely "Our names" 1989 [6]. Of particular interest for linguists studying Yakut anthroponymy are articles by Nikolayev addressing the issues of the history of Yakut anthroponyms' analysis, onomastic terminology in the system of Yakut language, word-forming affixes in Yakut anthroponyms, and borrowing from other languages in Yakut anthroponymikon [2].

3   Methods and Materials

Material of the study is the name-list of the Yakut personal names compiled by the Yakut historian, professor Fedor G. Safronov and later included in his work titled "Pre-Christian personal names of the peoples of East Siberia" published in 1985. The list presented a compendium of Yakut personal names collected from archive materials (among them being tax registers, household registers, income registers) from late 18th–early 19th centuries. In total, Safronov [6] collected around 5000 names in his pre-Christian Yakut name-list. He also succeeded to re-create the original Yakut spelling for 3337 of them. However, we succeeded in identifying and learning the original semantics of the appellative anthropo-stem for only 1210 of those. The part-of-speech affiliation analysis is performed using the methods of word-formation, statistic and definitional analyses.

4 Results and Discussion

4.1 Verbs as Anthropo-Stem

Shaykhulov states that "analysis and comparison of Turkic-origin names in structural and morphological aspects indicates that there are less than half of the total number of anthroponyms with the above-mentioned stem found in the name-list of ancient Tatar and Bashkir names included in the list. It apparently indicates at noun word-formation somewhat prevailing over the verbal one in the structure of anthroponyms. As for the verbal stems, they denote the most common every day and communication actions, states, necessities, and other things incorporated in an anthroponym" [11, p.40].

Out of 1210 anthroponymic units, 433 (35.75%) were formed from verbs. They are represented by both derivative and non-derivative names. Non-derivative names with verbal stems make up 169 units (39%). Among them there are units formed by appellative onymization with no derivation, as well as by names formed by clipping verbal stems.

Onymizated verbs:
Arbay – "arbay": to be disarranged, tousled, shaggy (of hair); *Akhchay* – "akhchay": straddled and slightly bent in knee (of legs or manner of walking); *Kylay* – "kylay": to gleam as to be visible or seen from afar, or to shine through something, to shine; *Chuguy* – "chuguy": to approach timidly, to flinch, to draw back; *Chochoy* – "cho-choy": to stick up.

The examples of anthroponyms formed by clipping verbal stems are presented below. Clipping in most cases occurs by omitting a verbal affix: *Kydaha* – "kydaharyi": to turn dark-red, to acquire the coloring likened to a blood clot, to turn crimson; *Dala* – "dalaadyi": to be, or seem very wide, or roomy from the bottom side; *Koemyuoekke* – "koetyoekkelee": 1. to hop, to bounce. 2. to constantly change place of work, living, to "bounce" from one place to another. 3. to hop from ones pot to next. 4. to be careless, airy.

Out of the above names with verbal stem, 97 anthroponymic units describe one's appearance, denoting "to look a certain way", and physiological characteristic, like, for instance, *Arbay* – to be disarranged, tousled, shaggy (of hair); *Daydaar*: to walk steadily, with a light walk, to float (as if floating above surface). There are six (6) verbs in the semantic group "personality traits", like, for instance, *Ayahalay* – to boast fully take on a difficult task or deed. Fourteen (14) verbs describe behavioural traits, like, for instance, *Kychchay* – to carefully, steadily look at something.

Derivative names having verbs as a stem are names formed by adding anthropo-formants to verbal stem. There are **263** verbs formed in this manner (60.4% from the total number of verbal stems).

Semantic distribution falls into the following groups:

1. in the group of names characterizing appearance there are 7 names (10% are derivative): the name *Kyuktyurche* was formed by adding the formant –che to verb Kyultyuryui (to be rounded, smoothed from all sides, to move in a corresponding manner as if rolling, or gliding). The name *Dyiaka* is formed by adding the formant

–ka to verb Dyiangnaa (to move clumsily, to slowly flap long limbs or wings of a lanky person, or bird with elongated body).

2. in the group of names characterizing physiological features of people there are 36 names (52% are derivative names). For instance, the name *Kyodjyoyos* was formed by adding the formant –es to verbal stem kyodjyoi (to look tall, heavy and slightly lopsided). Anthroponym *Dedeke* was formed by combining the verb dedey (to have disproportionately large, bulging belly (on a small person)) and formant –ke.

4.2 Adjective as Anthropo-Stem

Out of 1210 anthropo-units, 254 (20.99%) have adjective as stem. Non-derivative personal names with adjectival stem are represented by 155 units (62% of total number of adjectival stems): *Kihirges* – "kihirges": boastful; *Akhchahar* – "akhchahar": widely spread, or set (of legs or arms): *Kyltakh* – "kyltakh": too picky about food, having poor appetite; *Kichchey* – "kichchey": to gleam evenly from being completely smooth; *Kytaanakh* – "kytaanakh": 1. hard 2. tough in spirit, body; strong. 3. strict, tough in manner, severe; Other units were formed by clipping the adjectival stem: *Aka* – "aka-makaa": dumb, stupid, silly; *Kilchek* – "kilcherkey": having shaved head, bald.

95 derivative names formed by adding a formant to adjectival stem were identified, which makes up 38% of the total number of names with adjectival stem. Some examples are: *Keleheyeen* from "kelehey": severely stuttering person, with formant –ey added; *Ky'ylyka* from "kyhyl": red, with affix–ka added;

From the meaning of the appellative we distinguished the largest group denoted as "Appearance and physiological characteristics of an individual" represented by 86 (58.5%) adjectival personal names. For example, *Achchahar* (widely spread or set (of legs, arms)); *Kugas* (auburn, reddish-yellow, reddish-brown); 34 (23.12%) units of adjectival names represent the group "personal trait characteristics". For example, *Djebir* (1. Severe, strict, mean, cold, impassionate, cold in demeanor) 2. In dark tones, darkened, black); *Kyutyur* (1. Stingy, greedy. 2. Huge, gigantic. 3. Frightening, dangerous, cruel, threatening. 4. Nasty in character, mean).

Derivative personal names with adjectival anthropo-stem make up 32 units (33.6% of the total number of adjectival stems). Semantically, another largest group is a group of names describing physiological characteristics of (12 units, or 37.5%). For example, the name *Keleheyeen* was formed by composing the anthropo-stem Kelehey (a stutterer) and affix –en. Other semantic groups are smaller.

4.3 Noun as Anthropo-Stem

Despite the relatively small number of predicative anthropo-stems, nominal anthropo-stems are the most numerous; there are 467 of them (or 38.59 from the total number of analyzed anthroponyms).

Non-derivative personal names make up 282 units (or 62.9%). They are represented by simple two- and three-syllable nouns, as well as anthroponyms formed by clipping the noun-stem. Some examples are: *Alchakh* – "alchakh": a frog; *Akhta* – "akhta": a crotch; *Kynat* – "kynat": a wing; *Altan* – "altan": copper, gold; *Kulgakh* – "kulgakh": an ear. Clipping via omitting the ending formant: *Djelleng* – "djellenge": one of the morning stars which shines for a long period of time on the south-eastern part of the sky; Venus (ending "-e" was omitted); *Kuyuk* – "kuyukta": a maggot of an insect (usually of a gadfly) planted under animal's skin (formant "-ta" was omitted). Semantically, the larger portion of Yakut non-derivative personal pre-Christian names is comprised of the names belonging to the group "flora and fauna". There are 63 units (22.3%): *Kyohyon* (a mallard); *Kyhyyday* (a roach (fish)); *Kutuyakh* (a mouse); *Kuchukta* (name of a particular plant brought from the banks of the Uda river in order to treat various ailments).

Names belonging to the group "everyday life and hunting/craft objects" include 73 units (25%): *Adaha* (adaha – a wooden shoe put on horse's legs; shackes (archaic); an object which obstructs movement or anchors, weights; a burden, a disturbance); *Ketinche* (woolen or fur socks worn under boots); Inne 1. a needle. 2. a metal rod with sharpened end used for various purposes. 3. fir-needles. 4. hard, prickly formations, thorns found on bodies of certain animals and fish.

Derivative personal names having noun as a stem are presented by 45 units (31.4%). Names of this group are formed by addition of various formants. *Angyrchaan* from "angyr": a bittern with formant –chaan added; *Alchahay* from "alchakh": a frog with formant –ay added; *Kyunnyuk* from "kyun": the sun with formant –yuk added; *Kechchiikeen* from "kechchik/kerchik": 1. a piece, a stump (usually of something oblong). 2. a part, a fragment of something; a line between two points. 3. a sprig (of a plant), with formant –keen added.

The largest semantic groups in this category of names are "flora and fauna" (10 units) and "natural phenomena and materials" (6 units). For example6 *Altanyyr* was formed by adding formant –yr to the noun altan "copper".

4.4   Other Parts of Speech Serving as a Stem

Functioning for parts of speech like adverbs and numerals as anthropo-stems is not uniquely characteristic of the Yakut language. In Safronov's name-list 21 units (1.6%) out of total 1210 have adverbial stems, out of which 5 are non-derivative anthroponyms.

Formed by onymization with no derivation:
Balachcha – "balachcha": 1. for a long period of time; over a considerable period of time. 2. For long enough. 3. significant. 4. rather a lot.

Syugyun – "syugyun": expresses the demand to stop something; a call for order.

Semantically appellative stems, adverbial names are represented by names describing certain circumstances, like, for instance, birth: *Syugyun* "enough", as well as appearance and physiological characteristics of a person: *Byochyokh* in the meaning of "thickly, in thick clumps, flakes or batches, bales".

Names with numeral stems are comprised of 4 units, 3 of which are based on numeral 6 and 1 on numeral 7. Only the name *Alta* – "alta": six are non-derivative. Semantically these names could likely be linked to names describing certain birth circumstances (probably the order of the birth of the given individual).

5   Conclusion

The result of the conducted analyses shows that naming for the 18th–19th century Yakuts was not only significant in a way a person could be associated with a certain object or phenomenon. It was also an important indicator of actions, characteristic features and appearance. Thus, as the result of the analysis, it could be stated that the basis of the pre-Christian Yakut personal names was not composed of all parts of speech from verbs to numerals. It is believed that personal name possesses certain features characteristic of a noun. Therefore, its stem is based on noun, while imagery nature of ancient Yakut names determines a large number of predicative parts of speech acting as the stem.

The results obtained open a number of new perspectives of further analysis of Yakut personal names particularly in their structural aspect.

References

1. Bagdaryyn, S.: Dojdu surahtaah, alaas aattaah [Every place has the reputation, every alaas has its own name]. Yakutsk Book Publishing House, Yakutsk (1982). (in Yakut)
2. Nikolaev, E.: Ob izuchenii yakutskih antroponimov [On the study of Yakut anthroponyms]. North East. Humanit. J. **3**, 57–62 (2015). (in Russian)
3. Nikonov, V.: Imya i obshchestvo [Name and Society]. Nauka, Moscow (1974). (in Russian)
4. Safronov, F.: Dohristianskie lichnye imena narodov Severo-Vostochnoj Sibiri [Pre-Christian personal names of the peoples of North-Eastern Siberia]. Yakutsk Book Publishing House, Yakutsk (1985). (in Russian)
5. Shayhulov, A.: Tatarskie i bashkirskie lichnye imena tyurkskogo proiskhozhdeniya ["Tatar and Bashkir personal names of Turkic origin"]. Bashkir State University, Ufa (1983). (in Russian)
6. Samsonov, N.: Nashi imena [Our names]. Yakutsk Book Publishing House, Yakutsk (1989). (in Russian)
7. Skryabina, N.: Stanovlenie lokal'noj antroponimicheskoj sistemy (na materiale russkih delovyh dokumentov Yakutii) [Formation of a local anthroponymic system (based on Russian business documents of Yakutia)]. Yakut State University, Yakutsk (1983). (in Russian)
8. Spravochnik yakutskih lichnyh imen i familij [Directory of Yakut personal names and surnames]. APP, Yakutsk (2014) (in Russian)
9. Suvandii, N.: Tuvinskaya antroponimiya.[Tuvinian anthroponymy]. Tuvinian State University, Kyzyl (2011) (in Russian)
10. Vasileva, A.: K voprosu ob onomasticheskih issledovaniyah v Respublike Saha (Yakutiya) [On the issue of omomastic studies in the Republic of Sakha (Yakutia)]. Gramota Sci. J. **2**, 101–104 (2017). (in Russian)
11. Vasileva, A.: Antroponimicheskie issledovaniya v tyurko- i mongoloyazychnyh regionah RF [Anthroponymic researches in the turcic and mongolic-language regions of the Russian Federation(overview)]. Sociosfera Sci. J. **4**, 53–56 (2017). (in Russian)

Author Index

© Springer Nature Switzerland AG 2019
Z. Anikina (Ed.): GGSSH 2019, AISC 907, pp. 441–442, 2019.
https://doi.org/10.1007/978-3-030-11473-2

Printed in the United States
By Bookmasters